W9-ACM-270

The Facts On File

DICTIONARY
OF
ENVIRONMENTAL
SCIENCE

The Facts On File

DICTIONARY

OF

ENVIRONMENTAL

SCIENCE

L. HAROLD STEVENSON
BRUCE WYMAN

Facts On File

The Facts On File Dictionary of Environmental Science

Copyright © 1991 L. Harold Stevenson and Bruce C. Wyman

Facts On File, Inc.
460 Park Avenue South
New York, NY 10016

Library of Congress Cataloging-in-Publication Data

Stevenson, L. Harold, 1940—
    The Facts On File dictionary of environmental science / L. Harold
Stevenson, Bruce C. Wyman
        p.   cm.
    ISBN 0-8160-2317-4 (HC)
    ISBN 0-8160-3066-9 (PB)
    1. Environmental engineering—Dictionaries.   2. Environmental
protection—Dictionaries.   I. Wyman, Bruce C.   II. Title.
TD9.S74   1990
628'.03—dc20   90-28139

A British CIP catalogue record for this book is available from the British Library.

Facts On File books are available at special discounts when purchased in bulk quantities for businesses, associations, institutions, or sales promotions. Please call our Special Sales Department in New York at 212/683-2244 or 800/322-8755.

Composition by the Maple-Vail Book Manufacturing Group
Manufactured by the Maple-Vail Book Manufacturing Group
Printed in the United States of America

10 9 8 7 6 5 4 3 2

This book is printed on acid-free paper.

This book is dedicated to our wives,
Elizabeth Stevenson and Barbara Wyman,
without whose encouragement this effort
would not have been possible.

CONTENTS

ACKNOWLEDGMENTS

We would like to thank the following individuals for suggesting words for inclusion in this dictionary and/or contributing information used to prepare definitions: James Beck, George Fister, Gary Gaston, James Lane, Robert Maples, Davis Parker, Sandra Stephens, William Taylor, Aubrey Thompson and Reid Trekell. A special thanks is due our department head, Robert Maples, who not only supported us in this endeavor through contributing both words for inclusion and material for preparation of definitions, but also demonstrated tolerance while we delayed action on departmental duties to complete the project.

PREFACE

The field of environmental science examines the adverse effects of human activities on human health, wildlife, or ecosystems. The vocabulary of the science includes terms used in the definition, control, remediation, and prevention of harm to public health and the environment. Some of the many subject areas that fall within environmental science are the chemical contamination of air and water, species preservation, the environmental transmission of human disease, natural resources conservation, the stability and diversity of natural ecosystems, pesticide risks, workplace health and safety, waste management, and the effects and control of ionizing radiation. Given this broad scope, the terms in this dictionary are drawn from a variety of areas: chemistry, statistics, public health, biology, microbiology, ecology, engineering, toxicology, meteorology, geology, occupational health, soil science, mining, physics, computer modeling, epidemiology, and environmental law, and the electric power (fossil and nuclear), refining, and chemical process industries.

This dictionary is intended for high school and college students, members of environmental interest groups, environmental managers in government and industry, and, generally, the interested layman who wants to better understand an environmental concept or problem. Our goal has been to produce definitions that are accurate, but also clear and direct to the nonexpert. Some definitions will not completely satisfy the specialist.

We have selected terms for the dictionary to include those that may be used in a government report, in a newspaper or magazine article, in a regulatory compliance matter, or in a public debate or hearing. We strove to include the basic scientific and technical concepts applicable to environmental problems, the measurement methods and units employed, and the special terms and acronyms used in the intricate web of governmental statutes and regulations. A listing of the defined acronyms and abbreviations is included in the Appendix.

The terms in this volume are defined in their environmental sense only. For example, the term "runoff" appears in a water quality context and there is no mention of its political meaning.

Bruce Wyman   L. Harold Stevenson
Lake Charles, Louisiana

A

abandoned site A closed, hazardous waste storage or disposal site, the original owner or operator of which is normally no longer in business.

abiotic Nonliving components of the environment (or ECOSYSTEM), including chemicals in the air, water, and soil, the level and variability of solar radiation, and other aspects of the CLIMATE.

abrasive blasting The removal of coatings, rust, or SCALE from surfaces with a pressurized airstream containing silica sand, aluminum oxide, steel grit, nut shells, or other coarse material.

abscission The dropping of leaves from a plant. Excessive exposure to the air contaminants hydrogen fluoride, chlorine, and ethylene, among others, is associated with premature abscission in certain plant species.

absolute error The difference between a measured value and the true value, indicating the accuracy of a measurement.

absolute humidity The amount of water vapor in a unit mass of air, usually expressed in grams of water per kilogram of air.

absolute pressure The pressure exerted by a gas relative to zero pressure. Common units are millimeters of mercury (mm Hg) and POUNDS PER SQUARE INCH (ABSOLUTE). Compare GAUGE PRESSURE.

absolute temperature (T) A temperature expressed on the thermodynamic scale, measured from absolute zero. If Θ is the temperature in degrees Celsius, then:

$$T = \Theta + 273.15$$

absolute zero The zero value of thermodynamic temperature; 0 kelvin or $-273.15°C$. Theoretically, the volume of an ideal gas at this temperature becomes zero, according to CHARLES' LAW. In practice, however, substances become solids.

absorbance An expression of the amount of light that is absorbed by a solution; the measure is used to determine the concentration of certain ions or molecules in the solution. Specifically, the logarithm of the radiant power of light striking the solution minus the logarithm of the amount of light transmitted through the solution. See BEER-LAMBERT LAW.

absorbed dose The amount of a chemical substance that the body absorbs through the skin, gastrointestinal tract, or respiratory system. One can be exposed to a substance without absorption taking place. The substance may not penetrate the skin; it may pass out with body wastes unabsorbed; or it may be exhaled without surface or blood absorption in the lungs.

absorber A material capable of taking in a substance (such as oil) or radiation, as a sponge takes up water.

absorption The incorporation of a substance or energy into another substance or body; for example, the absorption of water by clay or the absorption of infrared radiation by atmospheric carbon dioxide.

absorption coefficient See SABIN ABSORPTION COEFFICIENT.

absorption factor The amount (expressed as a fraction) of a chemical that is absorbed into body tissues upon exposure.

absorption tower See PACKED TOWER.

absorptivity The ratio of the amount of radiation absorbed by a body to the maximum amount it can absorb. A surface that is a poor reflector is a good absorber. If no radiation is reflected, the surface acts as a BLACKBODY and has an absorptivity and emissivity of 1. Equal to the EMISSIVITY of a body, according to KIRCHHOFF'S LAW.

abyssal zone The bottom of a deep ocean. Compare to BATHYAL ZONE and EUPHOTIC ZONE.

acceleration (a) The rate of change of the velocity of a moving body with respect to time. Units of acceleration are velocity (e.g., ft s^{-1}, m s^{-1}) per time (e.g., ft s^{-2}, m s^{-2}).

accelerator An additive that increases the rate of a chemical reaction.

acceptable daily intake (ADI) The daily dose of a chemical substance determined to be without adverse effect to the general human population after long-term exposure. The dose is expressed in milligrams of the substance per kilogram body weight per day and is often used for the intake of PESTICIDES and PESTICIDE RESIDUES or for the amount of a food additive that can be consumed on a daily basis without the development of a health hazard. The U.S. Environmental Protection Agency has replaced acceptable daily intake with the REFERENCE DOSE for protection against noncarcinogenic health effects.

acclimation Synonym for ACCLIMATIZATION.

acclimatization The physiological adjustment by an organism to new, usually physical, environmental conditions. Often used in reference to the greater ability to tolerate increased air or water temperature following prior exposure to elevated temperatures.

accuracy The degree to which a reading given by an instrument or the calculation of a statistic agrees with the true value of what is being measured. Compare PRECISION.

acid Chemicals that release hydrogen ions (H^+) in solution and produce hydronium ions (H_3O^+). Such solutions have a sour taste, neutralize bases, and conduct electricity.

acid bright dip The cleaning of metal surfaces of corrosion and scale by immersion in acid.

acid deposition The introduction of acidic material to the ground or to surface waters. Includes WET DEPOSITION, from precipitation; DRY DEPOSITION, from particle fallout; and ACID FOG. See ACID RAIN.

acid dissociation constant (K_a) See DISSOCIATION CONSTANT.

acid aerosol Airborne particles composed of SULFATES, sulfuric acid, nitrates, and/or nitric acid. Moderate air concentrations have been shown to cause constriction of the respiratory passages in exercising asthmatics, and much higher levels have been linked to excess mortality in past air pollution EPISODES. Dry particle diameters are typically less than 1-2 microns. See ACID FOG.

acid fog Airborne water droplets containing sulfuric acid and/or nitric acid; typical diameters are 5-30 microns. See ACID AEROSOL.

acid mine drainage   Acidic water that flows into streams from abandoned coal mines or piles of coal mining waste. The acid arises from the OXIDATION of iron sulfide compounds in the mines by air, DISSOLVED OXYGEN in the water, and CHEMOAUTOTROPHS, which are bacteria that can use the iron sulfide as an energy source. Iron sulfide oxidation products include sulfuric acid, the presence of which has reduced or eliminated aquatic life in many streams in coal mining areas.

acid rain   Rainfall with a greater acidity than normal. Normal rainfall absorbs CARBON DIOXIDE from the air and forms CARBONIC ACID, which has a pH of about 5.6. Greater rainfall acidity is observed downwind (often at significant distances) from areas with many large sources of the air contaminants SULFUR DIOXIDE and NITROGEN DIOXIDE. These oxides react with water to form acids. Long-term deposition of these acids is linked to adverse effects on aquatic organisms and vegetation in geographical areas with poor BUFFERING CAPACITY within the sediments. Exposure to acid rain also deteriorates marble and limestone buildings and statuary. Because acidic materials also can be delivered to the earth by acidic particles, acid rain is a form of ACID DEPOSITION. See also WET DEPOSITION, DRY DEPOSITION.

acidification   Raising the acidity (lowering the pH) of a fluid by adding an acid.

acidity   The level of hydrogen ion concentration in a solution. On the pH scale, used to measure acidity, a solution with a concentration of hydrogen ions per MOLE greater than 1×10^{-7}, or a pH less than 7.

acoustic   Related or pertaining to sound or hearing. Excessive noise can effect hearing or otherwise impact negatively on individuals.

acoustical lining   A sound-absorbing lining inside air-conditioning ducts or electrical conduits that reduces the transmission of noise through walls of otherwise-soundproofed enclosures.

acoustical materials   Sound-absorbing materials, generally soft and porous in nature, such as acoustic tile or carpet.

acre-foot   A unit of water use or consumption; one acre-foot is the amount (volume) of water covering one acre to a depth of one foot. Equal to 43,560 cubic feet or 1234 cubic meters.

acrolein   An aldehyde (C_3H_4O) used widely in the manufacture of organic chemicals. The compound is a very strong contact irritant to humans.

actinic range   See ULTRAVIOLET RADIATION, ACTINIC RANGE.

actinides   The chemical elements from ATOMIC NUMBER 89 (actinium) or atomic number 90 (thorium) through atomic number 103 (lawrencium). The elements are grouped separately on the PERIODIC TABLE of the elements.

actinolite   A calcium-iron-magnesium AMPHIBOLE; a type of ASBESTOS.

actinomycetes   A group of soil bacteria that are moldlike in appearance and that are important in the ecology of soil. These organisms are involved in the decomposition and conversion of complex plant polymers into HUMUS. The characteristic musty smell of soil is derived from compounds released by this group of bacteria. Streptomycin and related antibiotics are produced by members of the actinomycetes group.

action level (AL)   Under regulations of the Occupational Safety and Health Administration, a measured concentration of certain airborne contaminants or a noise level in the workplace that requires medical testing of exposed workers. Action levels are usually set at one-half the permissible exposure limit.

activated carbon See ACTIVATED CHARCOAL.

activated charcoal A material produced by burning a carbon material such as wood or coconut shells and then increasing the adsorptive capacity by steam treatment, which creates a very large surface area. The product is used to adsorb organic materials, either as an air or water pollution control technique or as an air-sampling device. See ADSORPTION.

activated sludge The suspended solids, mostly microorganisms, present in the AERATION tank of a sewage treatment plant. See ACTIVATED SLUDGE PROCESS.

activated sludge process A form of SECONDARY TREATMENT of sewage, this process uses microorganisms suspended in well-aerated wastewater to degrade the organic material. A sedimentation tank placed after the activated sludge/aeration tank removes the SLUDGE, routing some of it back to the inflow side of the aeration tank. The rest is pumped outside the system for treatment and disposal.

activation In nuclear physics, the NEUTRON bombardment of stable atoms, which absorb the neutrons and become RADIOACTIVE.

activation energy The energy that must be put into a system in order to start some reaction or process. The activation energy is later recovered during the reaction or process. The simplest example is a match that is used to start a forest fire.

active Describing chemicals that readily combine or react with other chemicals. For example, oxygen gas is an active chemical because it reacts with other substances during the combustion process; metallic sodium is an active material because it reacts spontaneously with water to release hydrogen gas. Compare INERT.

active solar system A system used to capture solar energy as heat and to distribute that heat by a system of pumps and pipes. A typical system consists of solar collectors that use sunlight to heat water, which is stored in an insulated tank. The heated water is then pumped through a series of heat exchangers to provide heat for a building. Compare PASSIVE SOLAR SYSTEM.

activity Another term for RADIOACTIVITY, expressed in CURIE units. One Curie equals 3.7×10^{10} atomic decays per second.

actual cubic feet per minute (ACFM) The volumetric flow rate of air, uncorrected for STANDARD CONDITIONS of temperature and pressure. See STANDARD CUBIC FEET PER MINUTE.

acute Describing a sudden exposure to a significant dose of a dangerous chemical or radiation that usually results in a severe reaction.

acute exposure In toxicology, doses administered or received over a period of 24 hours or less. Compare CHRONIC EXPOSURE.

acute health effect A circumstance in which a chemical or physical agent results in the rapid development of severe symptoms in exposed humans or animals.

acute toxicity Toxicity resulting from an ACUTE EXPOSURE; adverse effects closely spaced in time to the absorbed dose of a toxic material.

additive effect The interaction of two or more chemical substances on a biological system that has a combined effect equal to the sum of the effects of each substance acting alone. Compare SYNERGISTIC EFFECT, ANTAGONISTIC EFFECT.

additive reagent A chemical element or compound that is included along with the normal REAGENTS employed in a pollution abatement procedure to facilitate the operation. A substance added in trace amounts to improve some standard process.

additives Chemicals added in small amounts to a product such as food in an effort to preserve freshness or quality.

adenoma A benign tumor of the epithelial (protective, membranous) tissue of a gland. An adenoma observed in a rat that has been administered a dose of a test chemical is taken to imply that the chemical may pose a human cancer risk.

adenosine triphosphate (ATP) An energy-storage compound common to all biological systems. The high-energy IN-TERMEDIATE is formed during photosynthesis or by the breakdown of energy-containing material, such as glucose. ATP supplies the energy for cell reactions and functions.

adiabatic lapse rate The constant decline in temperature of an AIR PARCEL as it rises in the atmosphere. The temperature change is due to the pressure drop and gas expansion only, and no heat is considered to be exchanged with the surrounding air through convection or mixing. The DRY ADIABATIC LAPSE RATE for air not saturated with water vapor is 0.98°C per 100-meter rise (0.54°F per 100 feet). The WET ADIABATIC LAPSE RATE is about 0.6°C per 100 meters (0.33°F per 100 feet).

adiabatic process A change involving no gain or loss of heat.

adjusted rate In EPIDEMIOLOGY, a rate that has been statistically modified to allow a proper comparison with other (adjusted) rates. For example, the death rate for a retirement community will be greater than the death rate for a typical suburb. The difference in the age distribution of the two areas is statistically eliminated when an adjusted death rate is computed. See AGE ADJUSTMENT, DIRECT METHOD and AGE ADJUSTMENT, INDIRECT METHOD.

administrative controls A system that reduces or controls exposure of workers to chemical or physical agents. The time workers spend in higher-exposure areas may be limited, and those workers with greater ABSORBED DOSES are rotated to lower-exposure areas. Activities with potentially high exposures are scheduled at times or in places that reduce the number of workers present in the area.

administrative law The legal process by which STATUTORY LAW is implemented. For environmental matters, the process pertains to administrative agency decisions and procedures, judicial review of agency actions, issuance of regulations, permission to establish facilities, and citizen suits, among other areas. See CITIZEN SUIT PROVISION.

Administrative Procedure Act (APA) A federal law enacted in 1946 that requires regulatory agencies to follow certain procedures when they make decisions, such as issuing regulations or standards and granting permits to facilities, among others. Judicial review of environmental regulatory agency actions follows the provisions of the APA.

administrative record The studies, reports, hearings, letters, internal memoranda, or other material used by an administrative agency to write a regulation or standard, make a permit decision, and so on. For example, the information used by the United States Environmental Protection Agency to select a response action under the COMPREHENSIVE ENVIRONMENTAL RESPONSE, COMPENSATION, AND LIABILITY ACT (Superfund); the information includes the REMEDIAL INVESTIGATION/FEASIBILITY STUDY, the RECORD OF DECISION, and any public comments.

Administrator, the The chief officer of the United States Environmental Protection Agency.

adsorbate The chemical(s) being adsorbed or taken up in a pollution control device used for the ADSORPTION of a chemical substance. See ADSORBENT.

adsorbent   The material to which a pollutant is being adsorbed in a pollution control device operating by ADSORPTION. See ADSORBATE.

adsorber   A solid or liquid that can hold molecules of another substance on its surface.

adsorption   The collection of solids, liquids, or condensed gases on the surface of another substance or body; for example, the adsorption of volatile organic compounds by activated charcoal.

adsorption isotherm test (AI test) See SOIL AND SEDIMENT ADSORPTION ISOTHERM TEST.

advance notice of proposed rule making (ANPR)   An announcement in the FEDERAL REGISTER, usually with an extended discussion, of the regulatory approach an agency (such as the United States Environmental Protection Agency) is considering in an area of concern; the notice is not an actual proposed rule.

advection   Transport by a moving fluid (liquid or gas). Differs from CONVECTION by being used mainly to describe the horizontal movement of airborne materials (pollutants) by wind.

aeolian soil   Soil transported from one area to another by wind.

aeration   In wastewater treatment, the addition of air (oxygen) to wastewater. This prevents the DISSOLVED OXYGEN content of the water from falling to levels insufficient for rapid degradation of the organic material in sludge or other wastewater.

aeroallergens   Airborne particles that can cause allergic responses (such as hay fever) in susceptible persons.

aerobic   Requiring oxygen; the usual application is to bacteria that require oxygen for growth. An aquatic system characterized by the presence of DISSOLVED OXYGEN. Compare ANAEROBIC.

aerobic bacteria   Single-celled, microscopic organisms that require oxygen to live. These organisms are responsible, in part, for the AEROBIC DECOMPOSITION of organic wastes. Also called DECOMPOSERS.

aerobic decomposition   The biodegradation of materials by AEROBIC microorganisms; the process produces CARBON DIOXIDE, water, and other mineral products. Generally a faster breakdown process than ANAEROBIC DECOMPOSITION.

aerodynamic diameter   A way to express the diameter of a (usually airborne) particle by using the diameter of a perfect sphere of UNIT DENSITY with the same SETTLING VELOCITY as the particle. For example, a nonspherical particle roughly 6 MICRONS across and with a density of 2 grams per cubic centimeter may have an aerodynamic diameter of 8 microns, meaning that the nonspherical particle has the same settling velocity as an 8-micron sphere having a density of 1 gram per cubic centimeter. Compare STOKES DIAMETER.

aerosols   A suspension of liquid or solid particles in air or gas.

afforestation   The establishment of a forest by human planting in an area where trees have not grown previously.

aflatoxin   Toxins that are produced during the growth of fungi or molds in grains or grain meals stored under moist conditions. Some of the aflatoxins are thought to be responsible for the development of cancer in animals that consume the contaminated grain or meal.

afterburner   An air pollution control device that incinerates organic compounds in an airstream to carbon dioxide and water. Also called a vapor incinerator.

age adjustment, direct method The application of AGE-SPECIFIC RATES for area or community populations to a STANDARD POPULATION to allow a proper comparison between two or more different populations. The method eliminates the effect of different age distributions in the populations being compared.

age adjustment, indirect method The application of standard (i.e., nationwide) MORBIDITY RATES or MORTALITY RATES to an area or community population to obtain an expected number of illnesses or deaths. This number is compared with the actual number of illnesses or deaths in the population.

age pyramid See AGE-STRUCTURE DIAGRAM.

Agency for Toxic Substances and Disease Registry (ATSDR) An agency in the United States Public Health Service that performs public health assessments in areas near hazardous waste sites, including health surveillance and epidemiological studies. The agency also sponsors research on the toxicology of specific chemical materials and develops health information databases for use in future health assessments.

Agent Orange The herbicide mixture used as a defoliant during the Vietnam War. The primary agent in the mixture was 2,4,5-trichlorophenoxyacetic acid (2,4,5-T). The chemical synthesis of this herbicide results in slight contamination of the agent with DIOXINS. Consequently, persons exposed to Agent Orange in Vietnam also were exposed to a small amount of dioxins. Dioxins are known to cause cancer and birth defects in experimental animals, but the hazard they pose to human health at small doses is uncertain.

age-specific rate In EPIDEMIOLOGY, a MORTALITY RATE that is calculated for a specific age group; for example, the death rate for persons 70 to 75 years old in the United States in 1988.

age-structure diagram A bar graph showing the number or percentage of individuals in a population within various age ranges: 0–5 years, 6–10 years, and so on. The graph is often divided horizontally to show the relative number of males and females within the age ranges.

agglomeration In wastewater treatment, the grouping of small suspended particles into larger particles that are easier to remove. See COAGULATION.

aggregate risk The sum of the estimated individual excess risks of disease or death in a population caused by an exposure to a chemical or physical agent.

aggregation The collection of smaller units into a larger single mass.

agitator/mixer Blades or paddles that slowly rotate within a tank to facilitate the mixing of reagents or suspended materials.

agricultural runoff The runoff into surface water of herbicides, fungicides, insecticides, and the nitrate and phosphate components of fertilizers from agricultural land. A NONPOINT SOURCE of water pollution.

air The gaseous mixture that makes up the earth's ATMOSPHERE. Four gases account for 99.997% (by volume) of clean, dry air: nitrogen (78.084%), oxygen (20.946%), argon (0.934%), and carbon dioxide (0.033%). The remaining components include neon, helium, methane, krypton, nitrous oxide, hydrogen, xenon, and various organic vapors. Under actual conditions, air contains up to about 3% water vapor (by volume) and many solid, liquid, or gaseous contaminants introduced by human activities and natural events such as wind erosion.

Air and Waste Management Association (AWMA) A professional organization, based in Pittsburgh, Pennsylvania, dedicated to research and management issues involving air quality and hazardous waste. Founded in 1907, it now has 10,000 members. Formerly known as the Air Pollution Control Association.

air classifier A device used in solid waste recovery operations that separates paper and other low-density materials from mixed waste by pumping air up through the waste.

air injection 1. In groundwater management, the pumping of compressed air into the soil to move water in the unsaturated zone down to the saturated zone, or water table. 2. In automobile emission control, the addition of air into the exhaust gas to complete the oxidation of CARBON MONOXIDE to CARBON DIOXIDE, now accomplished by the CATALYTIC CONVERTER.

air mover A fan, pump, or other device that induces the movement of air in a duct in workplace ventilation or air pollution control equipment or through hoses in the case of an air-sampling apparatus.

air parcel A theoretical volume of air considered in air pollution dispersion predictions to rise (or fall) without exchange of heat with surrounding air. An air parcel will rise if its temperature is greater than the air around it. See ADIABATIC PROCESS, ADIABATIC LAPSE RATE, MIXING HEIGHT.

air pollution The presence of gases or AEROSOLS in the AMBIENT air at levels considered to be detrimental to human health, wildlife, visibility, or materials. The air contaminants can have a human origin (smokestacks, tailpipes) or a natural origin (dust storms, volcanoes).

Air Pollution Control Association (APCA) Former name of the AIR AND WASTE MANAGEMENT ASSOCIATION.

air preheater In industrial and utility boilers, a device in which hot flue gas exhausting from the boiler flows past cool air flowing into the boiler burners and warms the incoming air. The use of preheated combustion air increases boiler fuel efficiency by saving the natural gas, coal, or fuel oil energy otherwise used to raise the incoming air temperature.

air-purifying respirator (APR) A device worn by workers when the air contains, or may contain, harmful levels of contaminants, such as during certain hazardous workplace tasks or during the cleanup of chemical spills outside facility property. A filter or cartridge removes specific air contaminants from the incoming air breathed by the wearer. The purifying filters or cartridges can become saturated or exhausted and must be replaced. Each APR is effective only against a certain chemical or chemical type; it must be chosen to match the airborne chemical in each situation. Compare SELF-CONTAINED BREATHING APPARATUS (SCBA). See also SUPPLIED-AIR RESPIRATOR.

air quality control region (AQCR) One of 247 air management areas in the United States designated by the Environmental Protection Agency. Each region contains at least two urban areas, often in adjoining states, that share actual or potential air quality problems.

air quality criteria The adverse effects on human health, human welfare, wildlife, or the environment used to set ambient AIR QUALITY STANDARDS. See CRITERIA DOCUMENT.

air quality dispersion model Any of a variety of mathematical simulations of the downwind diffusion and/or physical/chemical removal of air contaminants emitted by one or more sources. The inputs to the model include the type of emission, emission rate, release height, and appropriate weather conditions. The model output is an estimated air concentration at specified points in the vicinity of the source.

air quality maintenance area (AQMA) Administrative areas defined by the United States Environmental Protection Agency for the production of STATE IMPLEMENTATION PLANS. Ambient air quality data, emissions data, air pollution control strategies, and so forth, that are part of a state implementation plan are organized by AQMA.

air quality related value (AQRV) For areas designated as Class I under the PREVENTION OF SIGNIFICANT DETERIORATION of an air quality program, criteria other than the allowable INCREMENT CONSUMPTION that may be used to assess the impact of a new emission source. The value mainly refers to any reduction in the visibility that may be caused by the new air emissions. Class I areas include major national parks and wilderness areas, mostly in the American West.

air quality standard (AQS) The allowable amount of a material in the AMBIENT air for a certain AVERAGING TIME or, less often, the allowable emissions of air contaminants into the atmosphere. See NATIONAL AMBIENT AIR QUALITY STANDARD; STANDARDS, ENVIRONMENTAL. Compare EMISSION STANDARD.

air stripping Removing a dissolved contaminant from water to the atmosphere by pumping the water over a large surface area in order to provide as much contact as possible between the liquid and the air. A common method for the removal of VOLATILE ORGANIC COMPOUNDS from contaminated groundwater.

airshed An area with similar air mixing and affected by the same air pollution sources, thus experiencing roughly the same air quality. Large-scale air movements commonly void the geographical boundaries of an airshed.

airway resistance (Raw) The resistance to airflow in the respiratory system, measured by lung function tests, including SPIROMETRY.

air-to-cloth ratio (A/C) In computations involving FABRIC FILTERS for air pollution control, the ratio of the volumetric flow rate of the gas to be cleaned to the fabric area. The air-to-cloth ratio is typically expressed as a FILTERING VELOCITY, and is measured in meters per minute.

air-to-fuel ratio (AFR) In internal combustion engines, the ratio of the mass of air available for fuel combustion to the mass of the fuel. High AFRs are described as lean; low ratios are said to be rich. Exhaust emissions are affected by AFRs. A rich mixture will emit higher levels of CARBON MONOXIDE and VOLATILE ORGANIC COMPOUNDS (unburnt fuel); a lean (air-rich) mixture will cause poor engine performance. The theoretical AFR, called the STOICHIOMETRIC RATIO, is neither rich nor lean. When an engine operates with an almost ideal mixture, carbon monoxide and volatile organic compound emissions are low, but emissions of NITROGEN OXIDES greatly increase. The THREE-WAY CATALYST emission control device must operate within a narrow AFR to be effective at controlling carbon monoxide, volatile organic compounds, and nitrogen oxides.

air toxics Air pollutants that may pose chronic health risks to human populations if their ambient levels are excessively high for a prolonged period of time. The ambient level and the length of time required to cause harm vary by pollutant. The health risks of concern include cancer; neurotoxic, mutagenic, and teratogenic effects; and reproductive disorders. Most air toxics are metals, VOLATILE ORGANIC COMPOUNDS, or PRODUCTS OF INCOMPLETE COMBUSTION. EXTREMELY HAZARDOUS SUBSTANCES—those posing acute health risks—are sometimes included as air toxics.

Aitken counter A device for counting the number of CONDENSATION NUCLEI in an air sample. Sample air is cooled by rapid expansion, and the droplets that form on suspended dust particles (nuclei) are counted with the aid of a microscope.

albedo The fraction of electromagnetic radiation reflected after striking a surface. The average albedo of the earth to sunlight is about 30 percent, which is a combination of many different surface reflectivities: water (at noon), about 5 percent; fresh snow, about 80 percent; deciduous forest, about 20 percent.

aldehyde A type of organic molecule containing a carbonyl group in which a carbon atom is bonded to an oxygen atom with a double covalent bond and to a hydrogen atom with a single covalent bond. Aldehydes can be reduced to form alcohols, and oxidized to acids. Two common aldehydes are formaldehyde and acetylaldehyde. The basic chemical formula for aldehydes is given below. The R group represents any type of organic side chain. Excessive exposure to aldehydes can result in central nervous system depression, skin or mucous membrane irritation, or allergic reactions in sensitive groups.

$$R-C\overset{\displaystyle O}{\underset{\displaystyle H}{\Big\langle}}$$

aldrin An insecticide belonging to the chlorinated hydrocarbon class. The compound is a cyclodiene, meaning that the parent molecule is cyclic and contains many double bonds between the carbon atoms. The insecticide is persistent in the environment and considered dangerous; its use is strictly controlled when allowed.

algae A group of one-celled, free-floating green plants often found in aquatic ecosystems. The singular form is alga.

algal bloom Rapid growth of ALGAE caused by the addition of a LIMITING FACTOR, such as the nutrient phosphorus, to water. The extent of growth may result in a coloration of the water or the pro-

duction of toxins at levels that present problems for man or wildlife.

aliphatic Organic compounds with a straight-chain structure. One of the two major groups of HYDROCARBONS, the other being the AROMATIC hydrocarbons (closed-ring structures). Aliphatic hydrocarbons are classified as alkanes (carbon atoms connected by single bonds), alkenes (carbon atoms connected by double bonds), and alkynes (carbon atoms connected by triple bonds). See also STRAIGHT-CHAIN HYDROCARBONS.

aliquot The amount of a sample used for analysis.

alkali A chemical substance that can neutralize an ACID. Also refers to soluble salts in soil, surface water, or groundwater.

alkaline fly ash scrubbing A system for removing sulfur oxides from the flue gas of a coal-burning boiler by using as an absorbing medium the alkaline constituents of ash that remain after the coal is burned.

alkalinity Generally, a measure of the ability of water to neutralize acids. It is measured by determining the amount of acid required to lower the pH of water to 4.5. In natural waters, the alkalinity is effectively the BICARBONATE ion concentration plus twice the CARBONATE ion concentration, expressed as milligrams per liter calcium carbonate.

alkylbenzene sulfonate (ABS) An early synthetic detergent that, due to its branched molecular structure, was not decomposed effectively by bacteria in wastewater treatment plants, thereby resulting in foaming discharges. The detergent molecule was reconfigured as LINEAR ALKYLATE SULFONATE (LAS) in the 1960s, resulting in a chemical form of detergents readily decomposed by the bacteria.

alkylbenzenes Any member of a family of compounds consisting of a BENZENE ring (a single-ring aromatic compound, C_6H_6) containing one or more aliphatic (straight) side chains (e.g., methyl or ethyl groups) attached to the ring. The common solvents toluene and xylene are included in this group. The usual problems associated with inhalation of these substances in excessive doses are caused by irritation of mucous membranes in the respiratory system. Acute exposure leads to depression of the central nervous system. The adverse reactions experienced by "glue sniffers" are usually related to this class of compounds.

alkyl mercury A group of compounds consisting of aliphatic organic compounds (e.g., methyl and ethyl groups) bonded to mercury atoms. The most well-known member of this group is METHYL MERCURY, the most toxic form of mercury to people, which affects a variety of systems in humans, including the central nervous system and the reproductive process.

allelopathy The secretion by plants of chemical substances that inhibit the growth of competing species; a significant influence on the rate and type of plant SUCCESSION.

allergen A substance that acts as an antigen, causing the formation of antibodies that react with the antigenic substance. This particular form of allergic antigen-antibody reaction manifests itself only in certain persons for certain substances, and can be accompanied by skin or mucous membrane irritation, respiratory function impairment, and eye irritation, among other effects. Allergens can cause a reaction at extremely low doses in sensitive persons.

allergic sensitizers Chemical substances that act as antigens to produce an allergic reaction after repeated (sensitizing) exposures to the skin or respiratory system. The agents that can act as sensitizers to susceptible persons include

the epoxy resins, toluene diisocyanate, formaldehyde, chromium compounds, and the plant resins in poison ivy.

allochthonous Describing organisms that are non-native or transient members of a community in a specific habitat. They do not carry out a metabolic function in that location, and they do not reproduce and occupy the habitat permanently. The term is most frequently used to describe bacteria native to one environment (e.g., the human body) that get transported to a completely different habitat (e.g., a deep ocean) where they might survive but they cannot perform metabolic functions or colonize. Compare AUTOCHTHONOUS.

allogenic Exogenous, caused by external factors, such as a change in a habitat caused by flooding; the opposite of AUTOGENIC.

allogenic succession Predictable changes in plant and animal communities in which changes are caused by events external to the community, such as fire. Opposite of AUTOGENIC SUCCESSION.

alluvial Referring to ALLUVIUM, the type of soil found in floodplains.

alluvium The fertile sediment deposited in the floodplain of a river.

alopecia The loss of hair. The condition can result from exposure to toxic chemicals within the environment or from chemicals administered as anticancer agents.

alpha counter A type of radiation detector that has a selective sensitivity to ALPHA PARTICLES.

alpha decay The spontaneous decomposition of the nuclei of atoms that results in the emission of ALPHA PARTICLES.

alpha particle A form of IONIZING RADIATION ejected from the nucleus of a radioactive material. The particle consists of two protons and two neutrons, which is equivalent to a helium ion. Although alpha particles only travel, at most, inches in the air, the ingestion or inhalation and absorption of radioactive materials that emit these particles is a significant health hazard because the ionization occurs in intimate contact with body tissues.

alpha radiation   See ALPHA PARTICLE.

alpine tundra   The grassland area found above the TREE LINE on mountain ranges.

alternate concentration limits (ACLs)   One of the three types of standards that may be applied when a leak is detected at a TREATMENT, STORAGE, OR DISPOSAL facility and groundwater compliance monitoring is required. ACLs are set by the United States Environmental Protection Agency for specific hazardous waste constituents at levels that are designed to prevent a substantial hazard to human health or the environment. Groundwater compliance monitoring can use the following standards: (1) background concentrations, or the levels found in the area naturally; (2) specific values set by federal regulations in Title 40, Part 264.94, of the CODE OF FEDERAL REGULATIONS for eight metals and six pesticides and herbicides; or (3) alternative concentration limits.

alternate hypothesis   In a statistical test, the interpretation of the data that indicates an association between two variables. For example, in a laboratory test one group of test subjects may be treated or exposed in some way, and a second group serving as the control group may receive no experimental treatment. In this case, the primary statement or NULL HYPOTHESIS will be that there is no difference between the two groups relative to some measured outcome. The alternate hypothesis will be that there is a difference, and therefore the treatment or exposure had a significant effect. Compare NULL HYPOTHESIS.

alternative technology   See APPROPRIATE TECHNOLOGY.

alum   Aluminum sulfate, used as a coagulant in wastewater treatment. See COAGULATION.

aluminosis   A disorder of the lung resulting from excessive inhalation of aluminum-containing particulate matter. In some cases, pulmonary FIBROSIS is present.

alveolar macrophage   A scavenger cell that can engulf (phagocytize) inhaled microbes or particles deposited in the ALVEOLAR REGION of the lungs. These migratory cells constitute a major part of the natural resistance of the body to diseases related to bacteria and viruses that enter the body through the lungs.

alveolar region   The terminal parts of the respiratory system containing many millions of small air sacs, or alveoli; the area of gas exchange and, therefore, the entrance point for some air contaminants to the bloodstream. The alveolar region itself can be scarred by certain PARTICULATE MATTER air contaminants. See FIBROSIS.

amalgam   A mixture (alloy) of the metal mercury and one or more other metals.

ambient   Describing a natural outside environment. The term *ambient air,* for example, commonly excludes indoor and workplace air environments.

ambient air   See AMBIENT.

ambient noise   In sound surveys, the general noise level arising from all sources within the confines of a specific area. Compare SOURCE MEASUREMENT.

ambient quality standard (AQS)   See AMBIENT STANDARD.

ambient standard   The allowable amount of material, as a CONCENTRATION, in air, water, or soil. The standard is set to protect against anticipated adverse effects on human health or welfare, wildlife, or the environment, with a MARGIN OF SAFETY in the case of human health. See STANDARDS, ENVIRONMENTAL. Compare EMISSION STANDARD.

amebiasis   See AMEBIC DYSENTERY.

amebic dysentery   A disorder of the gastrointestinal tract caused by a protozoan parasite belonging to the genus *Entamoeba histolytica*. The disorder is commonly found in communities with poor sanitary conditions and is transmitted by contaminated water supplies, contaminated food, flies, and person-to-person contact. Infected individuals experience abdominal cramps, diarrhea, and blood and mucus in the feces. The parasite invades the liver in some cases.

amensalism   A two-species interaction in which one population is adversely affected and the other is not, such as the secretion of chemicals by certain aromatic shrubs, inhibiting the growth of herbaceous plants in their vicinity. Compare MUTUALISM.

American Conference of Governmental Industrial Hygienists (ACGIH)   A professional organization of individuals involved in occupational health activities. Based in Cincinnati, Ohio, the organization publishes recommended occupational exposure limits for chemicals and physical agents in the work place. See THRESHOLD LIMIT VALUE.

American Petroleum Institute (API)   A 200-member trade association of large companies involved in the production, refining, and distribution of petroleum, natural gas, and their associated products. Based in Washington, D.C., the organization promotes the petroleum industry, supports research in all aspects of the oil and gas industry, and collects and publishes worldwide oil industry statistics. Many of its publications concern environmental matters.

American Public Works Association (APWA)   A national organization of individuals and organizations involved in the management of municipal solid waste and in the design and operation of wastewater treatment plants. Founded in 1894, it now has 25,000 members and is based in Chicago, Illinois.

American Society for Testing and Materials (ASTM)   The largest organization setting standards for materials, products, and services. ASTM standards include methods for the sampling and testing of many physical and chemical agents that may be of environmental concern. Based in Philadelphia, Pennsylvania.

American Society of Civil Engineers (ASCE)   A professional organization in New York City that supports the practice of, and research in, environmental engineering. Founded in 1852, it now has 108,000 members.

American Water Works Association (AWWA)   The national organization in Denver, Colorado, of individuals involved in the design and operation of public water supplies. Founded in 1881, it now includes 45,000 members.

Ames test   A test of the ability of a chemical to cause mutations and thereby act as a carcinogen. The process involves the use of a histidine-dependent strain of the bacterial species *Salmonella typhimurium*. The organism used is actually a mutant strain that requires the presence of the amino acid histidine for growth. The chemical being screened is usually mixed with an extract of rat liver, which provides enzymes that convert the chemical being tested from an inactive to an active form. The suspension of bacteria and the suspected mutagen are allowed to stand for an appropriate period of time. The mixture is then spread on the surface of an agar medium that contains all the ingredients needed for growth of the test strain of *Salmonella* except histidine. If a mutation occurs that converts the test strain back to the wild type, which does not need to find histidine in the environment in order to grow, colonies will develop on the surface of the agar medium. If mutations do not occur, nothing will grow on the medium, and the test will be considered negative. Because carcinogens are also mutagens, the test is used as a screening technique to test for carcinogenicity. The test is inexpensive and produces overnight results, but misses perhaps up to 40% of carcinogens. Also, the appearance of mutants does not prove that the chemical tested is a carcinogen. The procedure was developed by Bruce Ames at the University of California, Berkeley.

amictic lake   A lake that does not experience mixing or turnover on a seasonal basis. See DIMICTIC LAKE.

amines   A family of compounds related to ammonia (NH_3). Subclasses of amines are named for the number of groups attached to the nitrogen atom: primary amines, i.e., methylamine [CH_3NH_2]; secondary amines, i.e., dimethylamine [$(CH_3)_2NH$]; and tertiary amines, i.e., trimethylamine [$(CH_3)_3N$]. Hundreds of nitrogen-containing ring compounds are related to the amines, some of which are toxic to humans.

ammonia stripping   A method for removing ammonia from wastewater. The pH of the water is raised (made more basic) to convert the dissolved ammonium ions (NH_4^+) to (dissolved) ammonia gas (NH_3). The water is then allowed to cascade down a tower, and the ammonia is lost to the atmosphere. A form of TERTIARY TREATMENT of wastewater.

ammonification   A process carried out by bacteria in which organic compounds are degraded or mineralized with the release of ammonia (NH_3) into the environment. The simplest example is the bacterial decomposition of urea with the release of ammonia, which occurs in cloth diapers before they are washed.

amorphous   Noncrystalline, as in amorphous silicates, which do not cause FIBROSIS in the respiratory system as crystalline silicates do.

amosite   A type of ASBESTOS, also known as brown asbestos.

amphibole   A metasilicate, a type of ASBESTOS.

amphoteric   Describing a substance capable of exhibiting properties of an acid or a base, depending on the environment. For example, tin oxide dissolved in water displays properties of a base when added to an acidic solution, but properties of an acid when added to a basic solution.

amplitude   For an electromagnetic wave or a sound wave, the distance from the crest (or trough) of the wave to its equilibrium value; the distance from crest to the trough of a wave is twice the amplitude.

AMPLITUDE

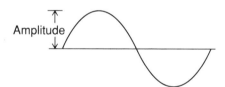

anabolism   The metabolic process involving the conversion of simpler substances to more complex substances or the storage of energy. Also called assimilation, biosynthesis, or constructive metabolism. Compare to CATABOLISM.

anadromous fish   A type of fish that spends its adult life at sea but returns to the upper reaches of a river to spawn (e.g., salmon).

anaerobic   An adjective that describes an organism that is able to live without oxygen. The term is frequently used to refer to microorganisms that possess this ability. Also used to describe environments that are devoid of gaseous or dissolved molecular oxygen.

anaerobic decomposition The degradation of materials by ANAEROBIC microorganisms living beneath the ground or in oxygen-depleted water to form reduced compounds such as METHANE or HYDROGEN SULFIDE. Generally slower than AEROBIC DECOMPOSITION.

anaerobic digester An airtight tank in which ANAEROBIC microorganisms decompose organic material and produce BIOGAS, mainly METHANE. SEWAGE TREATMENT PLANTS often use anaerobic digesters to reduce the volume of SLUDGE produced in PRIMARY and SECONDARY TREATMENT, and they sometimes use the methane as a heating fuel. In rural China and India, digesters are used to convert animal manure to methane.

analog Derivative of a naturally occurring compound that an organism cannot distinguish from the natural product; however, the uptake of this derivative results in the formation of a biological molecule that does not carry out its proper function. A common example is *para*-fluorophenylalanine, an analog to the amino acid phenylalanine. An organism may utilize the analog in place of the amino acid in the synthesis of protein; however, the protein containing the analog is not functional.

analysis In environmental testing of air, water, and soil, the separation of a mixed substance into its components and their quantification by type and amount.

analysis of variance (ANOVA) A statistical test that allows an investigator to determine if mean values observed for several REPLICATES of the same measurement are due to random chance or to real differences. For example, the mean concentration of hexachlorobenzene in the blood of three groups of 30 men each is determined to be 20.0 micrograms per 100 milliliters for group A, 17.8 for group B, and 26.9 for group C.

The analysis of variance will allow the testing of the NULL HYPOTHESIS, which states that there are no differences among the groups, and to make that determination with some specific degree of confidence—say 95 percent sure that there are no differences. The ALTERNATE HYPOTHESIS in this case would be that there is a difference among the groups.

analytical balance A highly sensitive device for determining the mass of a sample of material. The balance can measure mass to the nearest tenth of a milligram or even smaller units.

anechoic space A room in which the surfaces absorb all sound waves hitting them. Also called a free-field space or free-field conditions.

anemometer An instrument that measures air velocity. The cup-type anemometers often used at air-monitoring stations operate by measuring the rotation rate of three or four cups mounted on a vertical shaft. Hot-wire anemometers, used for measurements of indoor air velocities, operate by sensing the rate of heat loss from a heated wire to a moving airstream.

aneroid gauge An instrument that does not contain a liquid. An aneroid barometer, for example, contains a metal surface that reacts to differences in air pressure.

aneuploid In humans, having a chromosome number other than 23 or 46 (the normal complement in humans). An irregular number of chromosomes produces physical and chemical abnormalities, many of which can be severe. Opposite of EUPLOID.

Angstrom (Å) A unit of length measurement, equal to one ten-billionth (1×10^{-10}) of a meter. Wavelengths in the electromagnetic spectrum are often expressed in angstroms or nanometers $(1 \times 10^{-9}$ meter).

anion A negatively charged ION.

annual dose The amount of a substance or radiation to which an individual is exposed over a 12-month period.

annual outage A preset period of time set aside each year for the shutting down of a boiler or other piece of equipment to allow for routine maintenance and repair.

annulus For a well, the space between the pipe and the outer wall (casing) of the borehole, which may be a pipe also (the well casing).

annulus pressure The positive pressure maintained by a fluid introduced between the well piping and the outer wall (casing) of the borehole of an underground INJECTION WELL; an indication of the MECHANICAL INTEGRITY of the well.

anode The negative pole of an ELECTROLYTIC cell or a battery. Electrons flow from the anode to the CATHODE.

anoxic Without oxygen. Water with no DISSOLVED OXYGEN is said to be anoxic.

antagonistic effect The interaction of two or more chemical substances on a biological system that results in a combined effect less than the sum of the effects of each substance alone. Compare SYNERGISTIC EFFECT and ADDITIVE EFFECT.

anthophyllite A type of ASBESTOS.

anthracite The type of coal with the highest heat content or energy released per unit burned. Anthracite is the least common and usually most expensive type of coal and is used mainly in the steel industry. The other coal types, BITUMINOUS, SUBBITUMINOUS, and LIGNITE, are used for electric power production and vastly exceed anthracite in tonnage used annually.

anthracosilicosis A chronic lung disease associated with long-term overexposure to dust containing significant concentrations of FREE SILICA. First observed in anthracite coal miners; another term for BLACK LUNG DISEASE. Compare COAL WORKERS' PNEUMOCONIOSIS.

anthropocentrism A belief that considers man as the most important entity. All questions, including those related to environmental management, are resolved on the basis of value to humans.

anthropogenic Arising from human activities, as opposed to natural origin.

antibody A protein produced by the immune system in response to the presence of an ANTIGEN and that interacts with that specific antigen to eliminate or inactivate it.

anticyclone A clockwise air circulation associated with a high-pressure center. Anticyclones have descending air in their centers, which warms as it subsides and compresses. This can cause a SUBSIDENCE INVERSION, under which air contaminant levels can increase.

antidegradation policy Rules or guidelines, required of each state by federal regulations implementing the Clean Water Act, stating that existing water quality be maintained to protect existing uses of the water, even if the water quality in an area is higher than the minimum necessary, as defined by federal ambient water quality standards. Some (controlled) degradation is allowed for economic development. The comparable policy for air is found in the PREVENTION OF SIGNIFICANT DETERIORATION regulations.

antigen Any substance that causes the formation of an ANTIBODY. See ALLERGEN.

antiknock additive A chemical compound added to gasoline to prevent premature combustion in the engine cylinders, which creates a knocking sound. Tetraethyl lead was used as the primary antiknock additive from the 1920s until the mid-1970s, when United States Environmental Protection Agency regulations began a phaseout of the lead content of leaded gasoline. Oil refineries now use other antiknock additives in gasoline, such as methyl tertiary butyl ether (MTBE). See OCTANE NUMBER.

antiknock agent See ANTIKNOCK ADDITIVE.

antinatalist Describing public policies that attempt to discourage births. Opposite of PRONATALIST.

antioxidant A chemical substance that prevents or slows the OXIDATION of another material. Antioxidants are added to rubber products, such as automobile tires, to protect against the degradation caused by PHOTOCHEMICAL OXIDANT air pollutants. They are also used to prevent rancidity of fats in foods, as, for example, the food additives butylated hydroxyanisole (BHA) and butylated hydroxytoluene (BHT).

aphotic Without light. The aphotic zone of the ocean is the water deeper than about 800 meters, beyond which no light penetrates. Compare PHOTIC.

API gravity An expression of the density of liquid petroleum products; the arbitrary scale, as defined by the AMERICAN PETROLEUM INSTITUTE (API), is calibrated in degrees. Light crude oil is about 40 degrees API; heavy crude, about 20 degrees API.

API separator A device used to separate oil and oily sludges from water. A primary step in the wastewater treatment system of an oil refinery.

Appendix VII A listing of chemical substances, found in Title 40, Part 261, of the CODE OF FEDERAL REGULATIONS, that are considered to be hazardous constituents and therefore cause certain waste

streams to be listed by the United States Environmental Protection Agency as hazardous wastes. For example, spent HALOGENATED solvents used in degreasing are hazardous waste; Appendix VII gives the basis for this definition as the presence, in this waste, of tetrachloroethylene, methylene chloride, trichloroethylene, 1,1,1-trichloroethane, carbon tetrachloride, and chlorinated fluorcarbons.

Appendix VIII A listing of chemical substances which, if present in a waste, may cause the waste to be defined as a hazardous waste. The listing is found in Title 40, Part 261, of the CODE OF FEDERAL REGULATIONS.

Appendix IX list A list of over 200 chemicals for which groundwater from all monitoring wells must be analyzed if measurements of INDICATOR PARAMETERS/CONSTITUENTS at a hazardous waste TREATMENT, STORAGE, OR DISPOSAL facility show that the facility may be adversely affecting groundwater quality. The list is found in Appendix IX to Title 40, Part 264 of the CODE OF FEDERAL REGULATIONS.

applicable or relevant and appropriate requirement (ARAR) Under the SUPERFUND AMENDMENTS AND REAUTHORIZATION ACT, the standard used to define the minimum level of cleanup required at a contaminated site; the standard used to determine "how clean is clean."

appropriate technology The application of scientific knowledge in ways that fit the prevailing economic, social, and cultural conditions and practices in a country. The term also implies low-technology design, simplicity of use, and decentralized provision and maintenance. Usually used in assessing the transfer of agricultural technology from more- to less-developed countries. For example, food crops requiring heavy application of pesticides may be judged inappropriate for an area without sufficient resources to purchase, apply, and manage the pesticides. Not all less-developed countries agree on the virtues of appropriate technology.

aquaculture   The managed production of fish or shellfish in ponds or lagoons.

aquatic   Related to environments that contain liquid water.

aquiclude   A low-permeability underground formation located above and/or below an AQUIFER.

aquifer   An underground formation capable of storing and yielding significant quantities of water; usually composed of sand, gravel, or permeable rock.

aquitard   A low-permeability underground formation.

arable land   Land that can be farmed.

arbitrary and capricious   Shorthand for a standard used in judicial review of informal rule making by a regulatory agency. The actions of the agency should not be "arbitrary, capricious, an abuse of discretion, or otherwise not in accordance with the law," according to the Federal ADMINISTRATIVE PROCEDURE ACT (APA). The "arbitrary and capricious" standard results in judges generally deferring to informal agency actions. The APA holds formal rule making to a stricter standard; these agency actions must be supported by "substantial evidence." Environmental regulatory agencies maintain an ADMINISTRATIVE RECORD to support their decisions in case of subsequent judicial review.

arborescent   Resembling a tree in appearance or growth pattern.

arctic tundra   The grassland BIOME characterized by permafrost (subsurface soil that remains frozen throughout the year). Found in Alaska, Canada, Russia, and other regions near the Arctic Circle.

area method, landfill   A technique for depositing municipal solid waste in a landfill. The waste is spread out in narrow strips 2 to 3 feet deep and compacted. Then more waste is added until a wider compacted layer is 6 to 10 feet thick, after which at least 6 inches of cover material is applied. The width of the area receiving the cover material is usually 10 to 20 feet. The volume from one layer of waste and cover material to the next (adjacent or higher) layer is called a cell. Used in areas where the excavation required by the TRENCH METHOD is not practical.

area mining   The type of mining used to extract mineral resources close to the surface in relatively flat terrain. OVERBURDEN is removed in a series of parallel trenches to allow the extraction of the resource, and the overburden removed from one trench is used to fill in the adjacent trench after the removal of the resource. Also called area strip mining.

area of review   The area around an underground injection well that may be influenced adversely by fluid injection. The extent of the area of concern may be calculated by using the specific gravity and rate of introduction of the injected fluids; the size, storage capability, and hydraulic conductivity of the injection zone; and certain underground formation pressures. In other cases, the extent of the area is determined by a fixed radius around the well, not less than one-quarter mile in length.

area source   In air pollution, a geographic unit that combines many small sources of an air contaminant or contaminants. For example, part of an urban area could be treated as an area source for VOLATILE ORGANIC COMPOUNDS, which are emitted by hundreds of individual households, service stations, dry cleaners, and similar sources.

arithmetic growth   An increase in the number of individuals within a specific area by a constant amount per time period: for example, a population growth of 10 persons per year, year after year. Compare EXPONENTIAL GROWTH, SIGMOID GROWTH.

arithmetic mean The sum of a set of observations divided by the number of observations.

Aroclor Commercial preparation that contains a mixture of highly toxic POLY-CHLORINATED BIPHENYLS (PCBs) as well as traces of a variety of other dangerous compounds, such as DIBENZO-*PARA*-DIOX-INS.

aromatic One of the two main types of hydrocarbon compounds, containing one or more benzene rings or nuclei. Aromatic compounds are usually more difficult to decompose than are straight-chain hydrocarbons, and they present greater hazards to humans and the environment. Compare ALIPHATIC.

arrowhead chart A graphical display of ambient air quality data that plots the frequency of various levels of measured pollutant concentrations against different AVERAGING TIMES, usually for the observations made over a one-year period. The plotted frequencies are the percentages of the total number of measurements, for a given averaging time, that fall below a particular concentration. Longer averaging times result in fewer plotted points, until only one figure represents the annual average. The arrangement of the points on the graph, as they narrow down to the annual average, resembles an arrowhead. The arrowhead chart is used to display the variability and distribution of measured air concentrations at long-term sampling sites.

arsenic (As) A chemical element belonging to the nitrogen family. In pure form, arsenic is a gray crystal. Arsenic reacts readily with a variety of other chemicals, including the oxygen in moist air to produce an oxide. Arsenic is released into the environment as a byproduct from the smelting of copper ore

and through its use as an insecticide. The element undergoes BIOCONCENTRATION and is considered to be carcinogenic.

artesian aquifer See CONFINED AQUIFER.

artesian well A well producing groundwater at underground pressures great enough to force the water out of the ground without pumping; a free-flowing well.

artifactual An adjective that describes some compound, material, or process that is made by humans or influenced by human activities.

artificial recharge The deliberate addition of water to an AQUIFER. Two common recharge methods are injecting water through drilled wells and pumping water over the land surface and allowing it to infiltrate. The technology is employed to restore water to aquifers that have been depleted by excessive withdrawal.

asbestos A family of heterogeneous silicate minerals that have a variety of commercial applications. The mineral is fibrous, chemically inert, and nonconductive to heat and electricity. The fiber has been used extensively in insulating materials. A significant danger to individuals is present when airborne fibers of the mineral are taken into the lungs. The fibers are not subject to BIOTRANSFOR-MATION, nor are they excreted; rather, they remain in the body. Chronic overexposure can lead to ASBESTOSIS. There appears to be a strong association between the development of certain types of lung cancer in individuals who have both an occupational exposure to asbestos and a history of smoking cigarettes. The Asbestos Health Emergency Response Act of 1985 required that all elementary and secondary schools be inspected for the presence of ASBESTOS-CONTAINING MATERIAL and that an asbestos management plan be prepared, to include any necessary repair or removal.

The school guidelines are now generally used for all types of buildings. In 1989, the U.S. Environmental Protection Agency, by the authority of the TOXIC SUBSTANCES CONTROL ACT, issued a 10-year phaseout of the manufacturing, processing and distribution of asbestos in the United States.

asbestosis   A chronic lung disease produced by excessive inhalation of AS-BESTOS fibers. The fiber exposure reduces lung capabilities by causing extensive and progressive scarring of the lung tissue.

asbestos-containing material (ACM) Construction materials containing more than 1 percent asbestos. Federal regulations require the inspection of public and private elementary and secondary schools for such materials.

as low as reasonably achievable (ALARA)   The principle adopted by the INTERNATIONAL COMMISSION ON RA-DIOLOGICAL PROTECTION and the United States NUCLEAR REGULATORY COMMISSION for limiting exposures to IONIZING RADIA-TION to less than the allowable levels, if feasible.

aspect ratio   1. The ratio of the length to the width of an object.   2. In an ELEC-TROSTATIC PRECIPITATOR, the ratio of the COLLECTING SURFACE length to its height; it should be large enough to allow dislodged particles to fall into collecting hoppers before they are carried out of the precipitator.

asphyxiant   A substance causing suffocation by excluding oxygen, preventing oxygen absorption, or preventing oxygen use in the tissues. See SIMPLE AS-PHYXIANT, CHEMICAL ASPHYXIANT.

assimilation   See ANABOLISM.

assimilative capacity   The ability of air, water, or soil to effectively degrade and/or disperse chemical substances. If the rate of introduction of pollutants into

THE LAYERED ATMOSPHERE

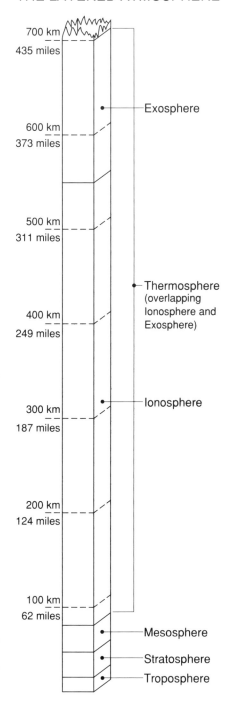

700 km — 435 miles

600 km — 373 miles — Exosphere

500 km — 311 miles

400 km — 249 miles — Thermosphere (overlapping Ionosphere and Exosphere)

300 km — 187 miles — Ionosphere

200 km — 124 miles

100 km — 62 miles

Mesosphere

Stratosphere

Troposphere

the environment exceeds the rate at which the environment can assimilate the materials, an adverse effect may occur to human health or welfare, wildlife, or other living components of an ecosystem.

association In EPIDEMIOLOGY, a relationship between two variables in which as one changes there is also a change in the other. For example, increasing age is associated with increased human cancer incidence. If both variables increase, it is called a positive association. Association does not necessarily imply a causal link between the variables.

asthenosphere The zone inside the earth beneath the LITHOSPHERE; the source of IGNEOUS ROCK (MAGMA).

atmosphere 1. The gaseous layer covering the earth. The regions of the atmosphere are the TROPOSPHERE, STRATOSPHERE, MESOSPHERE, CHEMO-SPHERE, thermosphere, IONOSPHERE, and exosphere. The atmosphere is one of the four components, together with the LITHOSPHERE, HYDROSPHERE, and BIOS-PHERE, that constitute the earth's ecosystem. 2. An expression of ATMOSPHERIC PRESSURE. One atmosphere is equal to about 14.7 pounds per square inch.

atmospheric pressure The force exerted by the atmosphere per unit area on the surface of the earth. Sea level atmospheric pressure is equal to about 14.7 pounds per square inch, or 101.33×10^3 newtons per square meter (pascals), or, in terms of the height of mercury in a barometer, 760 millimeters of mercury.

atmospheric stability An expression of the resistance of the atmosphere to vertical air motion, or dispersion. Stable air resists movement of air upward; unstable conditions result in good vertical dispersion. Atmospheric stability is important to the dispersion and dilution of air contaminants.

atom A unit of matter consisting of a nucleus and surrounding electrons; the smallest unit of an ELEMENT that retains the chemical characteristics of the element. See MOLECULE.

atomic absorption spectrophotometer (AA) An analytical instrument typically used to determine the concentration of various metals in an environmental sample. The instrument vaporizes the sample and measures the absorbance of the vapor at a wavelength of light that is specific for each element. The sample concentration is obtained by comparing the measured absorbance to an absorbance CALIBRATION CURVE prepared from samples of known concentration.

Atomic Energy Commission The predecessor agency to the NUCLEAR REGULATORY COMMISSION.

atomic mass unit (amu) A unit of mass equal to 1.6604×10^{-24} gram. One atomic mass unit is defined as one twelfth the mass of carbon-12, a carbon ATOM consisting of six PROTONS and six NEUTRONS.

atomic number The number of PROTONS in the nucleus of an ATOM. Each ELEMENT has a unique atomic number. For example, a material with six protons in its nucleus is the element carbon, atomic number 6.

atomic power The generation of electricity using the heat energy produced in nuclear reactors.

atomic waste Radioactive materials produced as by-products of activities such as the mining and processing of radioactive minerals, fabrication of nuclear weapons, and the operation of nuclear reactors.

atomic weight For each ELEMENT, the weighted sum of the masses of the PROTONS and NEUTRONS composing the ISOTOPES of that element. Approximately equal to the sum of the number of protons and neutrons found in the most abundant isotope.

atomizer An instrument used to produce a fine spray or mist from a liquid.

attainment area A geographical area designated by the United States Environmental Protection Agency that meets the NATIONAL AMBIENT AIR QUALITY STANDARD for a particular pollutant. The same area can be in attainment for one or more of the six national ambient air quality standards and be designated as nonattainment for the others. See NONATTAINMENT AREA.

attenuation Weakening; commonly applied to the reduction in noise level or level of IONIZING RADIATION with distance from a source or after passing through barriers or shields.

attractant A food, light, or sound that lures an insect to a poison, or a chemical sex attractant (PHEROMONE) that will disrupt insect propagation by sending false location signals to potential mates.

attributable risk The incremental increase in the risk that an individual will experience an adverse health effect that can be assigned to a particular source or activity.

audible range The sound FREQUENCY range accessible to human hearing, ideally about 20 to 20,000 HERTZ for young adults. The ability to hear higher-frequency sounds deteriorates normally with age or can be lessened by excessive noise exposure. See PRESBYCUSIS, NOISE-INDUCED HEARING LOSS.

audiogram A display of the HEARING THRESHOLD LEVEL, in DECIBELS, for an individual exposed to sound over a range of different frequencies; used to monitor hearing loss.

audiometer An instrument used for testing human hearing. With the person being tested isolated from other sounds,

the device generates a pure tone at selected frequencies, and the SOUND PRESSURE LEVEL is increased until the subject can just hear that sound, at a level called AUDIOMETRIC ZERO, or the hearing threshold, for that frequency. The test results are called an AUDIOGRAM.

audiometric zero The threshold of normal human hearing, usually defined as a SOUND PRESSURE of 2×10^{-5} newtons per square meter, or zero DECIBELS.

audit See ENVIRONMENTAL AUDIT.

Audubon Society See NATIONAL AUDUBON SOCIETY.

autecology The study of a single species within an ecosystem. Compare SYNECOLOGY.

autochthonous Describing an organism that is native, or indigenous, to a locale. The opposite of ALLOCHTHONOUS.

autoclave A sterilizing device that uses high-pressure steam to kill microorganisms.

autogenic Endogenous, arising from within an organism or a community. The opposite of ALLOGENIC.

autogenic succession The orderly and predictable changes of species composition within a community that are caused mainly by actions of the species present in the area. Compare ALLOGENIC SUCCESSION.

autoignition temperature The lowest temperature at which a gas or vapor that is mixed with air will ignite without the presence of a flame or spark.

autoradiography A technique employed to locate RADIOACTIVE substances in small, usually microscopic, samples. For example, the part of a cell that incorporates a particular substance can be determined by first incubating the cell in the presence of a radioactive form of the substance in question and then fixing and sectioning the cell. The sections are covered with a photographic emulsion and allowed to stand. The presence of the radioactive material is determined by developing the emulsion and examining the specimen for dark grains of silver.

autotroph An organism that has the capacity to satisfy its requirements for carbon-containing nutrients by the fixation, reduction, and incorporation of carbon dioxide from the environment. Green plants, which derive their energy from sunlight through the process of PHOTOSYNTHESIS, represent a major group of autotrophic organisms. Compare HETEROTROPH.

auxin A plant growth hormone. Synthetic auxins are used as weed killers.

auxotrophic A term employed in MICROBIOLOGY to describe certain mutant strains of bacteria that require for growth the presence of some organic compound or metabolic intermediate that is not required by the parent strain. The requirement generally reflects the loss by the mutant of its ability to produce one or more of the enzymes needed in the metabolic synthesis of the compound or INTERMEDIATE in question.

available water storage capacity The amount of water that soil can hold and that plants can use. A measure of water available to plants.

averaging time The length of time used for the time-weighted average of a measurement. For example, if the concentration of an air pollutant during an hour is 20 parts per million for 30 minutes and 10 parts per million for 30 minutes, the reported value for a one-hour averaging time would be 15 parts per million.

Avogadro's number (N_A) The number of molecules in one mole of a substance, equal to 6.02×10^{23}. Based on the assumption that equal volumes of different gases at the same pressure and temperature contain equal numbers of molecules. Named for the Italian physicist Amedeo Avogadro.

axenic Describing a culture of a microbe that contains only one kind of organism (i.e., a pure culture). The adjective is frequently used to describe cultures of algae or protozoa that consist of only one type of algae but that may also contain bacteria.

axial flow Fluid flow in the same direction as the AXIS OF SYMMETRY of the duct, vessel, or tank.

axial flow fan A mechanical device for moving air or other gases. The device consists of propeller-like blades rotating in a plane perpendicular to the flow of the gas stream. A fan inside a duct.

axis of symmetry An imaginary line passing through a figure or body around which the figure or body is symmetrically arranged.

B

back corona The buildup of a negative charge on the plates of an ELECTROSTATIC PRECIPITATOR, which reduces the collection efficiency of the plate for negatively charged particles. The condition develops if the collected dust has a high resistivity (resistance to the conduction of electricity). In this case, the poorly conducting, negatively charged particles caught on the positively charged plates do not drain their charge, which normally allows the efficient collection of additional dust. The retention of electrostatic attraction between the dust and the plate also makes the periodic removal of the dust more difficult.

back end system   The stage in a solid waste resource recovery operation in which recyclable materials are extracted from incinerator ash.

backfilling   In the reclamation of land on which improper contour STRIP MINING has occurred, the placement of spoils (waste soil and rock) in the notches cut in the hills and the restoration of the original slope. This process reduces soil erosion and allows for the reestablishment of vegetation.

background concentration   For a chemical substance, the level, or CONCENTRATION found in the air, water, or soil of a geographical area due to natural processes alone; the starting point for the determination of enhanced chemical concentrations caused by human activities.

background radiation   The IONIZING RADIATION level for a given geographical area produced by the natural radioactive materials in the soil and the cosmic radiation from outer space.

backwash   The residue that is removed from RAPID SAND FILTERS when clean water is employed to flush away accumulated particles.

backwashing   In a wastewater or drinking water treatment facility, the movement of clean water in a direction opposite (upward) to the normal flow of raw water through RAPID SAND FILTERS to clean them.

bacteria   Ubiquitous, one-celled, PROCARYOTIC organisms that, together with the fungi, serve as decomposers in aquatic and terrestrial ecosystems. A small number of them cause plant, animal and human diseases. See PATHOGEN.

bacterial plate count   A system employed to quantify the number of bacteria in a sample of solid or liquid material. The sample is mixed and/or diluted with a suitable sterile diluent and a measured, small portion of the mixture is placed on the surface of a solid substrate suitable for the growth of bacteria (a MEDIUM). Following incubation, the number of bacterial colonies that have developed on the medium is counted, and the necessary mathematical calculations are made on the assumption that each colony represents one bacterium that was present in the original sample.

bacteriostatic   A substance that is inhibitory to bacterial growth but not necessarily lethal. If a bacteriostatic material dissipates or lowers in concentration, bacterial growth may resume.

baffle   A wall, barrier, or screen placed in the path of a gas, liquid, light, or sound to disperse it or to regulate its passage.

bagassosis   A lung disease linked to the inhalation of dust from sugar cane residues.

baghouse   An air pollution control device that removes particulate matter from an exhaust gas by forcing air through large filter bags; the device is similar to a vacuum cleaner bag. Baghouses vary according to filter material and the method for dislodging the collected dust. See BAGHOUSE, PULSE JET; BAGHOUSE, REVERSE AIR; BAGHOUSE, SHAKER.

baghouse, pulse jet   A BAGHOUSE in which the air to be filtered flows through the bags from the outside to the inside. The bags are supported by internal cages and are cleaned by knocking the dust off with short pulses of high-pressure air.

baghouse, reverse air   A compartmentalized BAGHOUSE using reverse flows of clean air through the bags to dislodge the collected dust, allowing it to fall into a collection hopper. Each compartment is taken off-line before its bags are cleaned.

baghouse, shaker   Similar to a REVERSE-AIR BAGHOUSE but using a mechanical shaking device to dislodge dust collected on the bags.

bailer An instrument used to extract a water sample from a groundwater well or PIEZOMETER.

baler A mechanical device used to process solid waste into a large strapped bundle for convenient handling and efficient use of landfill space. The waste is mechanically shredded into smaller fragments, compacted, and tied to form a bale.

ball mill A device consisting of a rotating drum containing iron or steel balls and used to grind solid waste before compaction and disposal or as part of a recycling operation.

ballast Heavy material, often seawater, placed in the hold of a ship to gain stability. Routine ballast discharges from oil tankers account for a large portion of the total oil (including spills) introduced to the world's oceans each year.

banking The placement of EMISSION REDUCTION CREDITS (not a monetary sum) in a United States Environmental Protection Agency approved account, called a bank.

bar In meteorology, a unit of atmospheric pressure equal to 0.9869 atmosphere or 1×10^5 pascals. Sea-level pressure (one atmosphere) is equal to 1.013 bars or 1013 millibars.

bar racks The closely spaced rods (screen) that remove large solids from the wastewater entering a sewage treatment plant.

baritosis The presence in the lungs of small-diameter, chemically inert barium sulfate particles. Baritosis causes no impairment of lung function.

barometric pressure The force exerted on a surface by the acceleration of gravity acting on a column of air above that surface. It can be measured by a liquid barometer, which consists of a vertical tube sealed at the top and a mercury reservoir at the bottom, open to the atmosphere. Atmospheric pressure forces the mercury upward in the tube until the weight of the mercury column equals the weight of the air column above the mercury at the open end of the tube. The barometric pressure measured in this fashion is expressed as the height of the mercury, in millimeters or inches. See ATMOSPHERIC PRESSURE.

barrel A liquid volume, equal to 42 U.S. gallons, frequently used to describe the production or consumption of petroleum or petroleum products. Petroleum also is measured in tons. One ton of crude oil equals about seven barrels, whereas one ton of (less dense) petroleum products will be greater than seven barrels.

basal diet In a feeding study to determine the toxicity of a chemical or chemical mixture, food that does not contain the test chemical(s); in other words, an unaltered feed available from a commercial source.

basal metabolism The energy used by an organism while at rest.

basalt Rock of volcanic origin. Because of its strength and heat-dispersion capability, a desirable type of rock for long-term storage of HIGH-LEVEL RADIOACTIVE WASTE.

basalt aquifers Aquifers found in basalt rock in areas of past volcanic activity, such as the Pacific Northwest region of the United States and in Hawaii.

base Chemicals that release hydroxide ions (OH⁻) in solution. Such solutions have a soapy feel, neutralize acids, and conduct electricity.

base dissociation constant (K_b) See DISSOCIATION CONSTANT.

base flow The flow in a stream arising from groundwater SEEPS alone, excluding surface runoff into the stream.

base load facility   A generating facility that operates at a constant output to supply the minimum electric power demand of a system. It is supplemented by a CYCLING LOAD FACILITY as power demand increases.

baseline concentration   Under the PREVENTION OF SIGNIFICANT DETERIORATION program, the existing level of air quality in an area when the first permit application for a major source of certain air pollutants is submitted.

baseline data   Information accumulated concerning the biological, chemical, and physical properties of an ecosystem prior to the initiation of some activity that may result in the pollution of that ecosystem.

baseline emissions   For a particular pollutant, the level of air emissions below which the reduction in emission from a facility will be counted as EMISSION REDUCTION CREDITS. For example, if emissions of pollutant A, with a baseline of 1000 pounds per year, are reduced from 1100 pounds per year to 900 pounds per year, emission reduction credits of 100 pounds per year are earned.

basic   Describing a solution, sediment, or other material that has a pH greater than 7. See BASE.

batch method   Any industrial or laboratory procedure that involves adding all raw materials or reagents and allowing the process to be completed before the products are removed. For example, a batch method for composting involves adding all leaves, organic materials, or other solid waste and allowing the mixture to sit until the humic substances are removed from the holding vessel. Compare with CONTINUOUS ANALYZER and CONTINUOUS-FEED REACTOR.

bathing water   Water in swimming pools or natural fresh or marine waters used for swimming.

bathtub effect   The accumulation of LEACHATE in a landfill containing a good LINER, but not equipped with a leachate collection and removal system.

bathyal zone   The ocean stratum beneath the EUPHOTIC ZONE and above the ABYSSAL ZONE.

Bay of Fundy   A bay on the east coast of Canada located between New Brunswick and Nova Scotia. The bay experiences some of the highest natural tides in the world and is considered a likely location for the construction of facilities to utilize tidal energy for the generation of electrical energy.

Beaufort scale   A scale equating wind velocities and their effects. A Beaufort number, from 0 to 17, is assigned to each of 13 categories, the largest being a hurricane. (Numbers 12–17 describe six subcategories of hurricane winds.) For example, a wind of Beaufort number 2, velocity 4–7 miles per hour, can be felt on the face; Beaufort number 4, 15–18 miles per hour, causes small branches to move and raises dust. The scale was first devised in 1806 by Sir Francis Beaufort, who described the wind effects in terms of those on a fully rigged man-of-war. The inland and terrestrial effects are a later addition.

becquerel (Bq)   The SI unit of radioactivity, signifying one nuclear disintegration per second.

Becquerel rays   A general term for rays emitted by radioactive substances.

Beer-Lambert law   Law allowing the quantitative determination of the concentration of chemicals in solution by measuring the light absorbed upon passage through the solution. It is expressed as $A = \epsilon\, bc$, where A is the absorbance, ϵ is the MOLAR ABSORPTIVITY, b is the length of the light path in centimeters, and c is the chemical concentration in moles per liter. The absorbance A is expressed as $A = \ln(P_0/P)$, where ln is the natural logarithm, P is the radiant power of the light transmitted through the solution, and P_0 is the power of the incident light. See SPECTROPHOTOMETER.

behavioral standards Environmental, health, or safety standards that forbid conduct that is likely to have an adverse effect on human health or the environment or that significantly increase the risk of an accident. For example, a ban on outdoor burning of trash or leaves in an urban area, or the prohibition of smoking around fuel-loading facilities.

bel A dimensionless unit to measure sound. It is defined in two ways: (1) as the common logarithm of a measured power divided by a reference power or (2) as the common logarithm of the ratio of a measured SOUND PRESSURE to a reference sound pressure. One bel equals 10 DECIBELS, the more commonly used unit. See also SOUND POWER LEVEL, SOUND PRESSURE LEVEL.

benign See BENIGN NEOPLASM.

benign neoplasm A tumor that does not differ greatly from the tissue from which it arose and that is not likely to invade other sites (metastasize).

benthic Referring to bottom-dwelling aquatic organisms. Compare PELAGIC, LITTORAL.

benthos The organisms living on the bottom of an ocean, river, lake, pond, or other body of water. Compare NEKTON, PLANKTON.

benzene The simplest aromatic hydrocarbon. Its chemical formula is C_6H_6. The compound is illustrative of the structural features of all aromatic compounds. Benzene has a long history of use in the chemical industry as a solvent and as a starting compound for the synthesis of a variety of other materials, and is now also used extensively in the rubber, paint, and plastic industries. The liquid is volatile, and emissions are regulated as a toxic air pollutant. The major toxic effects of chronic exposure through inhalation, mostly an occupational hazard, concern the functioning of the bone marrow. A variety of blood disorders, such as aplastic anemia and leukemia, have been linked to excessive doses.

benzene hexachloride (BHC) Also known as hexachlorocyclohexane (HCH) and Lindane. An ISOMER of BHC is utilized as the insecticide lindane. Various forms of this compound have been implicated in exposures resulting in neurological symptoms associated with depression of the central nervous system. BHC has also been shown to have carcinogenic properties in mice and to be bioaccumulated (increased in concentration) in the fatty tissue of humans.

benzine A mixture of ALIPHATIC hydrocarbons, such as gasoline or certain cleaning solvents. Not related to BENZENE.

benzo(a)pyrene (BaP) A carcinogenic POLYCYCLIC HYDROCARBON found in soot emitted by coal-burning facilities, in automobile particulate emissions, and in broiled or smoked meat and fish.

benzopyrene See BENZO(A)PYRENE.

BENZENE

OR

OR

Berkson's fallacy In epidemiology, the incorrect application of findings from studies using hospital or clinic populations as representative of the general population. The error arises from the possible self-selection of the hospital and clinic patients. Characteristics of those seeking medical treatment may differ significantly from the general population, and the patients at a particular hospital or clinic can be unrepresentative.

Berl saddles A type of packing used in a gas cleaning device called a PACKED TOWER. The packing geometry is designed to maximize gas-liquid contact for pollutant removal efficiency. Other packing types include Pall rings, Lessing rings, Raschig rings, and Intalox saddles. See PACKING.

best available control technology (BACT) Technology-based emission standards required by the United States Environmental Protection Agency for all new or modified major sources of air pollutants in geographical areas meeting the NATIONAL AMBIENT AIR QUALITY STANDARDS (also called PREVENTION OF SIGNIFICANT DETERIORATION areas). BACT is applied on a pollutant-specific basis; for example, a new source of sulfur dioxide emissions must apply BACT for sulfur dioxide if the area is clean with respect to sulfur dioxide levels. But if the facility is also a major source of carbon monoxide and the area does not meet the ambient standard for carbon monoxide, then a more stringent emission standard, LOWEST ACHIEVABLE EMISSION RATE (LAER), would be applied to the carbon monoxide emissions.

best available demonstrated technology (BADT) The level of EFFLUENT LIMITATION technology required by the 1972 CLEAN WATER ACT to be used in setting NEW SOURCE PERFORMANCE STANDARDS for new industrial direct dischargers of water pollutants.

best available retrofit technology (BART) The level of air pollution control technology required on certain major existing sources (e.g., electric power plants) that may cause or contribute to visibility reduction in areas of the United States designated as Class I under the PREVENTION OF SIGNIFICANT DETERIORATION provisions of the CLEAN AIR ACT. Class I areas include major national parks and wilderness areas, mostly in the American West.

best available technology economically achievable (BAT) In water pollution control, the technological method required for all existing sources of toxic or NONCONVENTIONAL POLLUTANTS; it is set by industrial category or subcategory. The CLEAN WATER ACT specifies that the United States Environmental Protection Agency, in defining BAT, consider such factors as age of the facilities and equipment involved, process employed, engineering aspects of the control technique, process changes, cost of the reductions, and environmental impacts other than water quality, including energy requirements.

best conventional control technology (BCT) The level of water pollution control technology required of existing dischargers for the treatment of CONVENTIONAL POLLUTANTS by the 1977 CLEAN WATER ACT.

best demonstrated available technology (BDAT) The treatment standard for hazardous wastes that, if met, allows the wastes to avoid the LAND DISPOSAL BAN.

best engineering judgment (BEJ) See BEST PROFESSIONAL JUDGMENT.

best management practices (BMP) For facilities that manufacture, use, store, or discharge toxic or hazardous pollutants as defined by the 1977 CLEAN WATER ACT, a required program to control the potential spill or release of those materials to surface waters, such as dikes to contain tank overflows or heavy rainfall runoff.

best practicable control technology (BPT)   The level of EFFLUENT LIMITATION technology required by the 1972 CLEAN WATER ACT for existing industrial plants as of July 1, 1977; defined as the average of the best technology in each industrial category.

best professional judgment (BPJ)   Used by environmental regulators to choose TECHNOLOGY-BASED treatment requirements for a water pollutant discharge permit when no national industry standards (such as BEST CONVENTIONAL CONTROL TECHNOLOGY) are available. See NATIONAL POLLUTANT DISCHARGE ELIMINATION SYSTEM.

beta counters   A device used to measure IONIZING RADIATION. The counter is configured in such a way that the ionizing events caused by a BETA PARTICLE are selectively measured.

beta particle   Particles equivalent in mass to an electron emitted by certain substances undergoing RADIOACTIVE DECAY; a form of IONIZING RADIATION. Beta particles can carry a positive charge or a negative charge. Some emissions result when a neutron is converted to a proton in the atomic nucleus. This increase in the number of protons in the nucleus changes the ATOMIC NUMBER of the substance, and it therefore becomes a different element. For example, beta emission by a radioactive ISOTOPE of phosphorus (atomic number 15) converts the atom to sulfur (atomic number 16).

beta radiation   The emission of a beta particle by an unstable atomic nucleus.

bias   An error in data gathering or analysis caused by faulty program design, mistakes on the part of personnel, or limitations imposed by available instrumentation.

bicarbonate   A compound containing the HCO_3^- group, for example, sodium bicarbonate ($NaHCO_3$), which ionizes in solution to produce HCO_3^-. See CARBONATE BUFFER SYSTEM.

bimodal distribution   In statistics, a collection of observations with a large number of values found around each of two points. The distribution of the diameters of airborne particulate matter in urban areas, for example, is bimodal, with observations clumped around 0.2 and 10 microns.

binder   1. The chemical material that holds paint pigments together and attaches the paint film to a surface. It is nonvolatile—that is, not lost to the air as the paint dries.   2. The chemical material combined with the active ingredient in a pill for easier administration. Common binders in pills are starch, sugars, and gums.

binding energy   In an atomic nucleus, the energy holding the neutrons and protons (collectively, NUCLEONS) together. The atoms of *heavier* elements, those with larger numbers of protons and neutrons, have lower amounts of binding energy per nucleon because of the repelling force of the positively charged protons. Elements of ATOMIC NUMBER 83 and above are low enough in binding energy per nucleon to be UNSTABLE; in other words, they undergo RADIOACTIVE DECAY. In addition, FISSION, or splitting, of a heavy nucleus, such as uranium, creates lighter, stabler nuclei and releases the binding energy. Thus, the sum of the masses of the lighter fission nuclei is less than the mass of the heavier nucleus that splits; the difference in mass is represented by the binding energy released.

binding site   In toxicology, the location in the tissues where a foreign chemical will attach and possibly accumulate, such as the liver, kidney, and plasma proteins. Also refers to the location on a protein molecule where the actual chemical bonding occurs between a chemical and the protein.

bioaccumulation The increase in concentration of a chemical in organisms that reside in environments contaminated with low concentrations of various organic compounds. For example, fish living in aquatic environments contaminated with compounds such as the CHLORINATED HYDROCARBONS will absorb those compounds through the gills. Chemicals likely to be bioaccumulated are not readily decomposed in either the environment or in an organism and are likely to be stored in the fatty tissue. Bioaccumulation can also be used to describe the progressive increase in the amount of a chemical in an organism resulting from the uptake, or absorption, of the substance exceeding its breakdown or excretion rates. In the complementary process of BIOLOGICAL MAGNIFICATION, an increase in chemical concentration in organisms is a result of the passage of the chemical through the food chain, not directly absorbed from the air, water, or soil, as in bioaccumulation. Also called bioconcentration, biological amplification, and biological concentration.

bioactivation A process that takes place in animal or human tissues, during which a foreign substance is metabolized and transformed into a different INTERMEDIATE. In turn, the intermediate produced by the metabolic machinery of the body is more harmful than the original compound. Examples of compounds bioactiviated in the body are benzene, chloroform, and parathion.

bioassay The use of living organisms to assess the adverse effects of an environmental sample. The organisms are placed in the sample (commonly, aquatic species in a water sample), and observed changes in the activity of the test organism are used as indicators of sample toxicity.

bioavailability The magnitude and rate of availability of a DOSE of a chemical substance to body tissues. Some factors affecting bioavailablity are the material's absorption, body distribution, metabolism, and excretion rates.

biochemical oxygen demand (BOD) The use of (or demand for) oxygen dissolved in water during the decomposition or metabolism of biodegradable organic compounds by microbes. The greater the amount of waste material (organic carbon) added to water, the greater will be the requirement for dissolved oxygen needed to convert the organic material to the mineral state (CO_2). Since dissolved oxygen is required by the organisms native to the body of water, BOD is a measure of the ability of a waste to cause damage in a receiving stream or lake. The customary units of the BOD measurement, which are not frequently expressed, are milligrams of oxygen utilized by one liter of wastewater incubated under the proper conditions for five days. The BOD of untreated municipal sewage is usually about 250, that of unpolluted water about 5. See CARBONACEOUS BIOCHEMICAL OXYGEN DEMAND, NITROGENOUS BIOCHEMICAL OXYGEN DEMAND.

biocide A chemical substance that kills living organisms. The designation is usually used to include materials that can kill desirable as well as undesirable organisms. PESTICIDES are designed to kill undesirable organisms.

biocoenosis A community of animal and plant life.

bioconcentration The increase in concentration of a chemical in an organism resulting from the uptake, or absorption, of the substance exceeding the rate of metabolism and excretion. See BIOACCUMULATION.

bioconcentration factor (BCF) Describes the accumulation of chemicals in aquatic organisms that live in contaminated environments. The factor is calculated by dividing the micrograms of the chemical per gram of aquatic organism by the micrograms of the same chemical per gram of water constituting the habitat.

bioconversion The extraction of energy from a BIOMASS directly, as in the combustion of wood or organic solid waste, or indirectly, as in the production of BIOFUEL.

biodegradable Describes a substance that can be metabolized into simpler components. The degradation or decomposition is usually performed by bacteria or microbes, referred to as DECOMPOSERS.

biodegradation The metabolic breakdown of materials into simpler components by living organisms.

biodisc A large rotating cylinder containing surface features that allow for the growth of attached microorganisms. The cylinder revolves and contacts the wastewater along one side of the cylinder while the other side is exposed to the air. This promotes mixing of the wastewater and allows maximum oxygenation of the water, which stimulates decomposition of organic material dissolved or suspended in the water. The apparatus is used in some wastewater treatment plants as a type of SECONDARY TREATMENT.

biofuel Gaseous or liquid materials with a useful energy content produced from plant material or BIOMASS, for example methane gas (BIOGAS), captured during the ANAEROBIC bacterial decomposition of organic material or ethyl alcohol (ethanol) produced from organic material by yeast. Biofuels are seen as possible substitutes for natural gas and gasoline.

biogas Methane gas produced during the ANAEROBIC decomposition of the remains of plants or animal wastes by bacteria.

biogenic Describing changes in the environment resulting from the activities of living organisms.

biogeochemical cycling The flow of chemical substances to and from the major environmental reservoirs: ATMOSPHERE, HYDROSPHERE, LITHOSPHERE, and bodies of living organisms. As the materials move in the cycle, they often change chemical form, usually existing in a characteristic form in each reservoir. For example, carbon exists mainly as CARBON DIOXIDE in the atmosphere; mainly as CARBONIC ACID, BICARBONATE, or the CARBONATE ion when dissolved in water; and as more complex ORGANIC compounds in animals and plants. Essential materials that are recycled in the environment are carbon, nitrogen, phosphorus, and oxygen, among others. See CARBON CYCLE, NITROGEN CYCLE, HYDROLOGIC CYCLE.

biogeochemistry The study of the transformation and movement of chemical materials to and from the LITHOSPHERE, the ATMOSPHERE, the HYDROSPHERE, and the bodies of living organisms.

biogeography The study of the geographic distribution of plants and animals.

biohazard The presence of microorganisms, such as bacteria, viruses, or fungi, that are capable of causing infectious disease in humans.

bioindicator A living organism that denotes the presence of a specific environmental condition. For example, the presence of coliform bacteria identifies water that is contaminated with human fecal material.

biological additives Cultures of bacteria, enzymes, or nutrients that are introduced into an oil discharge or other wastes to promote decomposition.

biological amplification See BIOACCUMULATION.

biological concentration See BIOACCUMULATION.

biological control Elimination or reduction of undesirable species, usually insects, by deliberate human introduction of other living organisms that are PREDATORS, PARASITES, or PATHOGENS for the pest. The practice is usually considered an ecologically sound alternative to chemical PESTICIDES.

biological effects of ionizing radiation report (BEIR report) One of several reports issued by the Committee on the Biological Effects of Ionizing Radiation, established by the National Research Council of the United States to review the standards for IONIZING RADIATION exposure.

biological exposure index (BEI) Guidelines published by the AMERICAN CONFERENCE OF GOVERNMENTAL INDUSTRIAL HYGIENISTS (ACGIH) for assessing the hazard posed to healthy workers by chemical substances present in blood, urine, or breath. The guidelines are tied to the amount of the substance expected to be present in the body if inhalation exposure is 8 hours per day, five days per week, at the THRESHOLD LIMIT VALUE/TIME-WEIGHTED AVERAGE concentration. See BIOLOGICAL MONITORING.

biological half-life The time required for one half of an absorbed chemical substance to be biochemically degraded or excreted from the body of an animal.

biological magnification An increase in the concentration of heavy metals (such as mercury) or organic contaminants (such as chlorinated hydrocarbons) in organisms as a result of their consumption within the food chain of a particular habitat. Fish that consume large amounts of PLANKTON contaminated with mercury compounds retain the metal within their bodies as the greater portion of the food resource is metabolized; consequently, the amount of mercury in the fish gradually increases. Then, when a sea bird consumes a diet of contaminated fish, the metal in each fish is transferred to the tissue of the bird, and so on up the food chain. Chemicals likely to undergo biological magnification are not readily decomposed in the environment or metabolized by an organism and are usually stored in the fatty tissue of an organism. Compare to BIOACCUMULATION, which is the concentration of a chemical in an organism resulting from direct uptake from the environment (air, water, or soil), as opposed to through the food chain. Also called biomagnification.

biological methylation The addition of a methyl group ($-CH_3$) to elemental or inorganic mercury (i.e., a mercury atom or ion) by ANAEROBIC bacteria, usually occurring in sediments within a water body. The methyl mercury produced is more toxic to humans than other forms of mercury and much more subject to BIOLOGICAL MAGNIFICATION.

biological monitoring In occupational health, the sampling and analysis of the blood, urine, or breath of workers to detect actual chemical exposure and absorption. The level of the chemical substance or its METABOLITES is compared with a BIOLOGICAL EXPOSURE INDEX, if one is available for the material, to determine their acceptability. Biological monitoring overcomes the problems of whether workplace air sampling is representative of actual human exposure and can warn of chemical overexposure before adverse health effects arise.

biological variability The observable differences among the individuals that constitute a given SPECIES. All humans belong to the same biological species, yet one can easily observe differences among individuals. Also, the differences that are observed when members of the same species are exposed to toxic or dangerous chemicals.

biological wastewater treatment The use of bacteria to degrade organic materials in wastewater. See SECONDARY TREATMENT.

biomagnification See BIOLOGICAL MAGNIFICATION.

biomanipulation The use of biological, chemical, or physical approaches to control the accumulation of plant nutrients in a lake. The process can include the introduction of new animal species to consume the algae overgrowth or the removal of phosphate-containing sediments, which promote algae growth.

biomass Any biological material. In reference to alternative energy sources, mainly plants or parts of plants: harvested trees, leaves, limbs, and the like. In ecological studies, the DRY MASS of living organisms in a specified area, often expressed as grams of biomass per square meter.

biomes Major ECOSYSTEM types of the earth, defined by the climate and the (climate-related) dominant vegetation. Some examples are tropical rain forest, coniferous forest, deciduous forest, boreal forest, chaparral, grassland, desert, savanna, and tundra.

biometrics See BIOSTATISTICS.

biosphere The entire planetary ECO-SYSTEM, including all living organisms and the parts of the earth in which they live or that support them: the ATMOSPHERE, the HYDROSPHERE, and the LITHOSPHERE. Also called the ECOSPHERE. The term is also used to refer to only the living organisms on earth and not to their physical and chemical environments.

biostatistics The methods for the mathematical analysis of data gathered relative to biological organisms. Also called biometrics.

biosynthesis The use of chemical energy by plants or animals to make carbohydrates, fats, or proteins. Synonymous with ANABOLISM.

biota The types of animal and plant life found in an area.

biotic Of or relating to life, as in biotic components of an ECOSYSTEM. Compare ABIOTIC. See COMMUNITY.

biotic factors The influence or impact that other living organisms have on an individual.

biotic potential The upper limit of the ability of a species to increase in number; the maximum reproductive rate, assuming no limits on food supply or environmental conditions. A theoretical number that is not observed in nature for extended periods. See ENVIRONMENTAL RESISTANCE; LIMITING FACTOR.

biotransformation The metabolic conversion of an absorbed chemical substance, usually to a less toxic and more excretable form.

birth rate The number of births per 1000 persons in a population in a given year as calculated by dividing the annual number of births by the mid-year population and then multiplying by 1000.

bitumen The heavy oil found in TAR SANDS, usually having a high sulfur content.

bituminous coal A soft coal with a relatively high energy content; the most abundant and widely used type of coal in the United States. The chief fuel for electric power generation in the United States. See also ANTHRACITE, LIGNITE, and SUBBITUMINOUS.

blackbody A theoretical body with a surface that reflects no light and, thus, if not at a light-emitting temperature, would appear black in color. A blackbody is also a perfect emitter of ELECTROMAGNETIC RADIATION because it absorbs all incoming energy. INFRARED RADIATION emission calculations for the earth or the human body make the simplifying assumption that they have the characteristics of a blackbody.

black box   A part of a living or nonliving system that has a known or described function or role, but for which the details of its operation or parts are unknown or omitted.

black lung disease   See SILICOSIS.

black oil   A term for transported heavy oil or fuel oil, in contrast to transported refined products, such as gasoline, which are called clean oil.

blank   A quality control sample that is processed to detect contamination introduced during laboratory tests designed to measure specific chemicals in samples collected from the environment. For example, if the system being examined is a lake or stream, a quantity of deionized water of the same volume as the sample drawn from the environment is processed as if it were an authentic sample. See TRIP BLANK.

blanketing plate   Steel plates that are used to isolate or close off portions of ducts in exhaust systems.

blast gate   A metal damper that can move back and forth in an air duct to achieve a balanced, local exhaust, ventilation system.

bloodworm   A SLUDGE-eating worm typically found in water with a DISSOLVED OXYGEN content of less than one or two PARTS PER MILLION.

bloom   In aquatic ECOSYSTEMS, the rapid growth of ALGAE. See ALGAL BLOOM.

blowdown   The water removed from COOLING TOWERS to prevent the accumulation of excessive amounts of dissolved salts that result from evaporation. The water that is lost is replaced by MAKEUP WATER.

blowout   The uncontrolled release of oil, natural gas, or both from a well into the environment.

blow-by   In an automobile, the leakage of VOLATILE ORGANIC COMPOUNDS from the combustion chamber between the piston and the cylinder wall into the crankcase. To prevent the loss of these CRANKCASE BLOW-BY gases or vapors to the atmosphere, automobiles are now fitted with a POSITIVE CRANKCASE VENTILATION system that routes the volatile organics back to the carburetor for combustion.

body burden   The amount of a chemical material present in the body at a certain time.

body weight scaling   The method of extrapolating DOSE-RESPONSE data from experimental animals to humans that adjusts the data based on relative body weights. For example, if a rat is assumed to weigh 0.35 kilogram and a human 70 kilograms, then a dose of one gram for a rat is extrapolated to a human using a scaling factor of 200, or a dose of 200 grams for the human. The dose for each is therefore about 2.857 grams per kilogram body weight. Compare SURFACE AREA SCALING FACTOR.

boiler water   The water that is heated in an industrial device to produce steam.

boiling point   The temperature at which the VAPOR PRESSURE of a liquid equals the ATMOSPHERIC PRESSURE. Commonly used as the temperature at which a liquid boils.

boiling-water reactor (BWR)   A type of nuclear reactor power plant in which the water in contact with the core is allowed to boil. The water is thereby converted to steam, which is piped to drive a turbine, generating electricity. The other main type of nuclear reactor used for the production of electric power is the PRESSURIZED WATER REACTOR.

bole   The trunk of a tree.

bomb calorimeter   A laboratory device used to measure the energy value of a material (in calories) through combustion of the material.

bone seeker A radioactive material, such as strontium-90, that is metabolically incorporated into the bones after inhalation or ingestion and absorption.

boom A device deployed on a water surface to deflect or contain an oil spill. The booms come in sections, often 50 feet in length, that are connected as needed. One typically consists of an above-water freeboard (containment wall) and an underwater skirt attached to a float. A ballast weight is attached to the bottom of the skirt.

COMPONENTS OF A BOOM

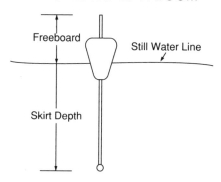

boreal forest Northern forest, as in the boreal forest biome, characterized by evergreen conifers and long winters. The boreal forest is found in the northern parts of North America, Europe, and Asia.

bottle bill A law that mandates the use of returnable beverage containers, usually requiring that a deposit be collected when the beverage is purchased. When the containers are returned, the deposit is refunded or credited toward the next purchase. Advantages include litter reduction, a reduction in solid waste volume, and the conservation of natural resources.

bottom ash The solid material left on the bottom of an incinerator or a boiler

after burning solid waste or a boiler fuel. Compare FLY ASH.

botulism A disease caused by the ingestion of food containing a TOXIN produced by the ANAEROBIC microorganism *Clostridium botulinum*. The toxin attacks the nervous system, and the fatality rate is about 60 percent. Primary risk is from home-canned food, especially improperly prepared beef and pork products, string beans, corn, olives, beans, and spinach.

bound water Water molecules that are held tightly to soil or other solids. This water is not easily removed by normal drying and is not available for other purposes such as plant growth.

box model The representation of STOCKS and flows of material or energy in and through systems by boxes for the stocks and arrows from box to box for the flow directions and amounts. At equilibrium, the stock in a box divided by the rate of flow in or out is called the RESIDENCE TIME. The model can be applied to ECOSYSTEMS, lakes, the BIOSPHERE, a human or animal body, and other systems.

Boyle's law The law, given by English scientist Robert Boyle, that the volume of a perfect gas is inversely proportional to the pressure exerted on the gas, given a constant gas temperature. Expressed as $pV=K$, where p is the pressure, V is the volume, and K is a constant.

brackish water Water with a salt content between 1000 and 4000 parts per million.

breakpoint chlorination A process used to guarantee the presence of sufficient chlorine concentrations in public supplies of drinking water for the protection of public health. The chlorine breakpoint is reached when a sufficient amount of chlorine is added to react with all OXIDIZABLE compounds in the water, such as ammonia. All additions after this breakpoint will increase the levels of residual or unreacted chlorine, available to eliminate dangerous microorganisms that may enter the water after treatment. Compare COMBINED RESIDUAL CHLORINATION.

breakthrough curve   A graph of the volume of gas flowing through a fixed-bed adsorber (such as an activated charcoal filter) versus the concentration of the chemical substance to be adsorbed measured in the outlet gas—that is, the amount of chemical that escapes collection. As a carboǐ filter becomes saturated with the chemical collected from the gas stream, the concentration of the chemical in the outlet gas begins to increase; the point at which the outlet gas concentration exceeds the EMISSION STANDARD for the gas or vapor being collected is called the breakpoint. When the breakpoint is reached, the adsorbing material must be replaced or cleaned.

breeder reactor   A nuclear reactor that converts nonfissionable uranium-238 to the FISSIONABLE material plutonium-239 by arranging the CORE materials so that NEUTRONS produced in the reactor are captured, through a series of steps, by the uranium-238. Natural uranium ore contains only 0.7 percent fissionable material, the ISOTOPE uranium-235. The other 99.3 percent is the nonfissionable isotope uranium-238. Liquid sodium metal is normally employed as the fluid used to cool the reactor. The liquid sodium is heated because it is in direct contact with the reactor core. Subsequently, the hot sodium is pumped to other units designed to remove the heat and thereby maintain the temperature of the reactor at a safe level.

Bremsstrahlung   ELECTROMAGNETIC RADIATION emitted by charged particles as they are slowed down upon impact or by passing through high-density material.

brine disposal   Removing water that contains high concentrations of salt.

British thermal unit (BTU)   The quantity of heat required to raise the temperature of one pound of water one degree Fahrenheit. One BTU equals 252 CALORIES or 1054.4 JOULES.

bronchioles   The air-conducting tubes in the lung; the two larger bronchi divide some 17 times into progressively smaller bronchioles where gas exchange beings. See ALVEOLAR REGION, MUCOCILIARY ESCALATOR.

bronchoconstriction   Narrowing of the BRONCHIOLES, with a resulting increase in AIRWAY RESISTANCE. The condition is caused by exposure to excessive concentrations of certain air pollutants, such as sulfur dioxide.

brown coal   A term for LIGNITE.

Brownian motion   The random, nondirectional motion of small particles suspended in a gaseous or liquid medium, caused by the movement of the molecules constituting the medium. Viewed with a microscope, small particles of nonliving matter in a liquid suspension will exhibit (random) Brownian motion, but bacteria or larger living cells will move in a specific direction. Described by Robert Brown, a British botanist, 1773–1858.

brown lung disease   See BYSSINOSIS.

Brundtland Commission Report   A 1987 report issued by the United Nations World Commission on Environment and Development, headed by Dr. Gro Harlem Brundtland, then-premier of Norway. The study concluded that solutions to the global problems of environmental pollution and natural resource depletion require extreme changes in agricultural, banking, and industrial practices, and reduced energy consumption and population growth. It urged movement toward SUSTAINABLE DEVELOPMENT, which emphasizes economic growth in an ecologically sound manner, especially in the less-developed countries.

bubble meter   A device for measuring GAS VOLUMETRIC FLOW RATE. The device consists of an inverted titration buret attached to a pump, with a squeeze bulb filled with soapy water at the end. The bulb is squeezed, producing a soap bubble that travels up the buret with the flowing air. The timed travel of the bubble between marks on the buret gives the gas flow rate.

bubble policy   The United States Environmental Protection Agency air quality management program that allows air emissions of a particular pollutant from a plant or group of plants to be treated as if all individual emission points were under an imaginary bubble and exhausted at one emission point. The total allowable emissions for a particular pollutant are then determined, and the facility manager can decide the most efficient mix of emission controls to use, as long as the overall limit is not exceeded. See also EMISSION REDUCTIONS CREDITS, EMISSIONS TRADING, OFFSET, and NETTING OUT.

bubbler   An air-sampling device that collects a gaseous air contaminant by passing air through a chemical solution that reacts with and captures the material to be collected.

buffer   A chemical substance in solution that neutralizes acids or BASES. Buffered solutions resist changes in pH. Surface waters and soils with chemical buffers are not as susceptible to ACID DEPOSITION as those with poor buffering capacity.

buffered solution   A solution containing BUFFERS.

buffering capacity   The amount of acid or base that a given liquid or soil can neutralize.

bulk liquids   Liquid hazardous waste not in specific containers. Along with waste containing FREE LIQUIDS, these liquids are subject to the LAND DISPOSAL BAN.

bulking sludge   SLUDGE that does not settle to the bottom of a CLARIFIER, causing a rise in the level of SUSPENDED SOLIDS and BIOCHEMICAL OXYGEN DEMAND in water leaving a treatment facility. This condition is typically controlled by adjusting the MIXED LIQUOR SUSPENDED SOLIDS concentration and the FOOD-TO-MICROORGANISM RATIO. See also SLUDGE VOLUME INDEX.

Bureau of Land Management (BLM)   A section within the United States Department of the Interior responsible for the management of public lands termed national resource lands. These holdings (about 244 million acres) constitute lands that are not included within the national parks, national forests, or national wildlife refuges. The lands managed by the BLM are available for mining, grazing, and timber removal.

Bureau of National Affairs (BNA)   A publishing company based in Washington, D.C., that issues timely reports on regulatory developments in several public policy areas, including environmental affairs. The reports are sometimes referred to as "banana" reports, because of the acronym for the company.

burning agents   Additives that improve combustibility of the material to which they are added.

burnup   1. The percentage of uranium-235 or plutonium-239 used as fuel in a nuclear reactor that is lost as a result of fission.   2. The amount of heat obtained per unit mass of fuel.

butterfly damper   A simple plate used to control the flow of gas through a duct by orienting it parallel to or perpendicular to the flow of the gas.

buy-back center   A facility that pays for waste paper, aluminum cans, plastics, and other recyclable items. The center delivers the material to a resource reprocessing operation or resells the material to a commodities broker for eventual reprocessing.

bypass To allow the release of exhaust gas or wastewater without passage through a pollution control device. Also, to allow drinking water through a treatment facility without adequate contaminant removal.

byssinosis A chronic lung disease resulting from long-term inhalation of high concentrations of cotton dust. Also called brown lung disease.

by-product A substance, other than the intended product, generated by an industrial process.

by-product material 1. A saleable substance that is recovered from a pollution control device. 2. Any material made RADIOACTIVE because of exposure to external radiation sources. 3. ISOTOPES formed through the process of nuclear FISSION.

C

cadmium A toxic metal used in galvanizing, in nickel-cadmium batteries, and as a pigment. Nonoccupational human exposure occurs from cigarette smoking, air and water industrial discharges, and improper disposal of cadmium-containing waste. Body accumulation occurs mainly in the kidneys, where the protein metallothionein protects against low-level exposure by chemically binding to the cadmium.

calcination A process in which a material is CALCINED.

calcine To heat a substance at a temperature below its fusion point, but high enough to cause the loss of water or VOLATILE components, making the material crumbly. The process is often performed in a rotating cylindrical kiln.

calibration The act of introducing known quantities to a measuring device to determine the response of the instrument.

calibration curve A set of responses given by a measuring instrument plotted against a set of known quantities introduced to the instrument. The responses of the instrument to unknown quantities are then translated to measured quantities using the calibration curve, either by the instrument itself or by the analyst.

California Air Resources Board (CARB) The California state air pollution regulatory agency; sets the California motor vehicle emission standards, which are generally lower than the federal limits.

California Aqueduct A channel built to transport water from the region of California north and west of San Francisco to the southcentral region of the state.

California list waste Certain liquid hazardous wastes banned from deep injection wells and, without proper treatment, from land disposal by regulations promulgated by the United States Environmental Protection Agency under the authority of the HAZARDOUS AND SOLID WASTE AMENDMENTS to the RESOURCE CONSERVATION AND RECOVERY ACT. See LAND DISPOSAL BAN.

calorie (cal) The amount of heat required to raise the temperature of one gram (one milliliter) of water one degree Celsius. When used to describe the energy content of food, is used with a capital "C" and equals 1000 calories (or 1 kilocalorie).

Canadian deuterium-uranium reactor (CANDU) A NUCLEAR REACTOR of Canadian design that uses naturally occurring, unenriched URANIUM (about 0.7 percent of the fissionable isotope uranium-235) as the fuel and HEAVY WATER (containing deuterium, the heavy isotope of hydrogen) as the MODERATOR and coolant.

cancer A class of more than 100 diseases characterized by the uncontrolled growth of cells. Cancers are divided into groups based on the type of cells from which they arise: CARCINOMAS, SARCOMAS, and LEUKEMIAS and LYMPHOMAS.

cancer potency factor (CPF) An estimate of the upper limit of the risk of contracting cancer per one unit of dose of a particular chemical; derived from the 95 percent upper-bound confidence limit of the slope of the DOSE-RESPONSE CURVE. This means that the true cancer risk is unlikely to be higher than the calculated risk, and is likely to be lower; i.e., a conservatively high estimate. Typical units are (milligrams per kilogram body weight per day)$^{-1}$ or (micrograms per cubic meter of air)$^{-1}$, the latter often called the UNIT RISK ESTIMATE. The CPF times the actual or modeled dose, for example in milligrams ingested per kilogram body weight per day, yields the lifetime cancer risk. Also called potency, slope factor, carcinogenic slope factor, potency slope factor, and q*.

cap In landfill management, the final, permanent layer of impermeable material placed on the surface as part of the CLOSURE process. A cap is commonly a compacted clay or synthetic material designed to resist liquid infiltration.

capacity factor A ratio used to describe the operation of a boiler. The actual or measured load on a boiler for a set period of time is compared with the theoretical rating of that boiler (that is, actual output per hour/theoretical output per hour × 100).

capture 1. A process whereby a particle is added to a nucleus, with the possible release of energy; for example, the capture of a neutron by a nucleus and the subsequent emission of GAMMA RADIATION. 2. In groundwater HYDROLOGY, the difference between natural groundwater discharge from an AQUIFER and the artificial increase in the amount of water in the aquifer due to pumping of water from surface sources.

capture velocity The air velocity at a point in front of a HOOD necessary to pull contaminated air from that point into the hood. Capture velocity varies with the release velocity of the airborne material to be captured and with the strength of competing airflows in the vicinity of the hood.

carbamates A group of NONPERSISTENT (degradable) synthetic insecticides that includes Sevin (carbaryl), Baygon, and Temik (aldicarb), among others. The carbamates are CHOLINESTERASE INHIBITORS and are generally selective in their toxicity.

carbaryl See SEVIN.

carbon (C) An element present in all materials of biological origin. The element contains six protons and usually six neutrons in the nucleus. Carbon atoms will bind with each other as well as with a host of other elements. Deposits derived from living sources, such as limestone, coal, oil, and natural gas, contain carbon as a principal element. Graphite and diamonds are composed almost exclusively of carbon.

carbon-14 (^{14}C) A RADIOACTIVE ISOTOPE of carbon that emits beta particles when it undergoes RADIOACTIVE DECAY. The nucleus of a ^{14}C atom contains eight neutrons rather than the six neutrons found in ^{12}C, which is the most abundant isotope in nature. Carbon-14 is formed naturally in the earth's atmosphere through the interaction of nitrogen gas in the air with cosmic rays. The atmospheric ^{14}C thus produced combines with atmospheric oxygen and is converted to CARBON DIOXIDE (^{14}CO$_2$). Because the radioactive form of carbon dioxide mixes with the normal, nonradioactive, carbon dioxide in the atmosphere and because the radioactive and nonradioactive forms of carbon dioxide undergo the same chemical reactions, both forms can be found in living plant tissue as a result of PHOTOSYNTHESIS. While the plant is alive, the ratio of ^{14}C to ^{12}C remains about the same as time passes. However, when the plant dies, carbon dioxide from the air is no longer incorporated into plant mass. The radioactive form of carbon decays with a half-life of about 5760 years; consequently, the ratio between the two isotopes in the dead plant changes as time passes and the radioactive carbon decays. The changes in this ^{14}C : ^{12}C ratio can be used to date materials. If a buried tree contains one-half of the radioactive carbon that one finds in a live tree (relative to the amount of normal carbon present), the tree is approximately 5760 years old. Materials older than 70,000 years cannot be aged by this technique since almost all of the radioactive carbon has decayed. Carbon-14 is also known as radiocarbon.

carbon black   A sooty material produced by incomplete combustion of hydrocarbons. Carbon black is made commercially under controlled conditions for use in pigments, tires, and inks.

carbon cycle   The movement of carbon in various chemical forms from one planetary RESERVOIR to another.

carbon dioxide (CO$_2$)   A normal gaseous constituent of the atmosphere, the product of aerobic respiration, decomposition, and carbon fuel combustion. Because carbon dioxide absorbs INFRARED RADIATION, rising levels of carbon dioxide in the global atmosphere over the past century have prompted concerns about climatic change. See GREENHOUSE EFFECT.

carbon filtration   The passage of treated wastewater or domestic water supplies through activated charcoal in an effort to remove low concentrations of dissolved chemicals.

carbon monoxide (CO)   An odorless, colorless, and tasteless gas that is an air contaminant formed by the incomplete combustion of carbon fuels, such as wood, coal, or gasoline. Classed as an asphyxiant gas, absorbed carbon monoxide combines with hemoglobin in the blood to form CARBOXYHEMOGLOBIN, which impairs oxygen delivery to the tissues. The major source for human exposure is the automobile.

carbon oil   A liquid fuel, derived from coal, used in lamps before the invention of electric lighting.

carbon polishing   The removal of residual dissolved organic substances from wastewater by ADSORPTION on ACTIVATED CHARCOAL (GRANULAR ACTIVATED CARBON). A form of TERTIARY TREATMENT.

carbon tetrachloride   A derivative of methane (CH$_4$) that is produced by substituting a chlorine atom for each hydrogen (CCl$_4$). Carbon tetrachloride is a common NONPOLAR SOLVENT that has significant industrial applications; it has also been used as a fumigant. Exposure to the HALOGENATED hydrocarbon has been linked to damage to the liver, kidney, and central nervous system. Carcinogenic properties have also been demonstrated.

carbon treatment   In a drinking water purification process, the removal of COLLOIDS by ADSORPTION on ACTIVATED CHARCOAL. This step often improves color, taste, and odor.

carbonaceouos biochemical oxygen demand   The incubation of a sample of water or wastewater for a relatively short time in order to determine the biochemical oxygen demand. The short incubation, usually five days, is sufficient to detect only the microbial utilization of carbon compounds. A longer incubation (15 to 20 days) would also detect the oxidation of inorganic nitrogenous compounds (ammonia and nitrite) and the subsequent demand for molecular oxygen by chemoautotrophic bacteria. See BIOCHEMICAL OXYGEN DEMAND for a description of the basic concept associated with the measurement. See also NITROGENOUS BIOCHEMICAL OXYGEN DEMAND.

carbonate   The CO_3^{-2} ION in the CARBONATE BUFFER SYSTEM. Combined with one proton, it becomes BICARBONATE, HCO_3^-; with two protons, CARBONIC ACID. The carbonate ion forms a solid PRECIPITANT when combined with dissolved ions of calcium or magnesium.

carbonate aquifer   An AQUIFER found in limestone and dolomite rocks. Carbonate aquifers typically produce HARD WATER, water containing relatively high levels of calcium and magnesium.

carbonate buffer system   The most important BUFFER system in natural surface waters and wastewater treatment, consisting of a CARBON DIOXIDE–water–CARBONIC ACID–BICARBONATE–CARBONATE ion equilibrium that resists changes in pH. For example, if acid materials (hydrogen ions) are added to this buffer system, the equilibrium is shifted and carbonate ions combine with the hydrogen ions to form bicarbonate; bicarbonate then combines with hydrogen ions to form carbonic acid, which can dissociate into carbon dioxide and water. Thus the system pH is unaltered, although acid was added.

carbonate hardness   See TEMPORARY HARDNESS.

carbonation, groundwater   The dissolving of carbon dioxide in surface water as it percolates through the ground. The carbon dioxide reacts with water to form carbonic acid, a weak acid that causes the water to have a slightly acidic pH.

carbonic acid   A mild acid formed by the dissolution of carbon dioxide in water. The carbonic acid content of natural, unpolluted rainfall lowers its pH to about 5.6.

carboxyhemoglobin (COHb)   The chemical compound formed by the combination of absorbed CARBON MONOXIDE and the blood pigment hemoglobin. Hemoglobin normally combines with atmospheric oxygen drawn into the lungs to transport it from the lungs to body tissues. When carbon monoxide is chemically bonded to hemoglobin, the oxygen-carrying capacity of the blood is reduced. The COHb level is expressed as a percent of all hemoglobin BINDING SITES occupied by carbon monoxide. Nonsmoking adults typically have an endogenous (formed within the body by the normal breakdown of HEME) COHb level of less than 1 percent. The United States Environmental Protection Agency NATIONAL AMBIENT AIR QUALITY STANDARD for carbon monoxide is set to prevent COHb levels exceeding about 1.5 percent.

carboy   A large glass jug or container, often with a wooden frame or case, used to store large amounts of liquid chemicals or to collect and transport large water samples.

carcinogen   A chemical substance or type of radiation that can cause cancer in exposed animals or humans. There is no single definition of the evidence necessary to classify a substance as a carcinogen. Four sources of evidence are used: EPIDEMIOLOGY, long-term animal testing, short-term tests (such as the AMES TEST), and STRUCTURE-ACTIVITY RELATIONSHIPS. Separate guidelines and policies defining carcinogens have been issued by the United States Environmental Protection Agency, the Occupational Safety and Health Administration, the Consumer Product Safety Commission, and the Food and Drug Administration. These agencies, and the INTERNATIONAL AGENCY FOR RESEARCH ON CANCER, classify chemicals by the degree of evidence available as to their carcinogenicity. Some 150–200 chemicals and several physical agents (ionizing radiation, ultraviolet radiation) appear on various lists as having sufficient or limited evidence of carcinogenicity; about 40 of these are classified as human carcinogens. Note that of the roughly 9 million known chemical substances, only about 7000 have been tested for carcinogenicity. Carcinogens vary in their estimated ability to induce cancer. See CANCER POTENCY FACTOR, NATIONAL TOXICOLOGY PROGRAM.

Carcinogen Assessment Group (CAG) A section of the United States Environmental Protection Agency uses risk extrapolation models to produce quantitative estimates of the risk of exposure to environmental CARCINOGENS. Risk extrapolation models use data from animal exposure to high doses of a chemical to derive the human risk of exposure to low doses of the chemical. The CAG is responsible for many UNIT RISK ESTIMATES, which result from risk extrapolation.

carcinogenesis The generation of cancer. Most cancers are derived from changes in the DNA or gene structure of a single cell (mutations). These changes are then passed on by inheritance mechanisms to the cells that descend from this single cell. The types of changes can be categorized into three primary areas: 1. those caused by chemical agents that bring about changes in the chemical structure of DNA; 2. those caused by ionizing radiation that result in changes in cellular chromosomes; and 3. those caused by viruses that introduce new DNA into cells. A single change produced by any of the three general mechanisms is not sufficient to produce a cancer cell. Several random changes are needed to convert cells from the normal to the cancerous state. As a consequence, cancer develops slowly from a population of mildly irregular cells. The agent or chemical that causes the first heritable change in the gene structure of the cell is termed the initiator (which is usually a mutagen). Following the action of the initiator, several exposures to other agents, termed promoters (which are not necessarily mutagens), are required to change one of the descendants of the first irregular cell

to a tumor cell. The initiators and promoters must act in the proper sequence and with proper timing in order to bring about the complete transformation of a normal cell to a tumor cell. The identification of initiators and promoters is not always possible with any degree of certainty. For example, agents such as tobacco smoke appear to act as mutagens and initiate various tumors, while reproductive hormones seem to function as promoters and stimulate the continued development of other tumors. The environment and life-style of an individual also play a significant role in the development of most cancers; however, the identification of the specific factors within the environment of an individual which are responsible for tumor development is not always possible. See MUTATION, MUTAGEN, CARCINOGEN, ONCOGENIC, and DEOXYRIBONUCLEIC ACID.

carcinogenic Describing a chemical substance or type of radiation that can cause cancer in exposed animals or humans.

carcinogenic activity indicator (CAI) A numerical cancer potency indicator for chemical substances, calculated by dividing the additional percentage of test animals found to have tumors by the lifetime dose of the tested chemical, in molecules per kilogram body weight. This calculation is done for each point on a DOSE-RESPONSE CURVE. CAIs are useful for comparisons of chemicals, such as pesticides, that have possible substitutes. The CAIs will reveal the lowest-risk material appropriate for the application.

carcinogenic potency factor (CPF) See CANCER POTENCY FACTOR.

carcinoma A tumor belonging to the most common group of cancers, those arising in epithelial tissues, such as the lining of the lung, gastrointestinal tract, or other organs, or the skin.

carnivore An animal that eats only meat, such as wildcats or wolves. Compare to HERBIVORE, OMNIVORE.

Carnot engine A theoretical device for converting heat into work (force), in which a piston moves in a cylinder without friction. The movement of the piston is performed by gas compression and expansion, which is caused by a heat source and a HEAT SINK, which absorbs the excess heat produced. The model illustrates that the efficiency of the engine depends on the absolute temperatures of the heat source and the heat sink.

carrier gas An inert gas, such as helium, that is employed as a medium for the transport of low concentrations of some active ingredient. The analysis of some chemicals requires that a carrier gas be used to transport the agent in question through the analytical instrument.

carrying capacity (K) The maximum number of organisms that an ECOSYSTEM can support over an extended time period without significant degradation of the ecosystem.

carryover Liquid particles introduced to a gas stream by the turbulent air-gas contact in a SCRUBBER. Carryover is removed by a MIST ELIMINATOR placed downstream from the scrubbing device. Also called ENTRAINMENT.

Carson, Rachel Marine biologist (1907–64) and author of *Silent Spring*, the 1962 book on pesticide (especially DDT) threats to the environment, which marked the beginning of the political ascendancy of the environmental movement in the United States. Other works include *Under the Sea Wind* (1941), *The Sea Around Us* (1951); and *The Edge of the Sea* (1954).

cascade impactor A device used to determine the size distribution of airborne particles. Air is drawn through a series of plates, each plate removing particles of a smaller diameter. The mass of the particles caught on each plate is measured to give a mass size distribution.

case-control study A type of epidemiological (disease-related) study that compares the past exposure of two groups to a particular environmental factor. The cases are a group of persons identified as having a certain disease, and the controls are members of a group similar to the cases in terms of age, sex, race, and other factors, that do not have the disease. Any significantly greater exposure to the factor in cases compared to the controls may indicate a causative link between the agent and the disease. Also called a retrospective study.

casing In water or oil and gas wells, a solid steel or polyvinyl chloride (PVC) pipe defining the outer diameter of the well. Other pipe may be run inside the casing.

cast-iron pipe (CI pipe) The type of pipe used in household water systems before copper and polyvinyl chloride (PVC) pipes were introduced.

catabolism The biological breakdown of materials into their simpler components; decomposition. Performed by DECOMPOSER organisms, mainly bacteria and fungi. Compare ANABOLISM.

catalyst A chemical substance that allows a chemical reaction to take place more readily, for example at a lower temperature or pressure, without itself being consumed or altered by the reaction. ENZYMES are biological catalysts, enhancing reactions within living organisms.

catalytic converter An air pollution control device, installed in an automobile exhaust system, that reduces the levels of HYDROCARBONS and CARBON MONOXIDE exiting the tailpipe. The CATALYST enhances the OXIDATION of hydrocarbons to carbon dioxide and water vapor and the oxidation of carbon monoxide to carbon dioxide. THREE-WAY CATALYSTS also include in the device a reducing catalyst that converts NITRIC OXIDE and NITROGEN DIOXIDE to nitrogen gas.

catalytic cracking A process for converting high-molecular-weight hydrocarbons found in crude oil to smaller hydrocarbon molecules by heating in the presence of a metal compound that serves as a catalyst (a substance that speeds up the rate of a chemical reaction without entering into the reaction itself). Through the use of a catalyst, the cracking process can be done at lower temperatures and pressures and with greater control than would be possible otherwise. The process increases the amount of gasoline and other light hydrocarbons that can be produced from heavy crude oil.

catchment area The area that draws surface runoff from precipitation into a stream or urban storm drain system; synonymous with WATERSHED.

categorical exclusion Types of actions determined by federal agencies to have no significant environmental impacts, and therefore do not need an ENVIRONMENTAL ASSESSMENT or an ENVIRONMENTAL IMPACT STATEMENT under the NATIONAL ENVIRONMENTAL POLICY ACT.

categorical variable A qualitative variable created by classifying observations into categories. For example, a series of temperature measurements could be classified into the categorical variables low, normal, and high, with low defined as less than $10°$ C, normal between 10 and $30°$, and high greater than $30°$. Many statistical techniques are inappropriate for categorical variables: an average, for example. Compare QUANTITATIVE VARIABLE.

cathode The positive pole of an ELECTROLYTIC cell or a battery. When the battery is connected in a circuit, electrons flow from the ANODE to the cathode.

cathodic protection A method to protect iron or steel tanks, pipes, or other structures from corrosion, to avoid leaks and spills. Iron or steel corrosion is caused by OXIDATION, which is a loss of electrons. A metal like zinc or magnesium is connected to serve as a source of replacement electrons for those lost by the tank or pipe. The zinc or magnesium ANODE is oxidized instead of the iron or steel CATHODE.

cation A positively charged ION.

cation exchange The displacement of one CATION for another, often on the surface of a soil or clay particle. Cations of sodium or potassium may be ADSORBED to clay, and if groundwater containing other metal cations (like lead, cadmium, or zinc) flows over the clay, the metal cations will displace the sodium and potassium ions, releasing them to the groundwater but removing the metals from the groundwater. This natural cleansing process is important in assessing the hazard of leaking waste sites.

cation exchange capacity (CEC) The ability of a soil to remove cations from groundwater. See CATION EXCHANGE.

caustic Alkaline, BASIC. Also used to refer to the heavily used industrial chemical sodium hydroxide, which is a strong BASE or source of OH$^-$ ions.

cell The unit of biological structure and function that can stand alone and carry out all the fundamental life processes. A cell is an independent unit that can obtain nutrients from the environment, derive energy from organic materials, reproduce exact copies of itself, and release waste products into the surrounding environment. No structure derived from a cell, such as the nucleus, mitochondria, or chloroplast, can carry out all of the life functions in an independent fashion. Viruses are life forms that are simpler than cells in structure and function; however, they are not capable of independent metabolism and cannot reproduce unless a cell is involved. All cells are bound by a plasma membrane, composed of lipids and proteins, and nucleic acids (DNA) necessary for inheritance functions. Bacteria and simple plants and animals consist of only one cell, which is capable of independent existence. In higher plants and animals, similar cell types form tissues, which in turn form organs.

cell culture The growing of animal or plant tissue in an artificial medium contained within a test tube or bottle.

cell, landfill A completed waste storage area in a landfill, separated from other cells by COVER material.

cementing Pumping of a cement SLURRY into a well or behind the casing of a well.

centerband frequency See OCTAVE BANDS.

centerline concentration In estimating the off-site impact of an air pollution source using a GAUSSIAN PLUME MODEL, the model's estimated concentrations along a line directly downwind from the source; estimated levels decrease with increasing distance on both sides of the centerline concentration.

Centers for Disease Control (CDC) An agency of the Public Health Service, United States Department of Health and Human Services. The lead federal agency in developing disease prevention and control, environmental health, health promotion, and health education as well as in the surveillance of disease patterns through epidemiological (disease-related) data collection, analysis, and distribution. CDC publishes and distributes the "Morbidity and Mortality Weekly Report," which is a major source of public health information. The agency is headquartered in Atlanta, Georgia, and was formerly known as the Communicable Disease Center.

centistoke A unit of KINEMATIC VISCOSITY equal to 0.001 square meter per second.

central nervous system The brain, spinal cord, and associated nerves. The functioning of this system is disrupted by many pesticides and other toxic contaminants added to the environment.

centrifugal collector An air pollution control device that collects particles in an airstream using CENTRIFUGAL FORCE. See CYCLONE.

centrifugal fan A fan commonly used in industrial exhaust ventilation systems. The fan blades turn inside a housing, as in vacuum cleaners and hair dryers.

centrifugal force An apparent force on a rotating object, directed outward from the center of rotation.

centrifugal pump A device that moves water with a rotating impeller (a series of metal vanes) surrounded by a casing. This compact, simple pump is the most common type used in water distribution and treatment.

centrifugation The separation of materials of different densities in a CENTRIFUGE.

centrifuge A device that employs CENTRIFUGAL FORCE to separate a mixture into components by their relative densities, especially to separate SUSPENDED SOLIDS from liquids.

cesium-137 (^{137}Cs) One of the important RADIOACTIVE alkali metals. It is a common product of nuclear weapons explosions and nuclear reactors. In humans, the isotope is absorbed rapidly and is distributed throughout the body where it enters into reactions that normally involve potassium. Both cesium-137 and its decay products release energetic BETA and GAMMA RADIATION, which can result in significant whole-body radiation damage. Deaths resulting from acute exposures usually result from dysfunction of bone marrow.

cesspool An underground cistern used for sewage disposal in areas not served by a community sewage collection system. This disposal method is generally prohibited in the United States.

chain of custody   The documented transfer(s) of an environmental sample from the person collecting it to any transporters until it reaches the person performing any required analysis, to ensure that no contamination or substitution occurs.

chain reaction   In nuclear processes, the sequence of atomic fission created by a neutron impacting an unstable nucleus, causing it to undergo fission, which produces additional neutrons, which can impact other unstable nuclei, causing more fission, and so on. The release of energy accompanying a controlled chain reaction is captured and converted to electricity by a nuclear power station. See FISSION, NUCLEAR REACTOR.

change of state   The change of a chemical substance from one physical state (solid, liquid, or gas) to another as a result of temperature or pressure changes; in the environment, the change is usually caused by a temperature change. The changes of state of water and the associated heat transfers are especially important in environmental heat regulation, including the moderation of ecosystem weather extremes and, on a larger scale, the movement of heat poleward at the surface of the earth. It is also vital for human body heat regulation and the removal of heat from industrial processes.

channelization   Flood control or navigation projects that straighten, widen, or deepen surface water channels such as streams, rivers, or canals. Adverse environmental consequences can include increased SEDIMENTATION, bank erosion, increased flooding, decrease in BIOMASS, and (surface) SALTWATER INTRUSION.

chaparral   A BIOME with hot, dry summers and rainfall mainly in the winter months. Vegetation consists of shrubs and small evergreen trees. Chaparral communities are found around the Mediter-

ranean Sea, in central and southern California, along coastal Chile, in southern Australia, and in southern Africa.

characteristic hazardous waste   Materials defined as hazardous waste by their possession of one or more of the following characteristics, as defined in regulations issued by the United States Environmental Protection Agency: IGNITABILITY, CORROSIVITY, REACTIVITY, or TOXICITY. Included in the CODE OF FEDERAL REGULATIONS, Title 40, Part 261, Subpart C. See LISTED HAZARDOUS WASTE.

charcoal filter   See CARBON FILTRATION.

charge rate   The rate at which material is introduced to an incinerator; typically expressed in pounds per hour.

Charles' law   The gas law stating that the volume of a given amount of gas is directly proportional to the absolute temperature of the gas, given a constant pressure. Expressed as $V = TK$, where V is the volume, T is the absolute temperature, and K is a constant. Charles' experimental law can be expressed as $V = V_0(1 + \alpha_v \Theta)$, where V is the volume at temperature Θ in °C, α_v is a constant equal to $1/273.15$, and V_o equals the gas volume at 0° C. [Also called the Gay-Lussac law.]

check dam   A small dam across a stream that decreases stream velocity and thereby reduces erosion of the banks and bottom while increasing sediment deposition.

chelating agent   Chemical compound that has the ability to bind strongly with metal ions. It is used to enhance the excretion of various toxic metals. For example, ethylenediaminetetraacetic acid (EDTA) is administered to treat lead poisoning. The EDTA chemically binds to the lead and the lead-EDTA complex is excreted.

Chemical Abstracts Service Registry Number (CAS number) A unique number assigned by the Chemical Abstracts Service, a division of the American Chemical Society, to each chemical substance and used for positive identification. A chemical may be known by several different names, but it has only one CAS number. The United States Environmental Protection Agency often uses CAS numbers to describe clearly the substances referred to in its regulations.

chemical asphyxiant A gas that deprives the body of proper oxygen absorption, such as carbon monoxide, or a chemical material that prevents oxygen *use* at the cellular level, like cyanide. See CYANOSIS and SIMPLE ASPHYXIANT.

chemical builders Inorganic phosphate compounds (such as tripolyphosphate) added to detergents to facilitate the COMPLEXing of calcium or magnesium ions that may be present in water supplies. The binding of the ions with the inorganic phosphate compound reduces the precipitation (solidification) of the detergent molecules in HARD WATER.

chemical energy The potential energy residing in the bonds of chemical compounds. For example, solar energy is converted by photosynthesis to chemical energy in the form of carbohydrates. The chemical energy originating in the sun is stored in the bonds holding the carbohydrate molecule together.

Chemical Hazard Information Profiles (CHIPs) A publication of the United States Environmental Protection Agency, Office of Toxic Substances, containing, for a particular chemical, estimates of occupational, consumer, and environmental exposure; human health and environmental effects; and pertinent standards and regulations. CHIPs are available for over 200 chemicals.

Chemical Hazard Response Information System (CHRIS) A set of manuals used for assessing the health, safety, and environmental hazards posed by chemical releases, emphasizing spills to surface water. The manuals were developed for the United States Coast Guard. The computer model version of CHRIS is the Hazard Assessment Computer System (HACS).

Chemical Manufacturers Association (CMA) A 175-member group of chemical producers, in Washington, D.C., that supports research, workshops, and technical symposia on the environment, health, and safety of chemical manufacturing and distribution. Operates the CHEMICAL TRANSPORTATION EMERGENCY CENTER (CHEMTREC) to support responses to chemical transportation accidents.

chemical oxygen demand (COD) A chemical measure of the amount of ORGANIC substances in water or wastewater. A strong oxidizing agent together with acid and heat are used to OXIDIZE all carbon compounds in a sample. Nonbiodegradable and recalcitrant (slowly degrading) compounds, which are not detected by the test for BIOCHEMICAL OXYGEN DEMAND, are included in the analysis. The actual measurement involves a determination of the amount of oxidizing agent (typically, potassium dichromate) that is reduced during the reaction.

chemical toilet A toilet facility in which waste is retained and chemicals, such as lime, are added to disinfect the sewage and/or to control odors.

Chemical Transportation Emergency Center (CHEMTREC) A service operated by the Chemical Manufacturers Association that provides timely information about chemicals involved in a hazardous material transportation incident. CHEMTREC personnel can send detailed information to first responders via the HAZARD INFORMATION TRANSMISSION (HIT) system and, if necessary, will contact the producer of the material(s) for additional guidance. The 24-hour telephone number of the Center is 800-424-9300.

chemical weathering The gradual decomposition of rock by exposure to water, atmospheric oxygen, and carbon dioxide.

chemiluminescent detector An analytical instrument that measures the level of a particular wavelength of light emitted by the chemical reaction occurring when a gaseous compound is introduced to a sample of ambient air. The intensity of the light emission is directly proportional to the concentration of a particular gas in the air sample, and can be calibrated to determine the gas concentration. This method is used for the determination of OZONE and NITROGEN DIOXIDE levels in the atmosphere.

chemoautotrophs Microorganisms that derive biologically useful energy from the oxidation of inorganic chemical substances, usually ammonia, sulfur, nitrite, and ferrous iron. The bacteria that oxidize iron sulfide to iron sulfate and sulfuric acid, the biological activity responsible for ACID MINE DRAINAGE.

chemodynamics The study of the transport, conversion, and fate of chemical substances in air, water, or soil, including their movement from one medium to another.

chemosphere The region of the upper atmosphere including the MESOSPHERE and upper STRATOSPHERE in which various sunlight-driven chemical reactions occur. See ATMOSPHERE.

chemotrophs See CHEMOAUTOTROPHS.

China Syndrome A popular but misleading term for the potential catastrophic result of a nuclear core overheating to the extent that the floor of the containment vessel melts and the now-molten core sinks toward the other side of the earth (China, for a reactor in the United States). Actually, the molten core would stop sinking a few meters below the surface. See MELTDOWN.

Chi-square test A method used to determine whether STATISTICALLY SIGNIFICANT differences exist in frequency data from separate groups, such as whether the number of persons in an exposed group that exhibit a certain adverse health effect is statistically different from the number of persons in an unexposed (control) group with the adverse effect.

chloracne An acne-like eruption of the skin caused by exposure to certain chlorinated aromatic organic compounds.

chloramines Compounds containing nitrogen, hydrogen, and chlorine, formed by the reaction between hypochlorous acid ($HOCl$) and ammonia (NH_3) and/or organic amines in water. The formation of chloramines in drinking water treatment extends the disinfecting power of chlorine. Also called COMBINED AVAILABLE CHLORINE.

chloramine-T A chlorine-containing material used to add chlorine to drinking water.

chlordane A CHLORINATED HYDROCARBON insecticide chemically related to ALDRIN and ENDRIN. The agent was widely used to treat soil around homes for termite control because it is persistent (long-lasting) in the environment. It is moderately toxic to mammals and has been shown to cause cancer and to alter fertility in some mammals. Currently, the use of this pesticide is severely limited in the United States.

chlorides Negative chlorine ions, Cl^-, found naturally in some surface waters and groundwaters and in high concentrations in seawater. Higher-than-normal chloride concentrations in fresh water, due to sodium chloride (table salt) that is used on foods and present in body wastes, can indicate sewage pollution. The use of highway deicing salts can also introduce chlorides to surface water or groundwater. Elevated groundwater chlorides in drinking water wells near coastlines may indicate SALTWATER IN-TRUSION.

chlorinated 1. Describing an organic compound to which atoms of chlorine have been added. 2. Describing water or wastewater that has been treated with either chlorine gas or a chlorine-containing compound.

chlorinated dibenzofurans (CDBF) A class of highly toxic aromatic halogenated hydrocarbons resembling the chlorophenols and dioxins.

chlorinated hydrocarbons Synthetic chemical substances containing chlorine, hydrogen, and carbon. The addition of chlorine to ORGANIC compounds causes these materials to break down slowly in the environment. The chlorinated hydrocarbons are also a class of insecticides, which includes DDT, mirex, ALDRIN, KEPONE, LINDANE, heptachlor, toxaphene, and many others.

chlorinated organics Synonym for CHLORINATED HYDROCARBONS.

chlorination The process of adding chlorine to water or wastewater in order to kill or inactivate dangerous microorganisms or viruses. Chlorine in various forms, such as chlorine gas, bleach, or solid chlorine-containing compounds, can be used.

chlorine (Cl_2) One of a group of elements classified as the halogens. Chlorine, the most common halogen, is a greenish yellow gas with an irritating odor. Chlorine is very reactive; it forms salts with metals, forms acids when dissolved in water, and combines readily with hydrocarbons. Various forms of chlorine are used to disinfect water. CHLORINATED HY-DROCARBONS are used widely as PESTI-CIDES and industrial chemicals. Freon is a synthetic material containing chlorine.

Chlorine is produced by the electrolysis of brine (concentrated salt solution).

chlorine demand The amount of chlorine that must be added to purify drinking water; the amount required to react with all dissolved and particulate materials and inorganic ammonia in the water.

chlorine residual Chlorine added to drinking water in excess of the amount needed to react with organic and inorganic materials suspended or dissolved in the water. This chlorine is available to eliminate microorganisms that enter the water distribution system after treatment.

chlorofluorocarbons (CFCs) A class of simple hydrocarbon derivatives in which chlorine and fluorine are substituted for some or all of the hydrogens (e.g., CCl_2F_2). They are commonly called freons. As a group, these compounds are VOLATILE, nonreactive, noncorrosive, and non-flammable. They are used widely in consumer products (propellants in aerosol sprays in some countries, and coolants in refrigerators and air conditioners) and in industrial applications (electronics manufacture and blown styrofoam). The chlorofluorocarbons have been implicated in the reduction of the ozone content of the stratosphere (OZONE LAYER), and are GREENHOUSE GASES; that is, they absorb outgoing infrared radiation from the earth. See OZONE LAYER DEPLETION, GREENHOUSE EFFECT.

chloroform A simple halogenated hydrocarbon ($CHCl_3$) obtained from the chlorination of methane (CH_4). Once used in human anesthesia, it remains an important industrial chemical. Chloroform is one of the more common halomethanes produced during the chlorination of water. Low concentrations in drinking water promote kidney and liver damage in animals.

chlorophenols See PENTACHLORO-PHENOL.

chlorophyll The green pigment in plants that absorbs a portion of incoming sunlight for use in PHOTOSYNTHESIS.

chloroplasts The structures within a plant cell that contain chlorophyll.

chlorosis The yellowing of plant leaves, indicating loss of CHLOROPHYLL or a reduction in the number of CHLOROPLASTS. It can be caused by a deficiency of iron, magnesium, sulfur, or nitrogen; disease; insufficient sunlight; or certain air pollutants. Air pollutants that can cause chlorosis to susceptible vegetation include sulfur dioxide, fluorides, and ethylene.

cholera An infectious waterborne disease that is characterized by severe diarrhea and its resultant dehydration and electrolyte imbalance. The disease is caused by bacteria belonging to the genus *Vibrio*. Outbreaks are associated with contamination of surface waters with human FECAL material.

cholinesterase inhibitors Chemical substances that inactivate the enzyme cholinesterase, resulting in nerve dysfunction. Normal transmittal of nerve impulses across synapses or the junctions connecting nerve fibers to each other and connecting nerves to muscles is accomplished by the chemical acetylcholine. After acetylcholine moves across the synapse to relay the nerve impulse, cholinesterase breaks down the acetylcholine, which is later reformed to carry another nerve signal. If acetylcholine is not broken down—that is, if cholinesterase has been inhibited—then the nerve stimulation is excessive, which can lead to twitching, convulsions, and death. The nerve gases developed for chemical warfare and their relatives, the ORGANOPHOSPHATE insecticides, are cholinesterase inhibitors.

chromatograms The pattern formed on or in an ADSORBENT material or on a printed output when closely related chemicals are separated by CHROMATOGRAPHY.

chromatography A process used to separate and/or identify similar compounds by allowing a solution of the compounds to migrate through or along a substance that selectively ADSORBS the compounds in such a way that materials are separated into zones. The simplest example of the process is the separation of the colors in ink by placing a drop of the ink on a napkin and dampening the cloth. As water migrates up the napkin, it moves and separates the different dyes in the ink.

chromosomal nondisjunction The failure of CHROMOSOMES to separate during cell division. As a result, daughter cells do not have the necessary number of chromosomes for normal functioning.

chromosome A threadlike structure in the cell nucleus, composed of DNA and protein, and the linearly arranged GENES. Each chromosome contains coiled DNA molecules. The number of chromosomes in the nucleus is characteristic of the species. Humans have 46.

chronic exposure In toxicology, doses that extend for long periods, from six months to a lifetime.

chronic toxicity Adverse health effects that are either the result of CHRONIC EXPOSURE or are permanent or long-lasting, as in scarring of lung tissue.

chrysotile In the past, the most widely used form of ASBESTOS in the United States. Also called white asbestos.

cilia See CILIATED MUCOSA and CILIATES.

ciliated mucosa The lining of the respiratory tract in which tiny, moving, hairlike projections called cilia move mucus upward. Inhaled particles are swept out of the lungs if caught by the ciliated mucosa.

ciliates A type of protozoa or single-celled animal that moves with the aid of short, hairlike projections termed cilia. These organisms are important members of the community of organisms that carry out mineralization, the conversion of an organic material to an inorganic form.

citizen suit provision A feature of many federal environmental statutes that allows private citizens or organizations to file suits involving enforcement of pollution control regulations, to challenge regulations of the United States Environmental Protection Agency, or to force the Administrator of that Agency to perform a nondiscretionary duty, i.e., a specific action required of the administrator by a federal statute. See also STANDING, LEGAL.

claims-made insurance policy See ENVIRONMENTAL IMPAIRMENT LIABILITY POLICY.

Clapeyron-Clausius equation See CLAUSIUS-CLAPEYRON EQUATION.

clarification The process of removing PARTICULATE MATTER from wastewater. Normally, the water is allowed to stand, which facilitates the settling of the particles.

clarifier A device or tank in which wastewater is held to allow the settling of particulate matter.

Class I, II, III, IV, and V injection wells Classifications of the United States Environmental Protection Agency that determine the permit requirements of an INJECTION WELL. Class I: A well into which liquid hazardous wastes or other fluids are pumped down, with the fluids being injected into an underground formation below the lowest underground source of drinking water that is within a one-quarter mile radius of the well. Class II: A well used to dispose of fluids produced by oil and gas wells, to introduce fluids for ENHANCED OIL RECOVERY, or for liquid hydrocarbon storage. Class III: A well used

to pump fluids underground for mineral extraction. Class IV: A well used to reinject treated fluid from a SUPERFUND cleanup site into or above an underground formation within a one-quarter mile radius of the well. Class V: Wells not included in Classes I–IV, mainly shallow industrial disposal wells or RECHARGE wells.

Clausius-Clapeyron equation The relationship used to calculate the change in the vapor pressure of a liquid as the temperature changes. The equation is also known as the Clapeyron-Clausius equation. It can take the form

$$\ln p_v = -\frac{\Delta H_v}{RT} + B$$

where $\ln p_v$ is the natural logarithm of the vapor pressure of the liquid, ΔH_v is the latent heat of vaporization, R is the universal gas constant, T is the absolute temperature, and B is a material-specific constant.

clay liner A layer of clay soil that is added to the bottom and sides of a pit designed for use as a disposal site for potentially dangerous wastes. The clay prevents or reduces the migration of liquids from the disposal site.

clay pan A tightly compacted layer of natural soil that restricts the migration of liquids into the underlying strata.

Clean Air Act (CAA) The basic federal air pollution control statute. First passed in 1963, following a 1955 federal statute authorizing research and technical assistance. The 1965 and 1967 amendments began automobile and stationary source standards. Major amendments in 1970 and 1977 provide for the NATIONAL AMBIENT AIR QUALITY STANDARDS, the STATE IMPLEMENTATION PLAN process, the PREVENTION OF SIGNIFICANT DETERIORATION program, emission standards for automobiles, NATIONAL EMISSION STANDARDS FOR HAZARDOUS AIR POLLUTANTS, and minimum technology standards for new or modified sources (NEW SOURCE PER-FORMANCE STANDARDS). The 1990 amendments include provisions for operating permits for stationary sources, a phaseout of ozone-layer-depleting chemicals, ACID RAIN controls, tradable emission credits, a system of ranking NONATTAINMENT AREAS by severity of air pollution, stricter auto emission standards, and new AIR TOXICS controls (MAXIMUM ACHIEVABLE CONTROL TECHNOLOGY).

Clean Air Scientific Advisory Committee (CASAC) A seven-member independent panel established by the CLEAN AIR ACT to review the basis (CRITERIA DOCUMENT) for each NATIONAL AMBIENT AIR QUALITY STANDARD every five years.

clean oil See BLACK OIL.

clean room A room maintained in a dust-free condition. A clean room is used to manufacture electronic components that would be damaged by contamination. Access usually requires special clothing and decontamination.

Clean Water Act (CWA) The basic federal water pollution control statute. The Water Quality Act of 1965 began setting water quality standards, and the 1966 amendments increased federal funding for sewage treatment plants. The 1972 amendments established a goal of zero toxic discharges and "fishable" and "swimmable" surface waters. Additional amendments passed in 1977 and 1987. The enforceable provisions include TECHNOLOGY-BASED effluent standards, administered through the NATIONAL POLLUTANT DISCHARGE ELIMINATION SYSTEM, for POINT SOURCES; a state-run control program for NONPOINT SOURCES; a CONSTRUCTION GRANTS PROGRAM to build or upgrade municipal sewage treatment plants; a regulatory system for spills of oil or hazardous waste; and a wetlands preservation program.

cleanup In hazardous waste management, the decontamination of water, soil,

or an aquifer that is determined to contain concentrations of a leaked or spilled substance that threatens the public health or the environment.

clear-cutting The removal of all trees in an area without regard to size or species. The process leaves large tracts of land without substantial vegetation, with a resulting increase in erosion.

clearwell An underground tank holding treated drinking water before it is distributed to consumers.

climate The weather patterns in a particular region, generalized over a long period of time. Humans are thought to be influencing climatic conditions through the addition of carbon dioxide to the atmosphere.

climax The last stage in ecological SUCCESSION. Relatively more stable, with a greater SPECIES DIVERSITY, than earlier, simpler stages.

clinker Solid residue formed in an incinerator from various noncombustible materials such as glass or metal.

closed system In physics, a system that does not exchange matter or energy with its surroundings. In ecology, a system exchanging energy, but not matter, with its surroundings. The planet earth is a closed system in the ecological sense, absorbing and radiating solar energy, but recycling matter within the biosphere. Compare OPEN SYSTEM.

closed water loop A process in which decontaminated wastewater is not discharged into a receiving stream but is reused. Any water lost during the process by evaporation or binding with some material is replaced by makeup water. Compare to OPEN WATER LOOP.

closed-cycle cooling A process in which cooling water used in industrial processes or in the generation of electrical energy is not discharged into receiving streams, where direct discharge can have adverse effects, but is circulated through COOLING TOWERS, ponds, or canals to allow the dissipation of the heat, and the water to be reused. Compare OPEN-CYCLE COOLING.

closure   The actions prescribed by regulations implementing the RESOURCE CONSERVATION AND RECOVERY ACT that must be performed at a hazardous waste facility if it will no longer receive waste for treatment or disposal. The actions include, among many others, the placement of a final cover on the buried waste, the establishment of a long-term groundwater monitoring program, and the filing of a notice in state property records that a hazardous waste facility has been closed at the location. The monitoring and property record notice are also termed postclosure actions.

closure plan   The written document, for a specific hazardous waste facility, outlining CLOSURE. See POST-CLOSURE PLAN.

Club of Rome   An informal international organization begun in 1968 by Dr. Aurelio Peccei, an Italian industrialist. The purpose of the group, limited to 100 members, is to foster understanding of the finite and interdependent nature of the world's resources. The organization sponsored the Project on the Predicament of Mankind, which led to the 1972 publication of the influential book, *The Limits to Growth*, followed by *Mankind at the Turning Point* (1974). Subsequently, many other reports on global problems and the future, including energy, waste management, education, and microelectronics, were issued.

coagulation   The grouping together of solids suspended in air or water, resulting in their PRECIPITATION. Coagulation is encouraged in wastewater treatment plants by the addition of ALUM, ferrous sulfate, and other materials, See COLLOID, FLOCCULATION, PRIMARY TREATMENT.

coal   A solid fossil fuel found in layers beneath the surface of the earth. The resource is mined and used primarily as a fuel to generate steam for the production of electricity. Coal is graded on the basis of heat content and classified as ANTHRACITE, BITUMINOUS, SUBBITUMINOUS, or LIGNITE. See SURFACE MINING and RECLAMATION.

coal gasification   The conversion of solid coal to a low-energy gas mixture containing mainly methane, hydrogen, and carbon monoxide. The process is not yet economically competitive with NATURAL GAS; however, the process provides a potential alternative fuel in the event of a shortage in crude oil supplies.

coal liquefaction   The conversion of solid coal to a liquid fuel. The process is not yet economically competitive with PETROLEUM; however, the process provides a potential alternative fuel in the event of a shortage in crude oil supplies.

coal tar   A crude mixture of aromatic hydrocarbons produced from the destructive distillation of coal. The mixture is usually a viscous material that can be used as a fuel and as a tar for roads and roofing. The mixture can also be further refined to produce a large array of chemicals, including creosotes, phenols, naphthalenes, and similar aromatic compounds. Products such as dyes, resins, perfumes, and flavoring agents are also prepared from coal tar.

coal workers' pneumoconiosis (CWP)   A chronic lung disease characterized by varying degrees of lung tissue FIBROSIS; the condition is caused by long-term overexposure to coal dust containing significant amounts of FREE SILICA, and is also associated with chronic inhalation of coal dust that has a very small free-silica content. See SILICOSIS.

coarse screen   See BAR RACKS.

Coastal Zone Management Act (CZMA)   A 1972 federal law, amended in 1980, that provides guidance and financial assistance to voluntary state and local coastal management programs. Goals of the programs include the protection of natural resources and the management of land development in coastal areas, along shorelines, and on shorelands (extending inland as far as a strong influence on the shore is expected). The state programs established under the Act vary widely in their approach and application.

cocarcinogen   In the two-stage model of carcinogenesis, a chemical substance that enhances the INITIATION stage by increasing the BIOAVAILABILITY of a CARCINOGEN, decreasing the metabolic detoxification of the carcinogen, or inhibiting DNA repair, among others. Cocarcinogens differ from PROMOTERS in that promoters act after the genotoxic initiation stage is complete to enhance the growth of a converted cell into a tumor.

cochlea   The spiral-shaped section of the inner ear connected to the auditory nerve. The nerve endings within this organ are damaged by excessive noise, with a resulting impairment of hearing.

Code of Federal Regulations (CFR)   The annual compilation of all current regulations that have been issued in final form by any federal regulatory agency; the publication is organized by subject titles. Environmental regulations are found under Title 40, Protection of the Environment. Occupational health and safety regulations are under Title 29, Labor.

coefficient   1. In mathematics, a number that multiplies another quantity or variable, such as the 7 in 7c.   2. In physics, a number or ratio that expresses the relationship between two quantities, given certain conditions.

coefficient of entry   In industrial ventilation, the ratio of the actual airflow rate into a HOOD to the ideal flow rate. The ideal flow rate would result if all hood STATIC PRESSURE were converted to VELOCITY PRESSURE, without losses. The coefficient is commonly calculated by dividing the measured hood velocity pressure by the hood static pressure and taking the square root of the result.

coefficient of haze (Coh)   A measurement of air visibility derived from the darkness of the stain on a white paper tape through which the air has been filtered. It is usually expressed as the number of Coh units per 1000 linear feet of air, with 1 per 1000 corresponding to clean air on a bright day and 5 per 1000 corresponding to significant visibility reduction due to smoke.

coefficient of permeability   See HYDRAULIC CONDUCTIVITY.

coevolution   Simultaneous EVOLUTION of two or more species of organisms that interact in significant ways.

coffin   A strong, shielded container used for transporting radioactive materials.

cogeneration   The use of steam or heat to process materials and generate electricity. For example, high-pressure steam may be routed through an electricity-generating turbine before its application in industrial processes, reducing the electricity demand from a central power station.

cohort   In EPIDEMIOLOGY, a group of people sharing one or more characteristics. A birth cohort consists of all persons born within a certain time period, usually a year. A group of persons exposed to similar levels of a toxic substance during a similar time period is a cohort.

cohort study   An epidemiological study that follows two groups, one exposed to a suspected disease risk factor, the other not exposed, and compares contagion rates.

coke   A solid carbon residue resulting from distillation of coal or petroleum. The product is used as a fuel and as a reducing agent in steelmaking. The volatile materials emitted to the air by coke ovens used in steel mills include known human carcinogens and are regulated by the United States Environmental Protection Agency.

cold-side ESP An ELECTROSTATIC PRECIPITATOR designed for flue gases less than 400°F (204°C) and located downstream from the AIR PREHEATER. The lower flue gas temperature means a smaller air volume and, thus, a smaller, less expensive unit. This type of precipitator is not as effective as a HOT-SIDE ESP on fly ash from low-sulfur coal combustion.

coliform bacteria Gram-negative, rod-shaped bacteria, including primarily FECAL COLIFORM, found in the digestive tract, but other forms are found in soil and water.

collecting surfaces The collection electrodes, either tubes or flat plates, that provide the attraction surface area in an ELECTROSTATIC PRECIPITATOR.

collection efficiency An expression of the performance of air pollution control equipment, or the percentage reduction in pollution concentration coming into the device compared with the pollutant concentration in the exhaust air. It is calculated by subtracting the pollutant concentration in the exhaust from the pollutant concentration in the incoming air and dividing by the incoming air concentration. This result is multiplied by 100 for expression as percent efficiency (by weight).

collection system 1. The underground pipe network that channels domestic sewage to a sewage treatment plant. 2. The underground pipes that capture and transport LEACHATE to the surface for treatment or disposal (LEACHATE COLLECTION SYSTEM).

colloids Particles with diameters of 1–1000 nanometers (10^{-9} meter) dispersed into a gaseous, liquid, or solid medium. Colloidal particles suspended in water or wastewater cannot be removed by filtration or sedimentation unless preceded by COAGULATION to increase their size.

colluvial Describing eroded material found at the bottom or on the lower slopes of a hill.

colony count A method for the quantification of bacteria in an environmental sample. A portion of a liquid sample or a dilution thereof is spread across the surface of a suitable solid nutrient medium and allowed to incubate. The number of bacterial colonies that develop on the surface are counted and the necessary mathematical calculations made to compute the number of bacteria per unit volume of the sample. The technique is based on the assumption that one bacterial cell will grow and divide to produce one colony on the surface of the nutrient medium.

colorimetry Methods of chemical analysis in which a change in color and/ or color intensity is the indication of the presence, concentration, or both of a particular material.

combined available chlorine Chlorine present in water as CHLORAMINES; produced by COMBINED RESIDUAL CHLORINATION. Compare FREE RESIDUAL CHLORINE.

combined residual chlorination The drinking water treatment method that involves the addition of chlorine to water at levels sufficient to produce, in combination with ammonia and/or organic amines, a COMBINED AVAILABLE CHLORINE residual. This chlorine residual maintains the treatment's disinfecting power throughout the water distribution system. Another approach to water chlorination is BREAKPOINT CHLORINATION.

combined cycle generation A system designed to increase the efficiency of a gas TURBINE. The otherwise-wasted heat energy from the hot gases used to drive the turbine is extracted with heat exchangers and used to produce steam for a conventional steam turbine that can generate electricity.

combined sewer   A water drainage pipeline that receives surface runoff as well as sanitary or industrial wastewater.

combustible liquid   A liquid with a FLASH POINT above 100°F and below 200°F.

combustible material   Any substance that will burn under ordinary circumstances.

combustion   A rapid chemical reaction of a fuel with oxygen that produces heat and light. The combustion of carbon fuels (wood, coal, natural gas, petroleum products) produces a mixture of exhaust gases that includes water vapor, carbon dioxide, nitrogen, and oxides of nitrogen.

combustion products   Gases, solids, or other material produced during the burning of some substance.

comfort chart   A graph of different combinations of air temperature, RELATIVE HUMIDITY, and air motion showing the percent of test subjects feeling comfortable under various conditions. The chart includes different comfort zones for summer and winter and is useful for the design and operation of air heating and cooling devices. The chart is applicable only to the culture and climate in which the test subjects' comfort zones were recorded.

command and control regulation   The use of detailed standards, regulations, permit provisions, penalties, and so forth, to meet a legislative mandate. Under this regulatory model, the legislature passes a law delegating to an administrative agency (such as the United States Environmental Protection Agency) the authority to write and enforce exacting rules for the regulated community to follow. This is the primary approach taken

in the United States to solve environmental problems.

commensalism   A form of species interaction in which one species is benefited but the other is unaffected; for example, shellfish may provide shelter to other, more mobile, organisms, but are neither harmed nor benefited as a result.

comminutor   A mechanical device that cuts and shreds solids as wastewater enters a treatment plant.

common ion effect   The decreased solubility of an ionized salt caused by the addition of a chemical that ionizes to form an ion that is the same as one formed by the salt. For example, if chemical AB forms A^+ and B^- ions and chemical BC, which forms C^+ and B^- ions, is added, then B^- is the common ion. The additional B^- ions from chemical CB increase the ion product $[A^+][B^-]$ such that it reaches the solubility product (the maximum amount of the two ions that can be present in a solution), after which the chemical AB will precipitate. The effect is important in the removal of ions from solution in water treatment.

communicable disease   A disease for which the causative pathogenic organism is readily transmitted by person-to-person contact, FOMITES, water, food, or air.

community   In ecology, the populations of plants and animals present in an ecosystem. Also called the biotic community.

Community Awareness and Emergency Response (CAER)   A program of the CHEMICAL MANUFACTURERS ASSOCIATION established to encourage, at the local level, planning for hazardous material emergencies by industrial facilities and nearby communities.

community water system   A public water system with 15 or more connections and serving 25 or more year-round residents and thus subject to United States

Environmental Protection Agency regulations enforcing the SAFE DRINKING WATER ACT.

compacted solid waste Solid waste after COMPACTION. The waste may be shredded first, compacted, and formed into bales. See SHREDDING, BALER.

compaction 1. The mechanical volume reduction (increase in density) of solid waste by the application of pressure. 2. The application of pressure to soil or clay, reducing its permeability to liquids.

compaction ratio The ratio of the volume of solid waste before COMPACTION to its reduced, compacted volume.

compartment model See BOX MODEL.

compensation point The point under water at which plant PHOTOSYNTHESIS just equals plant RESPIRATION. The water depth defines the lower boundary, where photosynthesis takes place, of the EUPHOTIC ZONE. Also called compensation level.

competing risks Causes of death of individuals that are not related to the particular risk factor being analyzed. For example, if one is studying the incidence of death in a human population from lung cancer, death of members of the study group from airplane accidents would represent a competing risk.

competition In ecology, the interaction among species or individuals of the same species in which they struggle to obtain the same food, space, or other essentials.

competitive exclusion The hypothesis stating that when organisms of different species compete for the same resources in the same habitat, one species will commonly be more successful in this competition and exclude the second from the habitat.

complete carcinogen In the two-stage model of carcinogenesis, an exposure containing materials that act as INITIATORS and PROMOTERS. Cigarette smoke is usually considered to be a complete carcinogen.

completed test The third, and last, part of the examination of water for the presence of bacteria of fecal origin. Cultures that are scored as positive in the earlier steps of the analysis (CONFIRMED TEST) are subjected to a verification by inoculating appropriate media (eosin methylene blue agar plates) and performing a GRAM-POSITIVE/GRAM-NEGATIVE stain on isolated colonies. See also PRESUMPTIVE TEST.

complex The incorporation into or combination of CATIONS with other molecules in such a way that the cations are no longer available to enter into reactions with other charged molecules.

compliance monitoring program The extensive follow-up groundwater monitoring required at a TREATMENT, STORAGE, OR DISPOSAL facility if the DETECTION MONITORING PROGRAM indicates a possible leak from the hazardous waste at the site. Data from UPGRADIENT WELLS are compared with data from DOWNGRADIENT WELLS for specific chemicals to help determine the source and extent of any groundwater combination.

compliance point The physical location with respect to a hazardous waste secure landfill from which groundwater samples will be taken to determine compliance with the United States Environmental Protection Agency groundwater protection standard.

composite sample A representative water or wastewater sample made up of individual smaller samples taken at periodic intervals.

compost The material produced by COMPOSTING, useful as a soil conditioner.

composting The controlled degradation by AEROBIC microorganisms of organic materials in solid waste to produce COMPOST, a soil conditioner and fertilizer. Wetted solid waste is stacked in piles or rows, which are turned periodically to ensure that sufficient oxygen is present for the DECOMPOSERS. The process is conducted on scales ranging from backyard heaps to tractor-using operations at municipal solid waste processing facilities.

compound A substance made up of two or more ELEMENTS, which are in a fixed proportion by weight. The various elements can only be separated by chemical reactions, and not by physical means. The physical and chemical properties of a compound are a result of the chemical combination of its elements and are not those of the individual elements. Compare MIXTURE.

comprehensive assessment information rule (CAIR) A chemical substance regulatory program implementing the TOXIC SUBSTANCES CONTROL ACT. The Act requires manufacturers to submit detailed information on a chemical, including data about its environmental fate (i.e., its persistence, transport, and distribution) and release, to the United States Environmental Protection Agency for use in determining the risk to human health and the environment. The CAIR is replacing the information submission requirements under the PRELIMINARY ASSESSMENT INFORMATION RULE (PAIR).

Comprehensive Environmental Response, Compensation, and Liability Act (CERCLA) The statute, also known as the SUPERFUND law, establishes federal authority for emergency response and cleanup of hazardous substances that have been spilled, improperly disposed, or released into the environment. The primary responsibility for response and cleanup is on the generators or disposers of the hazardous substances (see POTENTIALLY RESPONSIBLE PARTIES), with a backup federal response using a trust fund (see HAZARDOUS SUBSTANCES

SUPERFUND). The legislation was enacted in 1980 and significantly amended in 1986 (SUPERFUND AMENDMENTS AND REAUTHORIZATION ACT).

Comprehensive Environmental Response, Compensation, and Liability Information System (CERCLIS) A computerized system containing the basic information about and current status of a site being cleaned up under the NATIONAL CONTINGENCY PLAN, such as a Superfund hazardous waste site.

comprehensive general liability policy (CGL policy) An insurance policy covering a broad range of potential liabilities arising in a policy year, including claims resulting from sudden or accidental releases of pollutants, but not gradual leaks or normal pollutant emissions, which fall under the POLLUTION EXCLUSION CLAUSE. See also ENVIRONMENTAL IMPAIRMENT LIABILITY POLICY.

concentration The amount of a chemical substance in a given amount of air, water, soil, food, or other medium. The value can be expressed as mass of the chemical in a given mass of the medium, the volume of the chemical in a given volume of the medium, or the mass of the chemical in a given volume of the medium. For gaseous air contaminant concentrations, two expressions are appropriate: a volume/volume ratio and a mass/volume ratio. The volume/volume ratio units are typically PARTS PER MILLION (volume), equivalent to one liter of pollutant per million liters of air, or parts per billion (volume), one liter of pollutant in one billion liters of air. The mass/volume units are typically micrograms of pollutant per cubic meter of air. For airborne PARTICULATE MATTER, only mass/volume units are used, typically micrograms of particulate per cubic meter of air. In water, mass/volume and mass/mass ratios are used; the volume and mass of the aqueous medium are easily interchanged because one liter of water has a mass of one kilogram. Typical units are milligrams of pollutant per liter of water, which is the same as parts per million (mass), or micrograms of pollutant per liter of water, which equals parts per billion (mass). Soil and food concentrations are mass/mass ratios, in milligrams of a chemical per kilogram of medium, which is the same as parts per million (mass), or micrograms of a chemical per kilogram of medium, equal to parts per billion (mass). See Appendix for additional information.

concentration gradient   An expression of the change in the concentration of a material over a certain distance. Chemicals will diffuse from areas of higher concentration toward areas of lower concentration; the diffusion rate increases with an increase in the concentration gradient. This principle is used in the removal of pollutants from exhaust gases (SCRUBBERS) and water effluents (PACKED TOWER AERATION). The gaseous or liquid material into which the pollutant is diffusing is replenished rapidly to maintain its low concentration and, thus, a high collection efficiency. See FICK'S FIRST LAW OF DIFFUSION.

concurrent flow   The arrangement of material flow in systems designed to remove specific chemicals (such as sulfur oxides) from stack gases. The liquid containing the substance that absorbs or reacts with the undesirable gas enters the gas stream flowing in the same direction as the stack gas. Opposite of COUNTERCURRENT FLOW.

condensate   Liquid condensed from the vapor or gaseous state. In natural gas production, the liquid components sometimes present in the gas stream exiting a well or gases such as propane or butane that are readily condensed out to form LIQUEFIED PETROLEUM GAS (LPG) fuel.

condensation   The change of a gas or vapor to a liquid. At atmospheric pressure, the process is caused by the removal of heat from the gas or vapor. The amount of heat removed from (or released by) a unit of gas or vapor to cause it to become a liquid is called the heat of condensation, which is numerically equal to the HEAT OF VAPORIZATION of the liquid.

condensation nuclei   SUBMICRON particles naturally present in the atmosphere on which water vapor condenses to form droplets.

condenser   A heat-extraction device used in a steam engine or turbine to condense the steam to a liquid.

conductance   The ability of a material to transmit electricity; the opposite of resistance.

conduction   The transfer of heat by direct contact.

conductivity   See HYDRAULIC CONDUCTIVITY.

cone of depression   A drop in the WATER TABLE caused by a groundwater withdrawal rate exceeding the recharge rate. The resultant shape of the water table resembles an inverted cone.

confidence interval   A specified range of probability for a particular population statistic. For example, the statement "the 95% confidence interval for the mean dissolved oxygen (DO) level is 10.4–12.4 milligrams per liter" says that, based on information from a population sample, the true value for the mean DO level lies between 10.4 and 12.4 milligrams per liter, and that the chance that this range does not contain the true mean is 5 percent, or 5 in 100.

confidence limits   The endpoints of a CONFIDENCE INTERVAL.

confined aquifer   An AQUIFER located between two relatively impermeable layers of material with the groundwater confined under pressure significantly greater than atmospheric pressure. Also called an artesian aquifer.

confining layer A layer of underground rock or clay with low HYDRAULIC CONDUCTIVITY (permeability) that is located above and/or below an AQUIFER. Also called a confining bed.

confirmed test The second stage in the examination of water for the presence of bacteria of fecal origin. Cultures that are positive on the first portion of the testing procedure (PRESUMPTIVE TEST) are inoculated into tubes of brilliant green lactose bile broth and examined for fermentation when incubated at 35°C for 48 hours. If fermentation is present, a third stage, the COMPLETED TEST, is performed.

confluence The location at which two streams or bodies of water merge to become one.

confounding variable A characteristic, habit, exposure, and the like, that is linked to a disease risk and is found unequally in two groups in an epidemiological study, therefore confusing the comparison between the groups. For example, in a study of workplace lung cancer risk in coal miners compared to gold miners, if 75 percent of the sample population of coal miners were smokers, but only 25 percent of the gold miners smoked, a researcher finding excess lung cancer risk in coal miners could not conclude that coal mining is worse than gold mining in terms of lung cancer risk.

congener A chemical compound, person, or organism that resembles another in appearance or properties.

conifer One of the group of (mostly evergreen) cone-bearing trees or shrubs, such as pines, spruces and firs. They are softwoods. Compare HARDWOODS.

consent decree A binding agreement by both parties in a lawsuit that settles the questions raised by the case; no additional judicial action is required.

conservation Careful and organized management and use of some natural resource, emphasizing applied scientific principles. See also PRESERVATION.

constituent concentrations in waste extract table (CCWE table) A listing of hazardous waste treatment standards expressed as the concentration of certain hazardous waste substances in an extract of the treated waste, taken using the TOXICITY CHARACTERISTIC LEACHING PROCEDURE. If the extract concentration is less than the specified CCWE concentration, the treated waste may be disposed of in a landfill. See LAND DISPOSAL BAN. See also CONSTITUENT CONCENTRATIONS IN WASTES TABLE.

constituent concentrations in wastes table (CCW table) A listing of treatment standards expressed as substances present in a hazardous waste after treatment. If the results of an analysis of the treated waste (not a liquid extract, as in the CONSTITUENT CONCENTRATIONS WASTE EXTRACT TABLE) show the concentrations to be below the values in the CCW table, the waste may be disposed of in a landfill. See LAND DISPOSAL BAN.

construction grants program A program administered by the United States Environmental Protection Agency under the CLEAN WATER ACT that provides federal matching funds to build or upgrade PUBLICLY OWNED TREATMENT WORKS (municipal sewage treatment plants) to a level of SECONDARY TREATMENT, thus reducing municipal water pollutant discharges significantly.

constructive metabolism See ANABOLISM.

Consumer Product Safety Commission (CPSC) An agency of the federal government, established in 1973, responsible for issuing and ensuring compliance with regulations designed to protect the public from potentially hazardous household items, such as toys, lawn mowers, and flammable fabrics. The

CPSC has banned the following known or suspected carcinogens from certain consumer products: asbestos, benzene, vinyl chloride, and Tris (fire retardant in children's sleepwear).

consumers In ECOLOGY, organisms that gain energy by eating organisms. The place for a consumer in the FOOD CHAIN is defined by what it eats. HERBIVORES eat plants and are PRIMARY CONSUMERS; a human can be a primary consumer by eating plants or a SECONDARY CONSUMER by eating an animal that feeds on plants.

consumption, of water After water is withdrawn for industrial, commercial, or agricultural use, the degradation of the quality of the water or the loss of a significant portion of the water through evaporation, EVAPOTRANSPIRATION, or contamination by a substance such that the water cannot be reused. See WITHDRAWAL, OF WATER.

consumptive use of water See CONSUMPTION, OF WATER

contact chamber An enclosed vessel in which a gas is mixed with (and commonly absorbed by) a liquid, usually water.

contact hazard A chemical material with irritant or corrosive properties that, in case of a spilling accident, can be harmful to persons if it comes into contact with their skin or eyes.

contact inhibition A property of normal animal cells that results in the stopping of cellular growth when one cell touches or contacts another cell. A type of control that limits the growth and reproduction of normal cells. The inhibition does not occur in tumor cells, which continue to grow even when crowded.

contact process A basic process for the industrial manufacture of sulfuric acid. Oxidized sulfur, as sulfur dioxide, is cat-

alytically converted to sulfur trioxide, which is absorbed into a sulfuric acid liquid.

contact stabilization A version of the ACTIVATED SLUDGE PROCESS in which raw sewage is mixed with activated sludge in a small contact tank for about one-half hour, time enough for the microorganisms to absorb the wastewater organic material but not to decompose it. The wastewater and sludge mixture is then separated in a CLARIFIER, and the wastewater is discharged. The sludge settling in the clarifier is pumped to a stabilization tank, where about 3 hours of microbial decomposition of the absorbed organics takes place. The sludge is then returned to the contact tank for mixing with incoming raw sewage.

contact urticaria An allergic reaction of the skin caused by direct contact with some substance; the reaction is characterized by intense itching of red, swollen, fluid-filled patches. See also ALLERGEN.

containment building A term employed in the nuclear industry to identify the building housing the nuclear reactor and associated equipment. The building is constructed to confine any RADIOACTIVE substances that may be released from the reactor in the event of an accident.

containment vessel A pressurized steel vessel that houses the CORE of a nuclear reactor. This vessel contains the liquid used to cool the core and is designed to confine any RADIOACTIVE substances that may be released from the reactor core in the event of an accident.

continental drift According to the theory of plate tectonics, the movement of the earth's continents as the crustal plates on which they rest move across the surface of the semiliquid mantle of the earth.

continental shelf The seafloor sloping away from the continents at an angle of about 1°. Commonly defined as the shoreline area under water less than 200 meters deep and usually extending about 70 kilometers. At the outer edge of the shelf, the seafloor drops sharply.

continuity equation   The relation, based on the conservation of mass, that equates the VOLUMETRIC FLOW RATE Q of an incompressible fluid in a duct or pipe to the product of the fluid velocity V and the cross-sectional area A of the duct or pipe: $Q = VA$. If the area increases, then the velocity must decrease, and conversely. The equation is also applied to liquid flow through a system, stating that the flow in, Q_{in}, flow out, Q_{out}, and change in the storage volume for a given time must be in balance: $Q_{in} - Q_{out} = \Delta$ storage volume. For example, if water flow into a reservoir is 3000 cubic meters per day and the flow out is 1000 cubic meters per day, then during one day the reservoir volume must increase by 2000 cubic meters, excluding evaporation and seepage losses.

continuous analyzer   1. An instrument that samples and analyzes ambient air on a continuous basis, producing short-term (typically 5-minute) averages recorded on a computer disk or a paper tape.   2. Laboratory instrumentation that allows the operator to load a number of samples at one time. The instrument automatically draws the proper amount of material from each sample, adds the necessary reagents, promotes the required reaction between the material of interest and the reagents, monitors the result of the test, and provides the operator with a data printout of the results.

continuous emission monitoring (CEM)   The continuous sampling and analysis of gases, particulate matter, or OPACITY by monitors placed inside smokestacks. This type of arrangement provides more useful information than STACK SAMPLING, which may be conducted only every several years. Also called in situ monitoring.

continuous-feed reactor (composting)   A method of converting yard litter, sludge, or other solid refuse into humic, soil-like material. The solid waste is moved through the apparatus in such a way that the conversions are done rapidly, with the solid refuse added to the front end of the reactor and humic substances removed from the back in a continuous fashion. This contrasts to the BATCH METHOD, in which the proper composting mixture is prepared, placed in a vessel or static pile, allowed to compost for a specific period, and then removed.

continuous-flow microbiological system   An operation in which liquid wastes or media are added to a decomposition or growth vessel to allow for the growth and metabolism of bacteria on a continuous basis. The reactants are added to the vessel and the products are removed on a continuous basis. This system contrasts to the BATCH METHOD, in which the reactants are added at the start, the reaction is allowed to proceed, and the entire contents of the reaction vessel are discharged at the end of the growth period.

continuous-flow system   A system having an uninterrupted, time-varying input and output of material in one of two ways. The plug-flow model assumes the effluent from the system is discharged in the same order as the material entering the system, with little mixing. It is appropriate for systems that have an extended length and small cross-sectional area, such as small streams and pipes. The completely mixed–flow model assumes that material flowing through the system is uniformly mixed while in the system and thus the discharge is physically and chemically the same as the system contents. This model is appropriate for most lakes, reactors, tanks, and other mixing vessels.

continuous noise   A constant sound that varies less than five DECIBELS during a noise measurement or evaluation.

continuous variable   A quantitative variable that can take an infinite number of values in a set (for example, the mass of a soil sample), but practically the values are limited by the accuracy of the measurement method. See also DISCRETE VARIABLE.

contour ditch   An irrigation ditch that follows the elevation of the land.

contour mining   A type of surface mining of coal in which coal is removed from relatively shallow deposits by removing the earth and rock covering the deposit. The mining proceeds around the natural topographical features of a hill or mountain. See also SURFACE MINING.

control group   A group of humans, animals, or plants that is not treated or exposed in an epidemiological study or in an experiment conducted to determine the effect of some agent or condition. For example, in an experiment to determine the effect of high temperature on the activity of flies, the experimental group would be maintained at some elevated temperature while the control group is held at ambient temperatures.

control rods   Long cylinders of neutron-absorbing material, such as cadmium or boron, that are part of a nuclear reactor CORE. The rods are lowered or raised as required to control the rate of NUCLEAR FISSION.

Control Techniques Guidelines (CTGs)   A series of United States Environmental Protection Agency documents providing technical and economic information on the control of VOLATILE ORGANIC COMPOUNDS; the guidelines are used by state regulatory agencies to set limits for volatile organic compound emissions for existing sources in ozone NONATTAINMENT AREAS.

controlled area   In a facility using radioactive materials, a defined area within which worker exposure to IONIZING RADIATION is monitored and controlled by an employee specifically responsible for radiation protection.

controlling variable   The material or factor that determines the response of a process or experiment, given certain conditions. For example, the most important input controlling the yield of a crop in an area may be identified as the level of soil nitrogen. Note that if adequate nitrogen is added, then the controlling variable for crop yield will shift to some other factor, such as soil mixture. Related to LIEBIG'S LAW OF THE MINIMUM and LIMITING FACTOR.

convection   The transfer of heat by a moving fluid, such as air or water.

Convention on International Trade in Endangered Species of Wild Fauna and Flora (CITES)   A 1973 treaty that binds nearly 100 signing countries to establish a system of permits for the exporting and importing of endangered species or products of organisms that are endangered. The CITES protected species list contains about 1200 animals and plants. The list can be found at Title 50, Part 23, Section 23 of the CODE OF FEDERAL REGULATIONS.

conventional pollutants   Under the CLEAN WATER ACT, the following pollutants: materials exerting a BIOCHEMICAL OXYGEN DEMAND, FECAL COLIFORM, SUSPENDED SOLIDS, oil and grease, and pH. The Act specifies that the definition of a conventional pollutant is not limited to these five, but no additional categories have been added.

convergence   1. In meteorology, the flowing together of air masses. Air converges toward the center of low-pressure systems, where it rises. See DIVERGENCE. 2. Another name for CONVERGENT EVOLUTION.

convergent evolution   The development of plant or animal species that are similar in appearance and ecosystem function in separate, but environmentally similar, geographic regions. The species similarity is explained by independent adaptation to the same environmental conditions. Also called convergence. See ECOLOGICAL EQUIVALENTS.

coolant   Any gas (air) or liquid (water, antifreeze) used to conduct heat away from some source, such as a nuclear reactor or engine.

cooling pond   A holding pond into which hot water is discharged to allow cooling prior to reuse or discharge.

cooling tower   A large structure designed to transfer heat from industrial or electric power plant cooling water to the atmosphere. Wet cooling towers operate by introducing the water directly to the air and forcing evaporative cooling. Dry cooling towers retain the warm water within a piping system and allow heat transfer by CONDUCTION and CONVECTION only.

cooling   tower   blowdown   See BLOWDOWN.

cooling water   Water used to remove waste heat from industrial processes or turbines used to produce electricity.

coppicing   The cutting back of trees or bushes to cause shoots or sprouts to grow. The new growth is then harvested. Coppicing is an efficient method for the controlled harvest of firewood.

coprophagous   Describing organisms other than bacteria that feed on fecal matter, such as certain insects.

coprostanol   A LIPID soluble material, the presence of which indicates contamination with fecal material.

core   The part of a nuclear reactor where FISSION takes place; the location of the FUEL RODS.

core sample   A sample of soil or sediment taken with the aid of a pipe or tube that is pushed or driven into the soil or sediment. The soil that is removed is taken from the sampling device in a cy-

lindrical section in which the vertical positioning of layers of sediment is maintained.

Coriolis force   The apparent force, resulting from the earth's rotation, that deflects air or water movement. Winds seen from a moving frame of reference (the earth) are deflected to the right in the Northern Hemisphere and to the left in the Southern Hemisphere. The force influences the circulation of air in the atmosphere and the movements of water in the oceans.

corrosivity   A characteristic used to classify a hazardous waste. A waste is corrosive (and hazardous) if the pH is less than or equal to 2.0 or greater than or equal to 12.5.

cosmic radiation   See COSMIC RAYS.

cosmic rays   IONIZING RADIATION, both electromagnetic energy and particles, originating in outer space. Secondary cosmic rays, produced in the atmosphere by incoming cosmic rays, are the immediate source of human cosmic radiation exposure, which makes up about 30–40 percent of a typical BACKGROUND RADIATION dose and about 10–15 percent of the average American's annual dose of ionizing radiation from all sources.

cost-effectiveness analysis (CEA)   A study to determine the lowest-cost method of achieving a certain defined goal.

cost-benefit analysis (CBA)   A calculation of the present value of the costs and benefits associated with a project or action. Nonmonetary costs or benefits, such as illnesses or deaths avoided, are converted to monetary units for the analysis. Such an analysis is required of all major federal regulations by EXECUTIVE ORDER 12291.

cost-benefit ratio   The results of a COST-BENEFIT ANALYSIS expressed as costs divided by benefits or as benefits divided by costs (benefit-cost ratio).

coulomb (C)   The SI unit of electric charge corresponding to a current of one ampere in one second (1 C = 1 A s).

Council on Environmental Quality (CEQ)   A three-person panel established by the NATIONAL ENVIRONMENTAL POLICY ACT (NEPA). The panel administers, with staff, the ENVIRONMENTAL IMPACT STATEMENT process begun by the NEPA. The CEQ also issues an annual report on environmental quality.

countercurrent flow   The arrangement of material flow in a system designed to remove specific chemicals (such as sulfur oxides) from STACK GASES. The liquid containing the substance that absorbs or reacts with the undesirable gas and the stack gas move through the system in opposite directions, and collection of the contaminant occurs when the streams collide. See also CONCURRENT FLOW.

counting chamber   A special device used to facilitate the counting of cells by microscopic techniques.

covalent bond   A strong attraction between atoms that results from the sharing of electrons between two atoms. A single covalent bond results from the sharing of a pair of electrons by two atoms, each atom contributing one electron to the pair. The bonds between carbon atoms in organic molecules serve as examples. Covalent bonds are much more difficult to break than the weaker attractions between atoms illustrated by hydrogen bonds and ionic bonds.

cover   See COVER MATERIAL.

cover material   Soil required to be placed over solid waste at a SANITARY LANDFILL at the end of each day to control odors, flies, rodents, mosquitoes, blowing litter, fire. Usually a 6-inch depth is required. Also called cover and daily cover.

cracking   The utilization of heat and/ or a CATALYST to reduce high-molecular-weight hydrocarbons in crude oil to smaller molecules, which are the constituents of products such as gasoline, ethylene, or heating oil.

cradle-to-grave   A phrase used to describe the comprehensive management of hazardous waste under the RESOURCE CONSERVATION AND RECOVERY ACT. Substances meeting the definition of hazardous waste are regulated from their point of generation to their final treatment and/ or disposal.

crankcase blow-by   See BLOW-BY.

criteria document   A compilation of human health and environmental effects used by the United States Environmental Protection Agency to set NATIONAL AMBIENT AIR QUALITY STANDARDS for the CRITERIA POLLUTANTS.

criteria pollutants   The six air pollutants constituting the NATIONAL AMBIENT AIR QUALITY STANDARDS. The standards are set by the United States Environmental Protection Agency using CRITERIA DOCUMENTS. The criteria pollutants are carbon monoxide, nitrogen dioxide, particulate matter, lead, sulfur dioxide, and ozone.

critical   Describing a nuclear reactor or weapon in which a self-sustained fission chain reaction takes place. In the case of a nuclear reactor, the term describes normal operation and does not imply any emergency or dangerous condition.

critical aquifer protection area (CAPA)   A RECHARGE ZONE for certain SOLE SOURCE AQUIFERS designated for additional protection under the SAFE DRINKING WATER ACT.

critical mass The smallest amount of FISSIONABLE material that will allow a self-sustaining nuclear CHAIN REACTION without the aid of other MODERATORS.

critical organ An organ that an animal cannot live without, such as heart or brain, in contrast to organs that can be lost without loss of life, such as the spleen.

critical point The location downstream from a waste discharge at which the DISSOLVED OXYGEN of the water is lowest; the lowest point on an OXYGEN SAG CURVE.

critical reactor A nuclear reactor that is sustaining a spontaneous nuclear chain reaction. Such a reactor is engaged in normal operations, and the designation does not imply any emergency or dangerous condition.

critical wind speed In air pollution dispersion calculations, the wind speed (corresponding to the highest estimate of the ground-level, air concentration) downwind from a source, such as a smokestack. The critical wind speed depends on atmospheric stability and the height of release of the air contaminant.

crocidolite A form of asbestos-containing mineral, referred to as blue asbestos.

cross-media pollution The transfer of chemical contaminants from one environmental medium (air, water, soil) to another.

cross-sectional method In EPIDEMIOLOGY, a study designed to determine the prevalence of disease in a population along with existing population characteristics, such as the finding that people who watch more than 6 hours of television daily have a higher-than-average rate of heart disease. This snapshot approach is not useful for answering cause-and-effect questions; nor does it indicate which came first, the disease or the characteristic.

crown fire A forest fire among the tops of trees.

crucible furnace A furnace that raises the temperature of a small porcelain container (crucible), reducing the contents to ash. The device is used to process particulate samples for determination of the ash content (for example, the amount of metals) that will remain following burning at high temperatures.

crude ecological density The number of organisms per unit area or per unit volume of total space in an ecosystem. Compare with ECOLOGICAL DENSITY, which only includes the total HABITAT area.

crude oil A complex mixture of liquid hydrocarbons in the unrefined state as produced from underground formations.

crude oil fraction The hydrocarbon compounds in a CRUDE OIL that have boiling points within a certain range. PETROLEUM is separated into its components in an oil refinery by fractional distillation; i.e., the oil is heated, and the compounds that will vaporize within a certain temperature range are removed and condensed, resulting in various products such as gasoline and kerosene.

crude rate In public health statistics, a rate that includes the entire population in the population at risk. Specific rates narrow the population at risk to a certain age group, sex, race, or other category.

crystalline silica The mineralogical form of silica, or silicon dioxide, that causes SILICOSIS after prolonged exposure. Exposure to noncrystalline silica, or amorphous silica, does not cause silicosis. The three types of crystalline silica are quartz, tridymite, and cristobalite. Also called free silica.

cubic meter (m³) A metric volume equal to 1000 LITERS, 264.2 gallons, or 35.31 cubic feet.

cullet Mixed scrap glass.

cultural eutrophication The excessive addition of plant nutrients into aquatic ecosystems by human activities. A typical result is the growth of aquatic weeds and algal slime to such an extent that use of the water by humans is prohibited.

culture A term employed in MICROBIOLOGY to designate the growth and multiplication of bacteria or fungi in an artificial, closed container.

culture dish A shallow device for the cultivation of microorganisms on a solid nutrient medium (an agar) prepared in the laboratory. The most common device is a Petri dish measuring 100 millimeters by 15 millimeters.

culture media Solid or liquid substances prepared and sterilized in the laboratory to provide the nutrients needed for the growth of bacteria or fungi.

cumulative probability distribution A diagram of a cumulative distribution function $F(x)$, which is the probability that a random variable x will be less than or equal to a certain value of x. The figure shows the annual probability of an hourly ozone concentration being less than or equal to x, which could be created for a particular location based on past experience.

Cunningham correction factor A modification of STOKES' LAW that increases the terminal SETTLING VELOCITY for particles smaller than about 5 microns in diameter, accounting for the ability of the particle to slip between gas molecules in the air.

curie (Ci) A unit of RADIOACTIVITY corresponding to 3.7×10^{10} decays per second. RADON levels, however, are expressed as the radon decay rate per volume of air, usually picocuries per liter of air. A picocurie is 1×10^{-12} curie.

cut diameter The diameter of particles collected with 50 percent efficiency by a particulate control or air-sampling

device, or so that about half of the particles are captured.

cutting oil An oil or oil-water emulsion used to reduce heat and friction when operating metal-working machines. Prolonged skin contact can cause occupational acne.

cyanosis A condition in which the skin appears blue, caused by insufficient oxygen in the blood.

cycles per second The number of wave cycles passing a given point in one second; a FREQUENCY unit. The frequency of sound waves is sometimes still given in cycles per second. For electromagnetic radiation, one cycle per second is termed one HERTZ.

cycling load facility A power station that generates electrical energy to meet a fluctuating daily demand. See also BASE LOAD FACILITY.

cyclone In meteorology, a counterclockwise air circulation with a low-pressure center; tornadoes and hurricanes are intense cyclones. See also ANTICYCLONE.

cyclone collector A cylindrical or cone-shaped air-cleaning or air-sampling device designed to remove particles from an airstream by CENTRIFUGAL FORCE. The particulate-containing air enters the top of the cyclone and spins downward, throwing the particulate outward against the cyclone wall. The particles fall into a collection hopper, and the cleaned air exits the top of the cyclone.

cytotoxin Any toxic material that kills cells.

D

daily cover See COVER MATERIAL.

Dalton's law The law of partial pressures, discovered by English scientist John

Dalton, stating that a mixture of non-reacting gases or vapors exerts a pressure equal to the sum of the pressures that would be exerted by each gas or vapor alone. For example, the mixture of gases in the atmosphere exerts an ATMO-SPHERIC PRESSURE, which is equal to the total individual pressures of nitrogen, oxygen, argon, carbon dioxide, water vapor, and other components.

damage risk criterion   That noise level above which permanent hearing loss is likely to occur.

damp   In coal mining, a poisonous or explosive gas in a mine. Carbon monoxide is known as white damp, and methane is known as firedamp.

damper   In-duct movable plates used to adjust air VELOCITY PRESSURE and thereby balance the system airflow in the local exhaust ventilation system of a workplace.

dangerously reactive material   A substance that will react with itself or with air or water to produce a condition hazardous to individuals or the environment, for example metallic sodium.

daphnid   Small freshwater crustacean belonging to the species *Daphnia magna* or *D. pulex.* The organisms are used in the laboratory testing of pollutants for toxicity to aquatic biota.

Darcy's law   An equation stated by French engineer Henri Darcy in 1856 that relates groundwater velocity to the product of the HYDRAULIC CONDUCTIVITY (PERMEABILITY) of an AQUIFER and the slope of the WATER TABLE (the HYDRAULIC GRADIENT). The velocity V is given by

$$V = -K\frac{\Delta H}{L}$$

where K is the hydraulic conductivity and ΔH is the loss of HYDRAULIC HEAD over the distance L.

daughter   The material resulting from the RADIOACTIVE DECAY of a parent material. For example, radon-222 is a daughter of radium-226; the decay products of radon-222 are its daughters. Also called progeny.

day-night sound level (L_{dn})   The weighted equivalent sound level in DECIBELS for any 24-hour period. Ten decibels are added to the sound levels recorded from 10 P.M. to 7 A.M. to account for the greater disturbance typically caused by nighttime noise. A measure of community noise; not used for workplace evaluations.

de minimis   Quantities of pollutants or other chemical substances that are small enough to be exempt from environmental regulations. From *de minimis non curant lex,* "the law does not concern itself with trifles."

death phase   The terminal stage of growth of bacteria in laboratory culture. After the bacteria in the culture media have exhausted the supply of available nutrients, growth (cell division) stops. When the cells can no longer maintain viability, cell death results and the number of viable cells decreases.

death rate   The number of deaths per 1000 persons in a population in a given year. The rate is calculated by dividing the annual number of deaths by the midyear population and then multiplying by 1000.

decay product   An element or isotope resulting from radioactive decay; it may be stable or radioactive (undergo further decay). See RADIOACTIVE SERIES.

decay series   An illustration of the series of elements (DAUGHTERS) produced during the radioactive decay or decomposition of an unstable element until it reaches the point where it is no longer unstable or radioactive.

decibel (dB) A unit for expressing SOUND PRESSURE LEVEL or SOUND POWER LEVEL. The logarithmic decibel scale for sound pressure levels extends from 0 (the hearing threshold) to over 130 (causes pain); normal speech is at about 60 dB.

decibels, A-weighting network (dBA) The frequency-weighted SOUND PRESSURE LEVEL that best matches the noise sensitivity of the human ear. The overall measured sound pressure level is a sum of the sound pressures measured at the various frequencies composing a sound. Noise meters can be set to the A-weighting network, which is derived from the FLETCHER-MUNSON CONTOURS and reflects the relative insensitivity of the human ear to lower-frequency sounds. The meter adjusts downward the sound measurements at the lower frequencies to compute a total dBA. Noise standards are almost always in dBA units.

deciduous Describing vegetation that loses leaves seasonally. Compare EVERGREEN.

deciduous forest biome See TEMPERATE DECIDUOUS FOREST.

decommissioning Taking a power plant, industrial facility, machinery, or vehicle out of use. The monetary costs and environmental hazards of decommissioning a nuclear electric power plant are a feature of the debate over the expansion of the nuclear power industry.

decomposers A community of organisms, including bacteria and fungi, that metabolically break down complex organic matter into simpler materials. Used extensively to treat waste from domestic sewage.

decomposition The breakdown of a complex material into simpler materials. The complex material can be organic or inorganic, and the decomposition can be caused by heat, sunlight, water, chemicals, or metabolism. Metabolic decomposition is performed by DECOMPOSERS.

decontamination The physical or chemical removal of potentially toxic materials (chemical or radioactive substances) to an acceptable level.

decontamination factor The ratio of the amount of a given radioactive substance entering a cleaning process to the amount exiting the process.

deep ecology A perspective on environmental problems that emphasizes the interrelatedness of the earth and its biota, the equal importance of all species, and the need for radical social, economic, and political changes to improve and maintain environmental quality.

deep-well injection Pumping a waste liquid into geological strata below ground. See CLASS I, II, III, IV, V INJECTION WELLS.

defect action level (DAL) The allowable amount of insects or insect body parts in food, set by the United States Food and Drug Administration (for example, 50 insect fragments and 2 rodent hairs per 100 grams of peanut butter).

deficiency disease A disorder caused by the lack of some essential nutrient in the diet (such as scurvy, caused by a vitamin C deficiency).

definitive test A test of the acute or chronic toxicity of a chemical to an aquatic test organism, usually expressed in concentration-response units, such as a median lethal concentration (LETHAL CONCENTRATION: 50 PERCENT) or the percent of the test organisms that exhibited a certain response (EFFECTIVE CONCENTRATION: X PERCENT).

deforestation The removal of trees and other vegetation on a large scale, usually to expand agricultural or grazing lands. Global deforestation has contributed to the rise in atmospheric CARBON DIOXIDE levels over the past century. See GREENHOUSE EFFECT.

degenerative disease A disorder characterized by the gradual loss of some vital function, such as nerve function. For example, chronic exposure to low levels of pesticides may cause the gradual loss of motor nerve function.

degradation 1. A decrease in the quality of the environment. 2. The chemical, physical, or biological breakdown of a complex material into simpler components.

degreasing The removal of grease and oil from wastewater, machinery, or similar items.

degree-day A unit expressing the extent to which a daily mean air temperature is above (or below) a certain standard. For example, three daily temperatures of 78°F would be recorded as 18 degree-days over a standard of 72°F. Seasonal or annual degree-day totals for an area are used to estimate heating (or, at times, cooling) requirements for a building or other facility.

deionization The removal of all charged atoms or molecules from some material such as water. For example, the removal of salt from water involves the removal of sodium ions (Na^+) and chloride ions (Cl^-). The process commonly employs one resin that attracts all positive IONS and another resin to capture all negative ions.

deionized water Water that has been passed through resins that remove all IONS.

Delaney clause A 1958 amendment to the federal FOOD, DRUG, AND COSMETIC ACT that prohibits the addition to food products of any additive that is known to cause cancer in animals or humans.

delisting The formal removal of a chemical substance from a United States Environmental Protection Agency list of regulated materials, such as a list of hazardous wastes or hazardous substances.

This would be done, for example, if recent experimental or clinical evidence indicated that a substance is not as toxic as originally thought.

demister See MIST ELIMINATOR.

demographic transition The pattern of population growth exhibited by the now-developed countries during the nineteenth and early twentieth centuries. During the first of three stages, society experiences historically high birth rates and high death rates, especially in the young. Population growth is low. The second stage reflects the economic development of a country and the resulting higher standard of living; birth rates stay high, but death rates for infants and children decline sharply. This stage is a period of rapid population growth. The third stage arrives with a fall in birth rates in the developed country to match lowered death rates. Population growth returns to a low level.

demography The statistical study of human population size, growth, density, and age distribution.

denaturing In nuclear fuel management, the addition of NONFISSIONABLE material to nuclear reactor BY-PRODUCTS to make the material unfit for use in a nuclear bomb.

dendritic Treelike. See also ARBORESCENT.

denitrification The removal of nitrate ions (NO_3^-) from soil or water. The process reduces desirable fertility of an agricultural field or the extent of undesirable aquatic weed production in aquatic environments. See DENITRIFYING BACTERIA.

denitrifying bacteria Bacteria in soil or water that are capable of anaerobic respiration, using the nitrate ion as a substitute for molecular oxygen during their metabolism. The nitrate is reduced to nitrogen gas (N_2), which is lost to the atmosphere during the process.

density (ρ) 1. Mass per unit volume. Common units are kilograms per cubic meter ($kg\ m^{-3}$), grams per cubic centimeter, pounds per cubic yard, and pounds per cubic foot. 2. Number of individuals of a particular species per unit area of land.

density-dependent factor An influence on population growth that increases with size (density) of the population. Examples include PARASITISM, PREDATION, and COMPETITION. See also DENSITY-INDEPENDENT FACTOR.

density-independent factor An influence on population growth the strength of which is not affected by the size (density) of the population. The impact of seasonal cold weather on plant populations is an example. See also DENSITY-DEPENDENT FACTOR.

deoxyribonucleic acid (DNA) The macromolecule containing the genetic information governing the properties of individual cells and organisms; the genes of a cell. Exact copies of the genes of each cell are transferred to each daughter cell during cell division. Changes in the structure of DNA (mutations) can occur naturally or as a result of exposure to certain chemicals or radiation. Some changes result in the death of the cell, and others are passed on to future generations as an altered gene. Protection provided to individuals undergoing medical x-rays is designed to prevent alteration in the DNA structure of gametes. See MUTAGEN and MUTATION.

Department of Energy (DOE) A federal executive department, created in 1977, with responsibility for energy policy of the United States.

dependency load The ratio of nonworkers (children and retirees) to workers in a human population: the higher the ratio, the greater the dependency load.

The ratio is most useful when applied to industrial economies in which children are not employed and large numbers of older people depend on public pensions financed by taxes on the younger (working) generation.

dependent variable A characteristic or condition of an object, person, population, system, and so on, that changes its value or degree with changes in another, independent, variable. In epidemiology, the disease risk associated with an exposure, habit, or condition. For example, the increase in lung cancer risk with the number of cigarettes smoked daily. Compare INDEPENDENT VARIABLE.

depleted uranium Uranium that contains less than 0.7 percent of the fissionable isotope uranium-235.

deposition 1. The washout or settling of material from the atmosphere to the ground or to surface waters. 2. The absorption or ADSORPTION in the respiratory tract of inhaled gases, vapors, or particles.

depuration A process during which an organism, such as an oyster, eliminates dangerous chemicals or microorganisms when placed in uncontaminated water.

derived-from rule The United States Environmental Protection Agency regulatory provision that any waste derived from the treatment, storage, or disposal of a HAZARDOUS WASTE is itself a hazardous waste.

dermatitis Inflammation or irritation of the skin. The condition can be caused by allergic reactions associated with contact with some plants or excessive contact with irritant chemicals, such as strong acids.

dermatosis Any abnormality or disease of the skin, including DERMATITIS. The condition can be caused by overexposure to toxic chemical agents, bacterial or viral infection, or excessive exposure to sunlight.

desalination   The removal of salts from water to allow it to be used for drinking, irrigation, or industrial processing. The two main desalination techniques are DISTILLATION and OSMOSIS.

descriptive epidemiology   Studies of the distribution of human disease that describe disease rates or prevalence in various subpopulations based on age groups, sex, race, occupation, and so forth, without attempting to identify specific causes or environmental exposures that may increase disease risk in these groups. See also COHORT STUDY, CASE-CONTROL STUDY.

desertification   A process whereby land that is covered with vegetation is converted to desert. The term is generally applied to the production of artificial deserts where people have intensified the problems caused by droughts through overgrazing marginal land, repeated burning of natural vegetation, intensive farming of arid land, aggressive removal of trees, and prolonged irrigation of arid land for agricultural use.

design capacity   The number of tons that a solid waste burning facility is engineered to process in 24 hours of continuous operation.

design flow   The average flow of wastewater that a treatment facility is built to process efficiently, commonly expressed in millions of gallons per day (MGD).

desorption   The removal of a substance that has been absorbed or adsorbed by another material.

destruction and removal efficiency (DRE)   An expression of incinerator performance, in terms of the percent of a particular incoming chemical material that is destroyed. Calculated as $(I - O/I) \times 100$, where I is the input rate of the

material to be incinerated and O is the output rate of the material.

destructive distillation   The heating of organic matter, such as coal, oil, or wood, in the absence of air or oxygen. The process results in the production of volatile substances that are removed and recovered. The solid residue that remains is a mixture of carbon resembling COKE and ash.

desulfuration   A metabolic oxidation of a sulfur-containing organic compound within a biological system, resulting in the incorporation of molecular oxygen into the molecule and the concurrent elimination of sulfur.

desulfurization   The removal of sulfur-containing compounds, such as sulfur dioxide, from a gas exhaust.

detection limit   The smallest amount of a particular chemical that can be detected by a specific analytical instrument or method. See INSTRUMENT DETECTION LIMIT, METHOD DETECTION LIMIT, and QUANTITATION LIMIT.

detection   monitoring   program   Groundwater monitoring at the boundary of a TREATMENT, STORAGE, OR DISPOSAL facility (POINT OF COMPLIANCE) to detect any contamination caused by leaks from the hazardous waste at the facility. The materials for which the samples must be analyzed (the INDICATOR PARAMETERS/CONSTITUENTS) are specified in the facility permit.

detention basin   A relatively small storage lagoon for slowing stormwater runoff; it is filled with water for only a short time after a heavy rainfall. See also RETENTION BASIN.

detention time   The time interval for which stormwater is retained in a detention basin.

detergent   1. A synthetic, water-soluble compound capable of holding dirt in

SUSPENSION, emulsifying oils, or acting as a wetting agent; these compounds are commonly used in household cleaning agents. 2. An oil-soluble substance added to motor oils for the purpose of holding foreign substances in suspension.

detritivore   Animals (such as insects), that feed on particulate material derived from the remains of plants or animals. See DETRITUS FOOD CHAIN.

detritus   Dead organic matter derived from plant or animal body parts and excretions. Material that accumulates on the surface of the ground or the bottom of bodies of water.

detritus food chain   A feeding pattern in which an animal community survives through the consumption of decaying organic plant matter. Compare GRAZING FOOD CHAIN.

deuterium   An isotope of hydrogen that contains one proton and one neutron in the nucleus of the atom. Also called heavy hydrogen.

Deutsch-Anderson equation   The basic equation used in the design of ELECTROSTATIC PRECIPITATORS; the collection efficiency of the precipitator η is $\eta = 1 - e^{-w(A/Q)}$, where A is the total collection plate surface area, Q is the exhaust gas VOLUMETRIC FLOW RATE, w is the particle DRIFT VELOCITY, and e is the base of the natural logarithm. Named for W. Deutsch and E. Anderson.

dew point   The temperature at which, for a given water vapor content and constant pressure, condensation of water from the atmosphere commences. The temperature at which any gas begins to condense.

dewatering   Physical removal of water from sludge by pressing, CENTRIFUGATION, or air drying. The resulting product can be composted, landfilled, or burned.

diatomaceous earth   A geological deposit consisting of the remains of microscopic, unicellular plants called DIATOMS. The material is easily powdered and is often employed as a filter for treating public water supplies and for many other applications.

diatoms   A distinctive group of unicellular plants (algae) that produce cell walls composed of silicate minerals. The plates of the cell wall form overlapping halves termed frustules. Diatoms are distributed in almost every environment on earth and constitute some of the oldest known fossil records. See DIATOMACEOUS EARTH.

dibenzofurans   A family of toxic chlorinated organic compounds consisting of two benzene rings connected by two bridges, one a carbon-to-carbon link, the other a carbon-oxygen-carbon link. Along with dioxins, these compounds are present in minute amounts in the air emissions from solid and/or hazardous waste incinerators. Often called simply furans.

dibenzo-*para*-dioxin   See   TETRACHLORODIBENZO-PARA-DIOXIN

dichlorophenoxyacetic acid (2,4-D)   A chlorophenoxy herbicide that is one of the most familiar chemicals, used for control of broadleaf weeds and woody plants; an active ingredient, along with TRICHLOROPHENOXYACETIC (2,4,5-T), in Agent Orange, used as a defoliant during the Vietnam War. The chemical kills plants by acting as a plant hormone. Toxicity for animals requires relatively large doses.

dichlorodiphenyldichloroethane (DDD)   A major METABOLITE produced during the biological degradation of DDT. The material has some insecticide properties and is accumulated in organisms along with the parent compound. Not as toxic to mammals as DDT.

dichlorodiphenyldichloroethene (DDE)   One of the major products produced during the metabolism of DDT within biological systems. The material has a high lipid solubility and accumulates along with DDT in fatty tissue of animals. Like the parent compound, DDE is recalcitrant and demonstrates a persistence in the environment. Many of the adverse effects on birds ascribed to DDT are actually the result of DDE. DDE degrades the metabolism of birds, resulting in reproductive failure.

dichlorodiphenyltrichloroethane (DDT)   One of the best known of the chemical insecticides. First synthesized in 1874, the insecticidal property was discovered by P. Mueller, a Swiss chemist, in 1939. Used widely during World War II and since to control lice that spread typhus and to reduce other VECTOR-BORNE diseases, such as mosquito-carried malaria and yellow fever. DDT and its degradation products have been shown to be persistent in the environment and to be accumulated by organisms within adipose tissue. Wild birds are adversely affected by DDT, producing eggs with very thin shells, thereby resulting in a failure to reproduce. Use of the agent has been severely restricted in the United States since 1972, but DDT is still applied in many other countries. DDT is not as toxic to mammals as ENDRIN or the carbamate insecticides.

DICHLORODIPHENYL-TRICHLOROETHANE

dieback   A dramatic decline in the number of individuals within a population of organisms.

dieldrin   An insecticide belonging to the chlorinated cyclodiene class, which also includes aldrin and chlordane. Dieldrin is absorbed directly through skin contact and rapidly combines with blood serum proteins. The chemical accumulates in the fatty tissues of the body. Dieldrin is a neurotoxin that causes convulsive responses in high doses. Damage to both liver tissue and function has also been observed.

diethylstilbestrol (DES)   A hormone used as a medication from the 1940s until 1970 (when it was banned) to prevent miscarriages in problem pregnancies. DES has been found to cause unusual vaginal cancers in young women and abnormal development of the genitalia of both men and women born to mothers who took the hormone during early stages of pregnancy.

diffraction   A change in the AMPLITUDE or phase of waves that strike an object.

diffusion   The movement of a material within a supporting medium, such as the movement of molecules of a pollutant in a wastewater outfall into the surrounding water, or the movement of sugar molecules as a sugar cube dissolves in undisturbed water. Diffusion varies in response to changes in the concentration of the material. See GRAHAM'S LAW.

diffusion coefficient   See FICK'S FIRST LAW OF DIFFUSION.

diffusivity   The proportionality constant used to determine the mass movement of a chemical in a particular medium in response to changes in concentration.

digester   A water pollution control device used to enhance the microbial decomposition or MINERALIZATION of PARTICULATE ORGANIC MATTER. The device operates by way of ANAEROBIC DECOMPOSITION of the waste.

digester gas   The gas produced as a result of the microbial decomposition of PARTICULATE ORGANIC MATTER under anaerobic conditions. Methane and hydrogen are major components.

digestion   The process of MINERALIZ-ING or decomposing PARTICULATE OR-GANIC MATERIAL to lessen the impact of the addition of domestic waste to streams.

diluent   Substance used to dilute a solution or suspension.

dilution factor   The extent to which the concentration of some solution or suspension has been lowered through the addition of a DILUENT. The term is frequently used to describe the extent to which a sample must be diluted prior to the quantification of bacteria within the sample. Usually expressed as a negative exponent of 10 (10^{-3}) or a fraction (1/1000).

dimictic lake   A stratified lake that undergoes two OVERTURNS each year. The water in lakes layer in response to differences in the temperatures of surface and deep waters. The surface water will be warmer because of radiant heating by the sun, and the bottom water will be cooler and therefore denser. The waters in these two layers (termed the EPILIM-NION on the surface and HYPOLIMNION on the bottom) are separated by a boundary referred to as the THERMOCLINE. This layering is disrupted in response to variation in air temperature associated with changes in the seasons of the year. As the epilimnion cools, it sinks, mixing the water within the lake. Compare MEROMICTIC LAKE.

dinitrophenols   A family of chemical compounds that includes dinitrophenol (DNP), dinitro-ortho-cresol (DNOC), and 2-sec-butyl-4,6-dinitrophenol   (DINO-SEB). The compounds have been used for a variety of purposes ranging from weed control to a weight-control agent (DNP). The acute toxic effects on humans generally are related to the ability of the compounds to uncouple the metabolic conversion of carbohydrates to the chemical energy used in cellular metabolism within the body. The result is an increase in the rate of metabolism that leads to restlessness, sweating, flushing of the skin, and fever. Chronic overexposure may cause fatigue, anxiety, and weight loss, among other problems.

dioxin   See TETRACHLORODIBENZO-PARA-DIOXIN.

direct combustion reheat   The injection of hot combustion gases into a flue gas to increase the buoyancy of the exhaust exiting the smokestack.

direct photolysis   The transformation or decomposition of a material caused by the absorption of light energy.

direct plating   Method used to identify fungi present in soil or litter. A known weight of soil or litter is distributed over the surface of an appropriate medium and incubated to facilitate the growth of fungi. Following development of the fungal colonies, the organisms are counted and identified. The method is used in a similar manner to quantify specific bacteria present in water or wastewater.

directivity   A characteristic of a sound emission that is higher in some directions from the source than in other directions because the sound is emitted from one side of a machine, for example, or because the machine is placed against a wall.

directivity factor (Q)   A numerical expression of the DIRECTIVITY of a noise. Expressed as $Q = P_0/P_1$, the ratio of the sound power (P_0) of a small spherical source emitting sound equally in all directions to the sound power (P_1) of the actual source, where each sound power produces the same measured SOUND PRESSURE LEVEL at a certain distance and angle.

directivity index   Ten times the common LOGARITHM of the DIRECTIVITY FACTOR (DI = 10 log Q); the index is used in the calculation of expected noise levels from a source, such as in the vicinity of a machine.

discharge The volume of water flowing past a point in a stream or pipe for a specific time interval.

discharge monitoring report (DMR) The form filed by holders of NATIONAL POLLUTANT DISCHARGE ELIMINATION SYSTEM (NPDES) permits that contains the results of the periodically required pollutant analysis of their water discharges.

discount rate A factor applied to future monetary payments or receipts (and any environmental costs or benefits, if put in monetary units) to account for the delay in the payment or receipt. The discount rate applied is a function of the value of immediate receipt and use of the money, the interest rate that could be earned during the year, and the possible inflationary decline in the money's value.

The discount rate formula is a rearrangement of the equation to compute compound interest, $V_n = V_0(1 + r)^n$, where V_0 is the value at time period zero, r is the interest rate, n is the number of compounding periods, and V_n is the value at the end of n time periods. Thus, $V_0 = V_n/(1 + r)^n$ is applied to discount back to a present value V_0, where r is called the discount rate, V_n is the future value to discount, and n is the number of discount periods. For example, if $1.00 ($V_n$) is to be received one year from now ($n = 1$), and the sum of the above considerations implies a discount rate of 10 percent ($r = 0.10$), then the present value (V_0) of $1.00 to be received in one year is $1.00/(1 + 0.10)^1$, or 90.9 cents. The choice of discount rate can be the deciding factor when comparing current and future benefits with current and future costs of a proposed project.

discrete variable A quantitative variable that is limited to a finite or countable set of values, regardless of the accuracy of the measurement method. There can be no intermediate values between the numbers representing a discrete variable. For example, the number of persons residing in the state of Nevada must be represented by a whole number and exclude decimal fractions. See also CONTINUOUS VARIABLE.

disinfection The killing of dangerous bacteria or other microbes in water, in wastewater, in the air, or on solid surfaces.

disintegration A spontaneous change in the nucleus of a RADIOACTIVE element that results in the emission of some type of radiation and the conversion of the element into a different (DAUGHTER) element.

dispersal The breaking up, spreading out, or distribution of some material released from a concentrated source to a more diffuse distribution within the environment.

dispersants Chemicals added to a material, such as crude oil, to promote the formation of smaller aggregates (break up a concentrated discharge in the environment).

dispersion The act of breaking up concentrations of some agent into a more diffuse distribution, for example the spreading and mixing of a smokestack plume as it travels downwind.

dispersion coefficients Variables in the GAUSSIAN PLUME MODEL of horizontal and vertical dispersions of an air pollutant. Their value increases with increasing downwind distance from the air pollutant source and with increasing atmospheric turbulence. Larger values for the dispersion coefficients decrease the estimate of the average downwind pollutant concentration generated by the model.

dispersion model See AIR QUALITY DISPERSION MODEL, GAUSSIAN PLUME MODEL.

dispersion parameters See DISPERSION COEFFICIENTS.

disposal pond   A small, usually diked, enclosure that is open to the atmosphere and into which a liquid waste is discharged. See LAGOON.

dissociation   The partial or total separation of certain molecules into IONS when dissolved in water; for example, carbonic acid in water produces a positively charged hydrogen ion and a negatively charged bicarbonate ion. See DISSOCIATION CONSTANT.

dissociation constant   At equilibrium, the ratio of the molar concentrations of the ions produced by the DISSOCIATION (separation) of a molecule to the molar concentration of the undissociated molecule. For acids, the ratio is abbreviated K_a and is often expressed as pK_a, the negative logarithm of K_a. For bases the abbreviations are K_b and pK_b. For example, carbonic acid dissociates into hydrogen ions and bicarbonate ions. At 25°C, the K_a, the ratio of the molar concentrations of hydrogen ions and bicarbonate ions to the molar concentration of carbonic acid, is 4.47×10^{-7} and the pK_a is 6.35.

dissolved air flotation   A technique used to separate oil and SUSPENDED SOLIDS from water. Air is bubbled upward in a tank, bringing the oil and solids to the surface. This frothy top layer is removed by a skimmer.

dissolved gases   Gases that are in solution homogeneously mixed in water.

dissolved organic carbon (DOC)   A measure of the organic compounds that are dissolved in water. In the analytical test for DOC, a water sample is first filtered to remove particulate material, and the organic compounds that pass through the filter are chemically converted to carbon dioxide, which is then measured to compute the amount of organic material dissolved in the water. See TOTAL ORGANIC CARBON.

dissolved organic matter (DOM)   See DISSOLVED ORGANIC CARBON.

dissolved oxygen (DO)   The amount of molecular oxygen (O_2) dissolved in water. The units, often not expressed, are milligrams of oxygen per liter of water. DO is an important measure of the suitability of water for aquatic organisms. A level of 8 or 9 represents the concentration that one would expect to encounter in streams that have not been polluted with the organic waste common in domestic sewage. Waters with a dissolved oxygen value of 4 and below are not suitable for habitation by many forms of animal life.

dissolved oxygen sag curve   A graph of DISSOLVED OXYGEN levels downstream from an outfall through which wastewater containing a significant amount of organic compounds (domestic sewage) is discharged. Typically, the concentration of oxygen in the water falls precipitously close to the discharge point and gradually returns to normal levels farther downstream.

dissolved solids   Primarily, the inorganic salts in solution (homogeneously mixed) in water. These chemicals cannot be removed by filtration and must be recovered by evaporation of the water.

distillation   A process of separation and/or purification of the components of some substance based on differences in their boiling points. The components within a mixture of two liquids can be separated if the two liquids have different boiling points. The mixture is heated, and the material with the lowest boiling point is converted to a vapor first. That vapor can then be condensed to produce a liquid that consists primarily of the single liquid.

distillation tower   An apparatus used to separate the components of crude oil into different fractions, depending on their relative boiling points. See DISTILLATION.

distilled water   Water that has been purified by the DISTILLATION process. Water that contains various chemicals or ions in solution is heated to boiling and the water vapor is condensed. The process leaves behind various inorganic ions and results in a water that is free of dissolved salts.

distribution   In toxicology, the transport and diffusion of an absorbed material within the body. The body distribution of a compound is a function of the size, water or fat solubility, and ionization, among other factors, associated with the material.

dithiocarbamates   A group of compounds of low toxicity used as fungicides to protect seeds and vegetable products. The two most common are ferric dimethyldithiocarbamate (ferbam) and zinc dimethyldithiocarbamate (ziram).

diurnal   Daily; exhibiting a daily cyclical pattern.

divergence   In METEOROLOGY, the flow of air in different directions. In high-pressure systems, air diverges from the center, where other air masses sink to replace it.

divergent plate boundary   In the theory of plate tectonics, a boundary between two plates that make up the crust of the earth. The boundary is characterized by a chasm between the two plates, filled with molten rock from within the earth.

diversity   A measure of the number of different species, along with the number of individuals in each representative species, in a given area. Undisturbed environments tend to be characterized by high diversities, while polluted environments tend to exhibit low diversities. See RICHNESS, EVENNESS.

diversity index   A mathematical expression that depicts species DIVERSITY in quantitative terms. The Shannon-Weaver index is a widely used measure.

dolomite   A natural mineral consisting of calcium magnesium carbonate, $CaMg(CO_3)_2$. Also referred to as limestone, a form of marble, and dolomitic lime. Upon heating, carbon dioxide is released from the mineral.

domestic sewage   Wastewater and solid waste that is characteristic of the flow from toilets, sinks, showers, and tubs in a household.

domestic waste   See DOMESTIC SEWAGE.

domestic water   Water used within a household.

dominant gene   A gene that is expressed even though it may be paired with a matching recessive gene. When male and female gametes fuse in the reproductive process, individual gametes furnish one member of each pair of chromosomes that will determine the genetic makeup of the offspring. If the gametes are from plants and if plant height is governed by one gene, the possible genes may be designated *h* for the short variety and *H* for the tall variety. In the case where *H* is dominant and *h* is recessive, a plant with gene pair *Hh* would be tall, since *H* is dominant and would be expressed despite the presence of the gene *(h)* that codes for the short variety of plant.

Donora episode   The week-long STAGNATION period of high air pollution levels in Donora, Pennsylvania, during October 1948. See EPISODE.

donor-controlled flow   In BOX MODELS, a flow of a substance from one compartment to one or more others. The movement depends on the STOCK of the substance in the originating compartment.

dose   The amount of chemical agent or radiation to which an organism is exposed, or the amount the organism absorbs over a specific time period.

dose commitment The amount of radiation to which the body is exposed because of ingestion and retention of radioactive substances.

dose equivalent (DE) A measure of EFFECTIVE RADIATION DOSE, computed as DE = D(QF)(DF), where D is the absorbed dose, QF is a QUALITY FACTOR, and DF is a distribution factor. Typically expressed in REMS or SIEVERTS.

dose rate The amount of radiation to which a person is exposed per unit of time.

dose-distribution factor A consideration of the effects of some dangerous substance in the body when the substance is concentrated in specific areas rather than randomly distributed throughout the body.

dose response An adverse effect in an organism attributable to a particular physical or chemical agent.

dose-response assessment A description, using the DOSE-RESPONSE RELATIONSHIP between certain levels of exposure to a chemical or physical agent and the anticipated adverse effects, of the extent of disease, injury, or death resulting from an estimated exposure to a population.

dose-response curve A plot showing the increasing response in a group of organisms at various levels of exposure to a chemical or physical agent.

dose-response relationship The quantitative relationship between an exposure to a physical or chemical agent and a subsequent biological effect. The exposure to a chemical substance can be described as the mass ingested or inhaled or as the mass actually absorbed into the body tissues; the dose units can be in mass of the chemical per unit body weight on a one-time basis or mass per unit body weight per day, for an extended period or for a lifetime. The response can

be in terms of the severity of biological effect observed or the portion of a test group exhibiting a particular adverse effect, usually compared with an unexposed (control) group.

DOSE-RESPONSE RELATIONSHIP

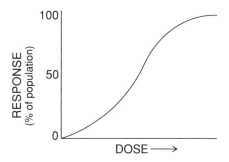

dosimeter A device to measure the amount of radiation to which a person is exposed. It can be a simple FILM BADGE or a complicated ionization meter.

dosimetry The measurement of the amount of radiation to which a person is exposed or of the amount of radiation absorbed by a person.

double recessive The circumstance in which recessive genes in an organism are expressed because they are not paired with a dominant gene that masks the recessive character. When male and female gametes fuse in the reproductive process, individual gametes furnish one member of each pair of chromosomes that will determine the genetic makeup of the offspring. If the gametes are from plants and if plant height is governed by one gene, the possible genes may be designated h for the short variety and H for the tall variety. In the case where H is the DOMINANT GENE and h is the RECESSIVE GENE, a plant with gene pair hh (double recessive) would be short, since the dominant H gene is not present to force the development of the tall variety.

doubling dose The amount of a specific radiation exposure needed to produce a doubling of the natural rate of appearance of genetic or somatic (cellular) abnormalities.

doubling time   The time required for a population to double in size.

downgradient well   One or more monitoring wells placed to sample groundwater that has passed beneath a facility with the potential to release chemical contaminants into the ground; one of a network of groundwater monitoring wells required by RESOURCE CONSERVATION AND RECOVERY ACT regulations for landfills or surface impoundments that treat, store, and/or dispose of hazardous waste. Results of testing downgradient well water are compared with data from an UPGRADIENT WELL to determine whether the facility may be contaminating the groundwater.

downwash   See STACK DOWNWASH.

draft   A flow of gas or vapor created by a pressure difference. See FORCED DRAFT, INDUCED DRAFT, and NATURAL DRAFT.

draft environmental impact statement (DEIS)   An unfinished version of an ENVIRONMENTAL IMPACT STATEMENT that is sometimes circulated for review.

drain field   See LEACHING FIELD.

drawdown   The vertical difference in the water table of an undisturbed aquifer and the water table in the same aquifer after the removal of water by pumping from wells.

dredge and fill permit   See SECTION 404 PERMIT.

drift   Mist that escapes from cooling towers used in CLOSED-CYCLE cooling systems.

drift velocity   For particles collected by an ELECTROSTATIC PRECIPITATOR, the rate at which the particles move toward the collection plates. The measurement is used in electrostatic precipitator design and to estimate particle collection efficiency. See DEUTSCH-ANDERSON EQUATION.

drill cuttings   Rock or other materials forced out of the borehole as a well is drilled.

drilling fluid   A dense fluid material, often containing bentonite clay and barite, used to cool and lubricate a well-drilling bit, seal openings in the wall of the borehole, transport DRILL CUTTINGS to the surface, reduce drill pipe friction, and control well pressure.

drinking water supply   Water provided for use in households. The most common sources are from surface supplies (rivers, lakes, and reservoirs) or subsurface supplies (aquifers). The distribution of water to households is regulated under the SAFE DRINKING WATER ACT of 1974 as amended.

droplet   A small airborne liquid particle that is larger than liquid AEROSOL and therefore settles out of the atmosphere relatively quickly.

dry adiabatic lapse rate   The ADIABATIC LAPSE RATE for air not saturated with water vapor, or 0.98°C per 100 meters rise (0.54°F per 100 feet). Expressed as

$$-\frac{dT}{dz} = \gamma_d$$

where T is air temperature, z is altitude, and γ_d is the dry adiabatic lapse rate. Compare WET ADIABATIC LAPSE RATE.

dry deposition   The introduction of acidic material to the ground or to surface waters by the settling of particles containing sulfate or nitrate salts. Compare to WET DEPOSITION.

dry gas meter   A device used in the field to measure gas flow rates in air sampling systems and in the laboratory to calibrate other flow meters. The gas being monitored should be free of moisture and corrosive material to avoid damage to internal parts of the meter.

dry mass The mass of a sample after any water has been removed.

dry scrubber 1. An air pollution control device that removes SULFUR DIOXIDE from stack gases by injecting a finely divided dry chemical reagent such as limestone, trona (Na_2CO_3), or nahcolite ($NaHCO_3$) into the flue gas. The sulfur dioxide reacts with the injected chemical to form a solid particulate, which is captured in a BAGHOUSE. 2. A gravel bed filter used to collect particulate matter from gas streams.

dry-bulb temperature The temperature reading from an ordinary thermometer; the reading is used with the WET-BULB TEMPERATURE to compute RELATIVE HUMIDITY.

duff Partially decomposed organic material found on a forest floor.

dump See OPEN DUMP.

dust Suspended solid particles in the atmosphere. Natural dust sources include wind erosion and volcanic eruptions; human sources are activities involving crushing, grinding, or abrasion of materials.

dust bowl A semiarid geographical area in which cultivation has led to extensive wind erosion of topsoil. It is often used to refer to the area of the United States that in the mid-1930s experienced soil loss and dust storms, including parts of Texas, New Mexico, Oklahoma, Colorado, Utah, and Kansas.

dustfall The amount of PARTICULATE MATTER that will settle out of the atmosphere at a particular location within a certain time. Measured with a DUSTFALL BUCKET. Dustfall is a crude measure of air quality and is no longer widely used.

dustfall bucket An open bucket or jar placed to collect larger-diameter (greater than 30 MICRONS) particulate matter settling out of the air. The mass of the dust collected during a certain time period is translated to dustfall units of mass per area per time, such as tons per square mile per month.

Dutch elm disease Fatal infectious disease of elm trees (genus *Ulmus*). The disease is caused by a fungus (genus *Ceratocystis*) and is spread from tree to tree by the elm bark beetle (genus *Scolytus*).

dynamic equilibrium state The condition of a system in which the inflow of materials or energy equals outflow.

dynamic viscosity See VISCOSITY.

dysentery A disorder of the gastrointestinal tract characterized by severe diarrhea with blood and pus in the feces. The disease frequently results from an infection by bacteria belonging to the genus *Shigella*.

dystrophic Describes the water in a lake that is high in humic substances and plant degradation products, resulting in water that is brown in color. Plant and animal life are typically sparse, and the water has a high OXYGEN DEMAND.

E

e The base of the NATURAL LOGARITHM, approximately equal to 2.718282.

Earth Day When first held April 22, 1970, the day was observed on American college campuses with demonstrations and "teach-ins" on environmental issues and was an early indication of the extent of political support for the environmental movement in the United States. Commemorated on a lesser scale in subsequent years until the widely observed twentieth anniversary in 1990, which coincided with a renewed worldwide concern for environmental issues.

Earthwatch A global environmental data system begun by the UNITED NATIONS ENVIRONMENT PROGRAM in 1972. See GLOBAL ENVIRONMENT MONITORING SYSTEM, INTERNATIONAL ENVIRONMENTAL INFORMATION SYSTEM, and INTERNATIONAL REGISTER OF POTENTIALLY TOXIC CHEMICALS.

ecesis The successful settlement of a plant or animal in a new habitat, through colonization, migration, or introduction by man.

Eckenfelder's equation One of several equations predicting performance of a TRICKLING FILTER wastewater treatment system. Given the filter area and depth, the volumetric flow of wastewater, and the BIOCHEMICAL OXYGEN DEMAND of the wastewater coming in to the treatment plant, the equation estimates the reduced biochemical oxygen demand of treated wastewater. One form of the equation is

$$S_t = S_0 \exp\left[-KD\left(\frac{A}{Q}\right)^x\right]$$

where S_t is the biochemical oxygen demand of the treated wastewater, S_0 is the influent biochemical oxygen demand, exp is the base of the natural logarithm, K is a rate constant, D is the trickling filter depth, A is the filter area, Q is the wastewater flow rate, and x is an empirical constant determined by the type of filter medium. Named for W. W. Eckenfelder.

ecologic fallacy In epidemiology, an erroneous association between a characteristic or exposure common to a group and a health effect or condition present in that group. For example, although most residents of a county might drink bottled water, an elevated rate of gastrointestinal illness in the county may not necessarily be associated with bottled water exposure. The ecologic fallacy is avoided by developing disease rates for subsets of a group, thereby preventing the potentially false assumption that average or majority group characteristics apply equally to individuals making up the group.

ecological density The number of organisms per unit area or per unit volume of a segment or compartment of a larger, complete ecosystem. This segment is referred to as a specific HABITAT within the larger system. An example would be the mud habitat on the floor of a lake ecosystem. The lake would include not only the bottom mud but also the water, shoreline, and other smaller habitats included within the complete lake ecosystem. The density is also reportable as the weight of BIOMASS within the specific habitat. Contrast with CRUDE ECOLOGICAL DENSITY, which includes the biota in all compartments of a complete lake ecosystem.

ecological efficiency The effective transfer of useful energy from a food to the animal consuming it. The term is normally applied to the feeding structure of animal communities in the natural environment, and the transfer is usually expressed as a percent.

ecological equivalents Unrelated species of plants or animals, living in different geographical locations, that are similar in appearance and perform similar ecosystem functions. For example, the zebra of Africa and the buffalo of North America both occupy grasslands and both are grazers. A product of CONVERGENT EVOLUTION.

ecological indicator An individual species or a defined assemblage of organisms that serves as a gauge of the condition of the environment. For example, the bacterium *Escherichia coli* indicates the presence of sewage in water, and the mussel, *Mytilus edulis* lives in polluted water.

ecological niche   The activities and relationships of an organism within its environment.

ecological pyramid   A construct depicting the relationships among organisms that belong to various producing and feeding groups within a specified area or location. The diagram attempts to show the transfer of useful energy from one feeding group to the next in a semiquantitative fashion. The amount of plant biomass in an area is normally placed at the bottom, and the mass of each succeeding feeding group is placed on top, with CARNIVORES occupying the highest level. The result is a figure resembling a pyramid. The diagram is also referred to as an energy pyramid when the caloric content of the organisms in each trophic (feeding) level is used to construct the diagram.

ecology   The study of the relationships among organisms and between organisms and their surroundings.

economizer   A device in a boiler that warms FEEDWATER by using the hot exhaust gases from the furnace.

ecosphere   Those environments that harbor living organisms. See BIOSPHERE.

ecosystem   A level of organization within the living world that includes both the total array of biological organisms present in a defined area and the chemical-physical factors that influence the plants and animals in it. The study of an ecosystem demands that the investigator consider all biological and nonbiological variables operative within a defined area.

ecotone   A zone of transition between two well-defined vegetated areas.

ectoparasite   A parasite that damages a host by attaching itself to the outside

of the organism, for example a leech attaching itself to a person's skin.

ectotherm   An organism that lacks the ability to regulate internal body temperature. The internal body temperature of this type of organism reflects the temperature of the environment. Fish are ectotherms. Also referred to as a poikilotherm. Compare ENDOTHERM.

edaphic   Soil characteristics, such as water content, pH, texture, and nutrient availability, that influence the type and quantity of vegetation in an area.

edaphic climax   The climax stage of an ecological SUCCESSION in which the COMMUNITY is in equilibrium with localized soil or climatic conditions.

eddy diffusion   The mixing of clean air and contaminated air by parcels (eddies) of air moving in a random, irregular manner. The eddies that cause this dilution are produced by MECHANICAL TURBULENCE and/or THERMAL TURBULENCE. Turbulent eddies that are roughly the same size as the pollutant mass are most effective at dilution. If the eddies are small relative to a plume of contaminated air, they will only dilute the outside edges of the plume; if the eddies are much larger than the plume, the entire plume will be moved without significant dilution.

edema   The accumulation of an excessive amount of fluid in body tissues.

edge effect   The observed increase in the number of different species along the margins of two separate communities of vegetation, the ecotone.

effective concentration dose   The amount of a chemical or physical agent that, after exposure, absorption, and metabolism, actually reaches a target organ or tissue and is able to elicit an adverse effect.

effective half-life   The time required for the amount of a RADIOACTIVE element or compound absorbed by a living organism to be reduced by a factor of 50 percent. The decreased radioactivity in the body is a combination of the natural decay of the radioactive element and the excretion of the element or compound by biological processes.

effective radiation dose See DOSE EQUIVALENT.

effective stack height In air pollution dispersion calculations, the height at which the pollutants begin their descent to the ground; the sum of the physical stack height and the length of the PLUME RISE. Gas exiting a stack rises above the stack exit by VERTICAL INERTIA due to the buoyancy of the stack gas. Higher exit velocities and higher exhaust gas temperatures cause a higher plume rise. The greater the effective stack height, the greater air mass for the pollutants to mix and disperse, which results in lower ground-level concentrations downwind of the stack.

effluent Wastewater that flows into a receiving stream by way of a domestic or industrial POINT SOURCE.

effluent limitation An amount or concentration of a water pollutant that can be legally discharged into a water body by a POINT SOURCE, expressed as the maximum daily discharge, the maximum discharge per amount of product, and/or the concentration limit in the wastewater stream, as a 24-hour or 30-day average. The applicable TECHNOLOGY-BASED standard is set by the United States Environmental Protection Agency by INDUSTRIAL CATEGORY or subcategory, but differs between new and existing sources and by broad types of water pollutants: CONVENTIONAL POLLUTANTS; toxic pollutants; nonconventional, nontoxic pollutants; dredge and fill wastes; and heat discharges.

effluent standard The maximum amounts of specific pollutants allowable in wastewater discharged by an industrial facility or wastewater treatment plant. The standards are set for individual pollutants and apply across industrial categories. The term can be contrasted with EFFLUENT LIMITATIONS, which are set for individual pollutants by INDUSTRIAL CATEGORY.

Eh An expression of REDOX POTENTIAL.

Ekman dredge A device for sampling bottom-dwelling organisms from lakes or streams with a slow current. A metal box, typically $6 \times 6 \times 6$ inches, the bottom of which can be opened and closed by the movement of spring-loaded jaws. The box, mounted either on a pole or line and with the jaws locked in the open position, is driven into soft sediments. The jaws are closed to trap the sediments within the sampler for removal to the surface. The Ekman dredge is most useful in the sampling of soft mud sediments for the examination of biota entrained within. Also called an Ekman grab.

Ekman spiral The change in wind direction with altitude caused by the varying effect of surface friction. At high altitudes (above about 700 meters), wind direction is unaffected by surface friction and is determined by a balance between PRESSURE GRADIENT and CORIOLIS forces. At lower altitudes, surface friction slows wind velocity, which reduces the Coriolis force, and the wind turns toward the lower pressure. The change in wind direction from the surface to the top of the friction layer can be represented by a spiral.

Ekman water bottle A tubular device used to sample water at selected depths. The tube is open at both ends as it is lowered by a rope, but flaps at the ends snap shut when a weight is dropped down the rope, enclosing the water sample derived from a specified depth.

electricity The form of energy arising from the movement or accumulation of electrons. The movement of electrons (called electric current) produces a mag-

Electromagnetic spectrum
(note: the figures are only approximate)

Radiation	Wavelength (m)	Frequency (Hz)
gamma radiation	-10^{-10}	$10^{19}-$
x-rays	$10^{-12}-10^{-9}$	$10^{17}-10^{20}$
ultraviolet radiation	$10^{-9}-10^{-7}$	$10^{15}-10^{18}$
visible radiation	$10^{-7}-10^{-6}$	$10^{14}-10^{15}$
infrared radiation	$10^{-6}-10^{-4}$	$10^{12}-10^{14}$
microwaves	$10^{-4}-1$	$10^{9}-10^{12}$
radio waves	1 $-$	-10^{9}

netic field, a phenomenon used to convert electrical energy to mechanical energy in electric motors. Conversely, generators use mechanical energy to move a magnetic field, producing an electric current in a conductor.

Electric Power Research Institute (EPRI) An industry organization, based in Palo Alto, California, that sponsors research and development in technologies for the production, distribution, and use of electricity, including the control of environmental impacts.

electrolysis The passage of an electric current through an electrolyte, causing the migration of the positively charged ions to the negative electrode (cathode) and the negatively charged ions to the positive electrode (anode).

electrolyte Any compound that dissociates into IONS when dissolved in water. The solution that results will conduct an electric current. For example, table salt (NaCl) is an electrolyte.

electrolytic Pertaining to electrolysis or an electrolyte.

electromagnetic radiation Wave functions that are propagated by simultaneous periodic variations of electric and magnetic field intensity. Types of electromagnetic radiation range from those of short wavelengths like x-rays and gamma rays to those of long wavelengths like radio waves. Also included are ultraviolet radiation, visible light, infrared radiation, and microwaves. The shorter the wavelength, the more dangerous is the radiation to humans. All types of electromagnetic radiation travel through a vacuum at the speed of light.

electromagnetic spectrum The range of types of ELECTROMAGNETIC RADIATION in wavelength (or frequency) order.

electromagnetic terrain conductivity A noninvasive survey of soil for the detection of buried metal objects and chemical contaminants that conduct an electrical current. The method is especially useful in the detection of electrically conductive leachate and contaminant plumes.

electron One of the elementary particles of an atom. Negatively charged electrons surround the positively charged nucleus of an atom and occur in numbers equal to the number of protons in the nucleus. An electron has a mass of 1/1837th that of a proton.

electron capture detector (ECD) A sensitive detector used in conjunction with a gas chromatograph, an instrument used to measure small amounts of compounds in samples collected from sources within the environment. The chromatograph employs a gas stream to carry and separate the organic compounds of interest. As the carrier gas (normally nitrogen) flows through the detector, the gas is ionized by a radioactive substance. The electrons resulting from this ionization migrate to an anode, creating a voltage. When the compounds in question are encountered in the carrier gas, the electric current is interrupted. The decrease in the current is proportional to the concentration of solute in the gas. The device can detect as little as 10^{-14} gram of organochlorine pesticide.

electron volt (eV)   A unit of energy equivalent to the amount of energy gained by an electron passing through a potential difference (the difference in the strength of an electric field between two points) of one volt in a vacuum. One electron volt is equal to 1.6×10^{-12} erg or 1.6×10^{-19} joule.

electrophoresis   A technique used to separate, identify, and quantify proteins and similar macromolecules as these molecules migrate within gel or cellulosic substrates under the influence of an electric current.

electrostatic precipitator (ESP)   A device to control the release of particles from an exhaust into the atmosphere. An electric charge is first applied to the particles, and the charged particles are then collected on the surface of oppositely charged plates.

element   A chemical substance that cannot be broken down to simpler units without changing its chemical properties. The atoms of an element all have equal numbers of protons and electrons. The list of 103 known elements includes oxygen, carbon, nitrogen, iron, silver, and gold. See Appendix.

elementary neutralization unit   A device for treating wastes that are defined as hazardous due to their CORROSIVITY.

elevation head   The potential energy in a hydraulic system, represented by the vertical distance between the hydraulic system (pipe, channel, etc.) and a reference level, and expressed in length units. The sum of the elevation head and PRES-

SURE HEAD is the HYDRAULIC HEAD. See also HEAD, TOTAL.

elutriation   The removal of particulate matter from a fluid stream by gravitational settling as the fluid moves upward.

emergency core cooling system (ECCS)   A complete system of pumps, piping, water reservoirs, and heat exchangers that serves as a safety device in nuclear reactors. The system is designed to remove excessive heat from a nuclear reactor should the normal core cooling mechanism fail.

Emergency Planning and Community Right-to-Know Act (EPCRA)   See TITLE III.

emergent plant   A plant that grows in shallow water with the root system submerged under the water and the upper vegetation rising above the water.

emigration   The departure of individuals from a POPULATION in a specific area or country. Compare IMMIGRATION.

emission factor   A numerical estimate of the mass of one or more air contaminants produced for a given amount of material processed by an industrial facility or, in the case of transportation sources, per mile driven. It is important to note whether the emission factor is for an uncontrolled source or one with properly functioning air pollution control equipment. This factor is used to arrive at a rough estimate of the total air emissions for a facility or geographical area. See EMISSION INVENTORY.

emission inventory   A compilation of estimated air pollutant emissions, by pollutant, from smokestacks, automobiles, and other sources in a given area. The inventory is usually conducted by an environmental regulatory agency.

emission inventory questionnaire (EIQ)   The form sent by an environmental regulatory agency to all facilities in a given jurisdiction for the production of an EMISSION INVENTORY. The questionnaire usually contains detailed questions about types, quantities, and locations of air emissions.

emission reduction credits (ERCs) Quantifiable permanent reductions in air pollutant emissions, beyond those legally required, placed in United States Environmental Protection Agency approved accounts (banks). The reductions can be used as part of a BUBBLE POLICY emission limitation or as an internal OFFSET, and, in some cases, traded or sold to other facilities. The use of the term "banks" does not imply monetary transfers.

emission standard The maximum legal amount of a particular pollutant that may be released into the air from a pollutant source. The standard applicable to a particular source depends on the type of pollutant, when the source began operation, the location of the source, and the quality of the air in the surrounding area.

emissions Pollutant gases, particles, or liquids released into the atmosphere.

emissions trading The internal exchange or sale to other companies of surplus air pollutant emission reduction credits in accordance with policies of the United States Environmental Protection Agency in regard to BUBBLE POLICY, the OFFSET program, the NETTING OUT for modifications, or the EMISSION REDUCTION CREDITS program.

emissivity The ratio of the amount of energy actually radiated by an object to the maximum amount of energy the body would radiate if it were a BLACKBODY. For blackbodies, emissivity and ABSORPTIVITY are equal. See KIRCHHOFF'S LAW.

empirical Based on experience or observation, as opposed to theory.

encapsulation A hazardous waste treatment process that permanently encloses the waste with materials such as special adhesives or concrete to prevent escape of dangerous substances into the environment.

endangered species A species of plant or animal that is present in such small numbers that the species is in danger of disappearing from either all or a significant part of its natural range. See ENDANGERED SPECIES ACT.

Endangered Species Act (ESA) The 1973 federal law establishing procedures for the listing of species of plants or animals as ENDANGERED or THREATENED. The law also prohibits federal agencies from engaging in projects that place in jeopardy the continued existence of species that are threatened or endangered. The prohibition extends to projects that involve either federal monies or the federal permitting process. The United States Fish and Wildlife Service, Department of the Interior, is responsible for placing and removing species on the endangered and threatened lists, which include species in habitats worldwide. About 830 animals and 270 plants are listed as endangered or threatened at Title 50, Part 17, sections 11 and 12 of the CODE OF FEDERAL REGULATIONS. See CONVENTION ON INTERNATIONAL TRADE IN ENDANGERED SPECIES OF WILD FAUNA AND FLORA.

endemic Characteristically found in a particular region or group of people; a disease constantly present at low levels in an area.

endoergic Describing a reaction or process that absorbs energy (same as ENDOTHERMIC). Compare EXOERGIC.

endogenous Originating within an organism. Compare EXOGENOUS.

endoparasite A parasite that lives within the host and damages the host because of its presence, for example a tapeworm living within a person's gastrointestinal tract. Compare ECTOPARASITE.

endotherm   An organism that has the ability to maintain a constant body temperature through physiological mechanisms. Mammals are endotherms. Also referred to as a homeotherm. Compare ECTOTHERM.

endothermic   Describing a reaction or process that absorbs energy or heat. Compare EXOTHERMIC.

endpoint   A biological or chemical change used as an indication of the effect of a chemical or physical agent on an organism. If the anticipated change occurs, the measured or estimated dose or exposure associated with the change is recorded. Endpoints range from the more objective and easily determined, such as the death or malformation of the organism, to the more subtle or subjective, like altered enzyme levels or behavioral abnormalities.

endrin   An insecticide belonging to the class of compounds referred to as chlorinated polycyclic hydrocarbons. Structurally similar to aldrin and chlordane, endrin is a neurotoxin that also has been demonstrated to cause birth defects and to adversely effect the immune system. The pesticide is persistent in the environment, and its use in the United States is banned.

energy   The capacity to do work or produce a change, for example, to move an object from one place to another or to change its temperature. Energy cannot be destroyed or created; however, the form of the energy can be changed. For example, the potential energy available in coal because of its chemical composition can be converted to heat energy through burning, and the heat energy released through burning can subsequently be converted to electrical energy by coupling the burning with the appropriate transfer technology associated with power generating facilities. Sources of energy used to generate electricity include the potential energy associated with fossil fuels (coal, oil, and natural gas), water held behind dams, and heat released in nuclear reactors.

energy cost   The amount of energy required to exploit a resource, for example, the quantity of energy required by an animal to obtain food or the energy used to remove coal from a deposit.

energy flow   The path of energy as it moves through the various components of a COMMUNITY of organisms. The path includes the input of solar energy, energy captured by photosynthesis, the utilization of energy by various animal groups, and the loss of heat from the community.

energy pyramid   See ECOLOGICAL PYRAMID.

Energy Research and Development Administration (ERDA)   The former federal agency charged with the responsibility of developing nuclear power and new energy sources. The agency was created in 1974 when the responsibilities of the ATOMIC ENERGY COMMISSION were divided into separate agencies, one (ERDA) dedicated to energy development and the other, the NUCLEAR REGULATORY COMMISSION, dedicated to safety and regulatory activities. ERDA's responsibilities were transferred to the DEPARTMENT OF ENERGY in 1977.

energy subsidy   The energy input required to produce one unit (calorie) of food energy in agriculture. Modern, high-yield agriculture requires fossil-fuel energy inputs for the production of pesticides and fertilizers and for the manufacture and operation of farm machinery. About 0.5 calorie of energy input is required to yield 1.0 calorie of corn in the United States. But to *deliver* 1.0 calorie of corn to the consumer may consume another 8 to 12 calories in processing and distribution.

enhanced oil recovery   Techniques for the removal of the remaining thick, heavy oil from reservoirs after PRIMARY RECOVERY and SECONDARY RECOVERY. A typical operation may involve steam injection into the reservoir to reduce the viscosity and provide pressure to force the oil into collection wells.

enrichment   A process by which the abundance of fissionable isotopes of uranium or plutonium is increased to create a nuclear fuel that will support a sustained nuclear reaction. Uranium recovered by the refining of uranium ore does not contain a sufficient abundance of the fissionable isotope (uranium-235) to be used in most nuclear power plants or in nuclear weapons. The relative abundance of uranium-235 must be increased before the material is suitable for use in reactors or weapons.

enteric organism   In the broadest sense, any microorganism whose normal habitat is the human gastrointestinal tract. The term is frequently used in a more narrow sense to signify the GRAM-NEGATIVE bacteria that inhabit the gastrointestinal tract. The term is also used to describe enteric viruses that are excreted in human feces and that cause disorders associated with the consumption of food contaminated with fecal material (hepatitis virus, for example).

enterococcus   Any   GRAM-POSITIVE, oval or round bacterium (coccus) that normally resides in the human gastrointestinal tract, frequently belonging to the genus *Streptococcus*. See FECAL STREPTOCOCCI.

enteropathogenic organism   Any GRAM-NEGATIVE bacteria capable of causing illnesses or dysfunction of the human gastrointestinal tract. The disorders range from mild upsets to diarrhea and dysentery to typhoid and cholera.

enterotoxin   Specific toxins or enteropathogenic organisms secreted by some bacteria, that cause dysfunctions of the human gastrointestinal tract. The agents are a type of EXOTOXIN.

entrainment   1. The capture of solid particles, liquid droplets, or mist in a gas stream.   2. In SCRUBBERS, the liquid aerosol in the gas stream exiting the control device.   3. The incidental trapping of fish and other aquatic organisms in the water used for cooling electrical power plants.

entrainment separator   See MIST ELIMINATOR.

entrainment velocity   The gas velocity that keeps particles of a given size and density suspended (dispersed) and causes deposited particles of that type to become resuspended; typically applied to airflow in ducts carrying airborne particles.

entropy (S)   The availability of energy to do work. When one form of energy is converted to another, for example, when coal is burned to produce heat, entropy increases. Also used as a thermodynamic measure of the randomness or disorder within a system. The higher the entropy, the more disordered is the system.

environment   An aggregate of the conditions that make up the surroundings of an individual or community. The components of an environment include climate, physical, chemical, and biological factors, nutrients, and social and cultural conditions. These influences affect the form and survival of individuals and communities.

environmental assessment   A preliminary study to determine the need for an ENVIRONMENTAL IMPACT STATEMENT.

environmental audit   1. An internal investigation of company compliance with environmental regulations.   2. Study of a site prior to a real estate transaction to uncover potential environmental liability associated with the property, such as the prior improper disposal of hazardous waste in the ground.

Environmental Defense Fund (EDF) An American environmental organization active in legal, economic, and scientific aspects of environmental issues; the fund has been responsible for a number of important environmental law cases coming to the attention of the courts in the United States, such as *Environmental Defense Fund, Inc. v. Ruckelshaus,* leading to the banning of the pesticide DDT. In 1989, membership numbered about 75,000.

environmental ethic A value system that judges human actions in terms of whether they harm, sustain, or improve environmental quality.

environmental fate The end result of the physical, chemical, and/or biological changes occurring in a chemical after its release into the environment.

environmental impact statement (EIS) A report required by the NATIONAL ENVIRONMENTAL POLICY ACT detailing the consequences associated with a proposed major federal action significantly affecting the environment.

environmental impairment liability policy (EIL policy) An insurance policy offering bodily injury and property damage coverage for gradual releases of pollutants for any claims made during a policy year, no matter when the release occurred. The EIL policy approach fills the gap left by POLLUTION EXCLUSION CLAUSES. Also called a claims made insurance policy.

environmental lapse rate The negative or positive change in air temperature with change in altitude. The rate varies with changes in solar insolation, high- or low-pressure systems, wind motion, and time of day. See also ADIABATIC LAPSE RATE, MIXING HEIGHT.

Environmental Protection Agency (EPA) An independent executive agency of the federal government, established in 1970, responsible for the formulation and enforcement of regulations governing the release of pollutants and other activities that may adversely affect the public health or environment. The agency also approves and monitors programs established by state and local agencies for environmental protection.

environmental resistance The forces of nature (predators, drought, etc.) that tend to maintain populations of organisms at stable levels.

environmental response team (ERT) A group of individuals with special training and equipment who provide assistance in the event of spills or releases that threaten human health or the environment. They provide special decontamination equipment, hazard evaluation, sampling and analysis, cleanup technologies, water supply decontamination, and the removal of contaminated material, among other services.

environmental tobacco smoke (ETS) Indoor emissions of a combination of gases and particles from smoking cigarettes, pipes, or cigars. These emissions can occur at nuisance levels or unhealthy concentrations. See also PASSIVE SMOKING.

enzyme A protein compound that acts as a CATALYST for biochemical reactions. Enzymes are sensitive to changes of temperature, pH, and other substances in the environment.

EP toxicity See EXTRACTION PROCEDURE TOXICITY TEST.

epidemic A condition or disease that affects many individuals within a community at about the same time. An epidemic is usually characterized by a sudden onset and rapid spread of disease throughout the community. An epidemic requires a sufficiently large number of individuals susceptible to the condition.

epidemiology The science that deals with the incidence and distribution of disease or other medical problems within a defined area.

epidermal cells   The cells forming the outer layer of the skin or the thin outer covering of plant leaves, stems, and roots.

epifluorescence   A type of microscopy in which the specimen under examination is illuminated by the projection of ultraviolet light through the microscope from above the microscope stage. It differs from standard microscopy, in which the specimen is illuminated from below the slide with white light. The observer sees the object as light emitted from the specimen by fluorescence mechanisms (glowing) stimulated by the ultraviolet light.

epigenetic   1. Describing a cancer mechanism that does not involve the direct interaction of a chemical with cellular genetic material (DNA). 2. Describing a geological deposit that has been changed from its original state because of outside influences, for example the conversion of accumulated shell deposits to limestone.

epilimnion   The upper layer of water in a lake in which the water is stratified by temperature. Large standing bodies of water tend to form layers due to non-uniform heating by the sun. The water at the surface, the epilimnion, is warmer, whereas the water at the lower depths, the HYPOLIMNION, is colder. The boundary between these two layers is the THERMOCLINE. The water in these distinct layers is of different densities, and the layers remain separate until either the water temperatures change or they are forced to mix by physical forces such as wind. See also FALL TURNOVER.

epiphyte   A plant that grows on the surface of another plant without damaging it. For example, single-celled diatoms grow on the surface of marsh grasses without apparent damage to the grass.

episode   A period of extremely high air pollutant concentrations, lasting for sev-

eral days to a week or more, associated with an abnormally high incidence of respiratory disease and, in the worst cases, an increased death rate in the affected area. The condition is caused by air STAGNATIONS, or persistent INVERSION conditions, that allow pollutant concentrations to increase. Sites of historic air pollution episodes include Meuse Valley, Belgium, in December 1930; Donora, Pennsylvania, in October 1948; and London, England, in December 1952. Major episodes have not occurred in the United States or Western Europe since the 1960s thanks to the installation of air pollution controls on major stationary sources (smokestacks) and the widespread switch in the energy source for domestic space heating from coal to natural gas or electricity generated at a central power station.

epithelium   A membranous, protective tissue layer that covers the surface of body organs and that lines various body cavities.

equal-loudness   contours   See FLETCHER-MUNSON CONTOURS.

equation of state   An equation derived from a combination of Boyle's law and Charles' law that describes the relationships among the pressure, density, and temperature of a gas: pressure = gas constant × density × absolute temperature. $P = R\rho T$, where R is the UNIVERSAL GAS CONSTANT, ρ is density, and T is the ABSOLUTE TEMPERATURE.

equilibrium   A STEADY-STATE condition where flow in equals flow out. For BOX MODELS, the condition in which there is neither net gain nor loss of the amount of matter or energy in each box or compartment.

equilibrium ethic   An outlook that attempts to establish a middle ground between the total development of a potential resource by economic interests and the absolute preservation of that resource in its natural condition.

equivalent method An analytical procedure deemed by the United States Environmental Protection Agency to be equivalent to the standard, or official, method for the determination of pollutant concentrations, as approved by the agency. An individual or organization must demonstrate the accuracy and reliability of the new method before it can be certified as an equivalent method.

equivalent weight (EW) For dissolved compounds, the MOLECULAR WEIGHT divided by the number of hydrogen or hydroxyl ions in the undissolved compound. For elements, the ATOMIC WEIGHT divided by the VALENCE.

erosion The process of wearing away of a surface by physical means. The term is usually applied to the loss of soil through the agency of wind or water.

erythema An abnormal redness of the skin due to irritation or tissue damage.

Escherichia The genus name of a type of bacteria whose normal habitat is the colon in man and other warm-blooded animals. The organism is GRAM-NEGATIVE, ferments lactose at 37°C, and can grow with or without molecular oxygen. If members of this genus, referred to as fecal coliforms, are found in water, the water is considered to be contaminated with fecal material.

estuary Coastal waters where seawater is measurably diluted with fresh water; a marine ecosystem where fresh water enters the ocean. The term usually describes regions near the mouths of rivers and includes bays, lagoons, sounds, and marshes.

ethylene dibromide (EDB) A substance used for fumigating foodstuffs for protection against insects and nematodes. EDB is found as a residual in some treated grains and has been implicated as a carcinogen in test animals.

etiology The cause of a disease or abnormal condition.

eucaryotic Describing a cell that has a well-defined nucleus confined by a nuclear membrane, or a multicellular organism comprised such cells. The cell type for all organisms except the bacteria and blue-green algae. Compare PROCARYOTIC.

euphotic zone The upper layer of a body of water that is penetrated by sunlight. Photosynthesis or primary production takes place in this layer. Compare BATHYAL ZONE and ABYSSAL ZONE.

euploid A cell that has the correct number of chromosomes. If a normal cell contains 22 chromosomes, a euploid cell would have either that number or some exact multiple of that number, like 44. See ANEUPLOID.

eury- A prefix that means wide or broad. A euryhaline organism, for example, is able to live in an environment with large changes in salinity. *Eurytherm* is used to describe an animal that can live in an environment with large changes in temperature.

eutrophic Describing a river, lake, stream, or other body of water enriched with excessive amounts of plant nutrients, such as nitrates and phosphates. Such environments are characterized by the excessive growth of aquatic plants. Compare MESOTROPHIC, OLIGOTROPHIC.

eutrophication The addition of excessive plant nutrients to a river, lake, stream, or other body of water. The nutrients in excess are usually nitrates or phosphates, and the process leads to prolific growth of aquatic plants.

evaporation A change of state from liquid to gas. Some molecules in the liquid have enough energy to escape to the gas phase. The rate increases with temperature.

evaporative cooling Cooling of a liquid, such as water, by allowing a portion to evaporate. The process is important in the operation of COOLING TOWERS used to cool heated effluents from power plants as well as in the cooling of the human body through the evaporation of perspiration. The process is more effective than convection cooling. See HEAT OF VAPORIZATION.

evapotranspiration The combined action of evaporation (a physical process that converts liquid water to a gas) and transpiration (the loss of water vapor from plants).

evenness A mathematical expression that describes the distribution of individuals in a COMMUNITY among the species represented. A community has a high evenness if there is an equal number of individuals among the species present. Conversely, a community is described as having a low evenness if most of the organisms are members of one species while the other species are represented by one or a few individuals. One calculation method for evenness is

$$\frac{1}{\sum (N_i/N)^2}$$

where N_i is the number of individuals in each species and N is the total number of individuals.

event tree A graph of all possible outcomes and their probabilities following an initiating event. The probability of an outcome, such as an accidental death, can be estimated by combining all of the event probabilities that lead to a fatality. The representation is used in accident and risk analysis. Compare FAULT TREE.

evergreen Plants that retain their leaves throughout the year. The term is most commonly applied to trees such as spruces and firs. Compare DECIDUOUS.

evolution A scientific theory proposing that higher forms of life have descended from lower forms by way of natural mechanisms. The process by which the characteristics of a POPULATION of organisms change over time in response to natural genetic variation within the population and to selection forces of nature acting on that population. The selection forces include any of a multitude of natural factors that have the potential to impact in a negative fashion on the survival of an organism, for example, predators or parasites, nutrient supply, climactic changes, or physical isolation from a suitable habitat.

exception report Under the MANIFEST SYSTEM, a notification that must be sent to the regional administrator of the United States Environmental Protection Agency by a hazardous waste generator if the manifest accompanying waste shipped off-site by the generator for treatment and/or disposal has not been returned to the generator within 45 days of the waste shipment. The manifest system documents hazardous waste in transit to its ultimate treatment and/or disposal site.

excess air The air supplied for a combustion process over the amount theoretically required for complete burning. See also STOICHIOMETRIC RATIO, CONDITIONS.

exchange reaction A process to soften water by altering its ionic composition through the trading of ions. A resin is saturated with sodium ions; then, as water containing divalent ions, such as calcium or magnesium, flows over the resin, the sodium ions are released into the water and the calcium or magnesium ions are adsorbed by the resin.

exchangeable cation A positively charged ion loosely bonded to soil particles. The ion is not removable by pure water; however, it is easily exchanged for the sodium ions in a neutral salt SOLUTION. The soil particle adsorbs the sodium ion and releases the exchangeable cation.

Executive Order 12291   An order issued by President Ronald Reagan in 1981 requiring administrative agencies to prepare a REGULATORY IMPACT ANALYSIS for all major regulatory actions.

Executive Order 12498   An order issued by President Ronald Reagan in 1985 requiring all federal regulatory agencies to submit to the OFFICE OF MANAGEMENT AND BUDGET an annual report summarizing agency objectives for the coming year, including the issuance of any major regulations. Together with EXECUTIVE ORDER 12291, it is an attempt to reduce uncertainty and costs associated with federal regulations and standards.

exfiltration   The movement of air out of a building through the agency of wind or temperature differences. Compare INFILTRATION.

exhaust gas recirculation (EGR)   An automobile emission control system that mixes a small amount of the exhaust gases with the fuel-air mixture entering the engine via the intake manifold. This lowers the combustion temperature, and reduces the formation of OXIDES OF NITROGEN (NO_x).

exoergic   Describing a reaction that gives off energy or heat (same as EXOTHERMIC). Compare ENDOERGIC.

exogenous   Originating outside an organism. Compare ENDOGENOUS.

exothermic   Describing a reaction that gives off energy or heat. Compare ENDOTHERMIC.

exotoxin   A protein produced and released by the bacteria that cause certain diseases in humans, for example botulism and tetanus. These proteins are extremely toxic in microgram quantities.

exp   When followed by a number or expression that represents an exponent, the base of the NATURAL LOGARITHM; the same as e.

expedited removal action (ERA)   The cleanup of a hazardous waste disposal site without first preparing a REMEDIAL INVESTIGATION/FEASIBILITY STUDY; performed when a release or a threatened release of a hazardous substance requires immediate efforts to protect public health or the environment.

experimental concentration–percent (ECx)   In a test of the toxicity of a chemical to aquatic life, the experimentally determined concentration of the chemical that is calculated to affect X percent of the test organisms.

explosive limits   See LOWER EXPLOSIVE LIMIT, UPPER EXPLOSIVE LIMIT.

exponential decay   The decline in the number of a population, amount of a pollutant, level of radioactivity, and so forth, according to the exponential function $N = N_0 e^{-kx}$, where N is the amount left after decay, N_0 is the initial amount, k is a constant, and x is a variable such as time, altitude, or water depth.

exponential growth   1. Growth in the size of a population in which the number of individuals increases by a constant percentage of the total population each time period. When the number of individuals in the population are plotted against time, the increase appears as a J-shaped curve; when the logarithms of the numbers of individuals are plotted against time, the increase appears as a straight line.   2. The growth of bacteria in culture where division is by binary fission. Represented by the equation $N(t) = N_0 e^{kt}$, where $N(t)$ is the population size at time t, e is the base of the natural logarithm, and k is a constant. This stage is followed by a STATIONARY GROWTH PHASE. See LOG PHASE.

exposure   Contact between a chemical, physical, or biological agent and the outer surfaces of an organism. Exposure to an agent does not imply that it will be absorbed or that it will produce an effect.

exposure assessment An estimate of the actual or anticipated contact of a chemical, physical, or biological agent with a population, including the route(s), frequency, concentration, and duration of the exposure, and the number of individuals exposed at various levels.

external radiation The exposure of the body to IONIZING RADIATION where the source of the radiation is located outside the body.

externality The cost or benefit of some activity that affects persons not involved directly with the activity. Called a *negative externality* if costs are imposed, for example a decrease in value of residential property near a new industrial facility, or *positive externality,* as in the aesthetic and economic benefits accruing to the neighbors of a person repainting his or her home.

extinction The elimination of every individual within a particular species.

extinction coefficient A variable used in KOSCHMIEDER'S RELATIONSHIP to determine visual range in the atmosphere. The extinction coefficient b_{ext} is expressed as $b_{ext} = b_{rg} + b_{ag} + b_{scatp} + b_{ap}$, where b_{rg} is scattering by gas molecules (RAYLEIGH SCATTERING), b_{ag} is absorption by gas molecules, b_{scatp} is scattering caused by air particulate matter, and b_{ap} is absorption by particles.

extractable organic halogens (EOX) Organic compounds combined with any members of the chemical halogens (mainly chlorine, bromine, and fluorine) that can be removed from a soil or sludge sample with the solvent ethyl acetate. The organic halogens represent a class of unusually toxic materials.

extractable organics Organic chemical compounds that can be removed from a water sample by the solvent methylene chloride under conditions of pH greater than 11 or less than 2. Organic compounds in water represent a class of pollutants that are potentially toxic materials.

extraction procedure toxicity test (EP toxicity test) A laboratory test in which a solid waste material is treated to leach out certain toxic metals and/or pesticides. Formerly the official toxicity characteristic test method for a hazardous waste, the EP toxicity test has been replaced by the TOXICITY CHARACTERISTIC LEACHING PROCEDURE.

extrapolation The prediction of outcomes, in particular circumstances, beyond known experience or experimental observations but based on existing empirical data.

extremely hazardous substance (EHS) One of about 400 chemicals listed by the United States Environmental Protection Agency under the Emergency Planning and Community Right-To-Know Act of 1986. The Act called for, among other things, community planning for the accidental release to the atmosphere of toxic materials from industrial facilities or along transportation routes and their possible adverse effects in the nearby community. Chemicals were placed on the list based on acute toxicity and annual production volume. See also TITLE III.

F

F factor The ratio of the volume of gas produced by the combustion of a fuel to the energy content of a fuel. The gas volume used can include all combustion gases (wet F factor), all gases excluding water vapor (dry F factor), or only carbon dioxide (carbon F factor). The wet F factor is in wet standard cubic meters per joule; the dry F factor is in dry standard cubic meters per joule, and the carbon F factor is in standard cubic meters per joule. The appropriate factor is used in calculations of air pollutant emission rates for particulate matter, sulfur dioxide, and nitrogen dioxide.

fabric filters   Filter bags made of Teflon, nylon, cotton, or glass fibers used to remove PARTICULATE MATTER from industrial exhaust gases. The material collected on the filters is periodically shaken or blown off and falls to a hopper for disposal or recycling. See BAGHOUSE.

facepiece   The part of a respirator covering all or part of the wearer's face. The facepiece, which contains connectors for inflowing and outgoing air, must make an airtight seal to operate properly.

facultative bacteria   Microorganisms able to grow with or without molecular oxygen (O_2).

fall turnover   The exchange of top and bottom waters of a stratified lake promoted by the cooling of the surface water during the fall of the year. The surface water (the epilimnion) cools in response to the falling air temperature and becomes denser as a result. When the surface water cools to a temperature lower than that of the bottom water (the hypolimnion), the upper layer of water sinks, forcing the bottom water to the surface. Compare SPRING TURNOVER.

fallout   1. Radioactive particles introduced to the atmosphere by a nuclear accident or explosion. The particles can be transported thousands of miles before settling to the ground. 2. Any solid matter emitted to the atmosphere by human activities (such as smokestack or tailpipe particulate matter) or natural processes (like dust storms, volcanoes, forest fires) and then returning to the surface by gravitational settling.

fallow   Describing arable land that is left uncropped for a growing season; a part of the rotation practiced under non-irrigated agriculture. The land stores water during the fallow period, and the next

crop has available a two-year supply and more nutrients.

false negative   An erroneous test result that labels a chemical or individual as not having a certain property or condition when in fact the property or condition is present. For example, the false determination that a chemical is a noncarcinogen when it actually is a carcinogen. Compare FALSE POSITIVE.

false positive   An erroneous test result that labels a chemical or individual as having a certain property or condition when in fact the property or condition is not present. For example, a false test result stating that a person is infected with the AIDS virus when in fact the virus is not present. Compare FALSE NEGATIVE.

fan curve   A graphical depiction of the volumetric airflow of a fan for different STATIC PRESSURES at a given fan turning rate in revolutions per minute (rpm). A family of curves is typically generated with a separate curve for each of a series of rpm values.

far field   The area away from a sound source in which the SOUND PRESSURE LEVEL is inversely proportional to the square of the distance from the source, according to the INVERSE SQUARE LAW. Compare NEAR FIELD.

fast breeder reactor   A nuclear reactor that produces a significant quantity of fissionable material (plutonium-239) when uranium-238 absorbs FAST NEUTRONS produced during the fission process. See BREEDER REACTOR.

fast neutron   A neutron with energy exceeding 1×10^5 electron volts as it is ejected from a nucleus that undergoes fission. Compare SLOW NEUTRON.

fat-soluble   Describing a material that dissolves (is stored) in fat. Fat solubility is a characteristic of a compound that is more likely to undergo BIOACCUMULATION or BIOLOGICAL MAGNIFICATION.

fault tree A graph of all the events or failures necessary for an accident or other adverse effect to occur. The representation is developed by starting with the undesired outcome, such as loss of radiation in a nuclear power plant accident, and working backward to describe what events or sequences of events can lead to the outcome. The technique is used in accident risk analysis and to develop strategies for risk reduction. Compare EVENT TREE.

fauna A general term for the animal life of an area or region. Compare FLORA.

feasibility study A detailed technical, economic, and/or legal review of a specific proposed project at a particular location to outline all potential costs, benefits, and problems. If the project is a hazardous waste site cleanup, the process is called a REMEDIAL INVESTIGATION/FEASIBILITY STUDY (RI/FS).

fecal bacteria Any type of bacteria whose normal habitat is the colon of warmblooded mammals, such as man. These organisms are usually divided into groups, such as FECAL COLIFORM or FECAL STREPTOCOCCI.

fecal coliform A type of bacteria whose natural habitat is the colon of warm-blooded mammals, such as man. Specifically, the group includes all of the rod-shaped bacteria that are non-spore-forming, Gram-negative, lactose-fermentating in 24 hours at 44.5°C, and which can grow with or without oxygen. The presence of this type of bacteria in water, beverages, or food is usually taken to mean that the material is contaminated with solid human waste. Bacteria included in this classification represent a subgroup of the larger group termed coliform.

fecal material Solid waste produced by humans and other animals and discharged from the gastrointestinal tract. Also referred to as feces or solid excrement, it is a component of domestic sewage and must be treated to avoid the transmission of fecal bacteria and other organisms or disease.

fecal streptococci A type of bacteria whose natural habitat is the colon of warm-blooded mammals, such as man. The group includes those bacteria that are Gram-positive, spherical, and form chains. This type of bacterium is catalase-negative and capable of growth in media containing 6.5 percent sodium chloride. Most of these organisms belong to the species *Streptococcus faecalis*.

Federal Energy Regulatory Commission (FERC) An independent federal administrative agency, founded in 1977. Governed by five commissioners, the FERC is responsible for regulating the sale of electricity or natural gas in interstate commerce, construction and operation of oil and gas pipelines, and licensing nonfederal hydroelectric power plants.

Federal Environmental Pesticide Control Act (FEPCA) The 1972 amendments to the FEDERAL INSECTICIDE, FUNGICIDE, AND RODENTICIDE ACT that gave the United States Environmental Protection Agency authority to register pesticides and to suspend or cancel registration of any pesticide found to unduly threaten human health or the environment.

Federal Food, Drug, and Cosmetic Act See FOOD, DRUG, AND COSMETIC ACT.

Federal Insecticide, Fungicide, and Rodenticide Act (FIFRA) The federal law that regulates the manufacture and use of pesticides. The law was first passed in 1947 and amended in 1972, 1975, and 1978. Under FIFRA, pesticides must be registered with and approved for use by the United States Environmental Protection Agency before they are sold. Marketing approval may require the pesticide container label to include directions for proper use and disposal, and the purchase and/or use of some chemicals is restricted to certified (trained) applicators.

Federal Land Policy and Management Act (FLPMA) A 1976 federal law that, together with the TAYLOR GRAZING ACT, outlines policy concerning the use and preservation of public lands in the United States. The FLPMA gave the Bureau of Land Management, in the Department of the Interior, the responsibility to manage all public rangelands not within national forests or national parks. It also granted the federal government power to control the environmental consequences of mining on public lands.

Federal Power Commission (FPC) The predecessor federal agency to the FEDERAL ENERGY REGULATORY COMMISSION (FERC), established in 1920. Its authority was transferred to the FERC in 1977.

Federal Register A daily publication of the United States Government that contains federal administrative agency proposed RULES, final rules, and other executive branch documents. Final regulations and standards are annually codified by subject in the CODE OF FEDERAL REGULATIONS.

Federal Test Procedure (FTP) The methods prescribed by the United States Environmental Protection Agency for the stationary testing of the efficiency of emission control devices or the fuel economy (miles per gallon) of an automobile.

Federal Water Pollution Control Act (FWPCA) See CLEAN WATER ACT.

feedback Corrective information or a signal generated within a self-regulating system or process that is intended to induce a change in that system or process.

feedlot A confined area in which cattle are held and fed to promote maximum weight gain prior to marketing. The large

quantities of animal waste produced can sometimes lead to pollution problems in nearby water bodies.

feedwater Boiler water that is converted to steam during the operation of a fossil-fuel or nuclear power plant. The water is boiled by heat transferred from the furnace or reactor core, and the steam is employed to drive turbines for the generation of electricity.

feldspar Rock formed by the slow cooling of molten magma, constituting about half of the earth's crust. Crystalline rock composed primarily of aluminum silicates.

fenceline concentration The concentration of an air contaminant measured just outside the perimeter of the property of an industrial facility, the closest possible site of exposure to the public.

Fenton's reagent A mixture of iron (Fe^{+2}) and hydrogen peroxide (H_2O_2) that can be used to chemically oxidize or degrade toxic organic chemicals in soil. Soil contaminated with a variety of chlorinated aromatic compounds can be mixed with the two ingredients to effect the conversion of the organic material to carbon dioxide and chloride ions. The reagent is useful in the cleanup of contaminated soil.

feral animal A domesticated animal living in the wild.

fermentation A type of bacterial or yeast metabolism (chemical reaction) characterized by the conversion of carbohydrates to acids and alcohols, usually occurring in the absence of molecular oxygen.

fermentation tube method A technique for the examination of water or wastewater for the presence of fecal bacteria. Portions of a water sample are inoculated into culture tubes containing a growth medium that has lactose. Coliform bacteria will produce acid and gas by the fermentation of lactose when in-cubated at certain temperatures.

fertile atoms Nonfissionable isotopes that absorb neutrons and then decay to a fissionable material. The most common fertile atom is uranium-238, which de-cays to fissionable plutonium-239.

fertile isotope See FERTILE MATERIAL.

fertile material A material that is not capable of undergoing nuclear fission but can be converted to a FISSIONABLE MA-TERIAL by irradiation in a nuclear reactor. The most common fertile isotopes are uranium-238 and thorium-232, which are converted to the fissionable materials plutonium-239 and uranium-233, re-spectively.

fertility rate The average number of offspring born to females within an ani-mal population. Within human popula-tions, only those females of childbearing age are considered.

fetotoxic Describing a chemical sub-stance or other agent that has adverse effects on a developing fetus.

fibers per cubic centimeter (f/cc) Common units for the expression of air-borne asbestos concentration. The as-bestos in air is present in the form of countable fibers. A cubic centimeter is equal to one-millionth of a cubic meter, or a cube about 0.4 inch on a side.

fibrosis The formation of an excessive amount of fibrous tissue in the lung re-sulting from overexposure to certain in-soluble particulate material. Fibrotic par-ticulates include rock and coal dust, fine sands, and asbestos, among others. The accumulation of the fibrous tissue re-duces the efficiency of oxygen transfer in the lungs. See SILICOSIS, ASBESTOSIS, CRYSTALLINE SILICA.

Fick's first law of diffusion A law stating that the rate of diffusion of one material through another is proportional to the cross-sectional area of diffusion, the concentration gradient, and a diffu-sion coefficient. The value of the diffusion coefficient depends on the size and elec-tric charge of the diffusing substance, the type of material it is moving through, and the absolute temperature. It is expressed as

$$\frac{M}{A} = -D\,\frac{dC}{dX}$$

where M is the mass transfer rate, A is the cross-sectional area, D is a diffusion coefficient, and dC/dX is the CONCENTRA-TION GRADIENT.

field blank A sample container car-ried to and from the sample collection site but not used for taking environmental samples. This filter, collection tube, or other container is analyzed along with the actual samples to detect contamina-tion that may occur during sample col-lection and transport.

film badge A personal monitoring de-vice used to determine an individual's radiation exposure level. The small badge is worn on the clothing and may contain one or more layers of photographic film, which is developed after an appropriate time interval to determine exposure.

filter cake 1. The solids or semisolids deposited on a filter as a fluid is moved through it. 2. The remaining solids or semisolids on a filter after the fluid in a material is extracted by a negative pres-sure.

filterable Of particles that are suffi-ciently small to allow their passage through filters capable of retaining most particles. For example, a filterable virus is one that will pass through a filter that will normally retain bacteria.

filtering velocity The speed at which air or fluid moves through a paper, syn-thetic, or sand filter for the purpose of removing particulate material from the air or liquid. The units used are volume measure (such as cubic meters, gallons, or liters) per unit of time (such as days, seconds, or hours).

final clarifier   A gravitational settling tank installed as part of some wastewater treatment plants and placed after the biological treatment step. The tank functions to remove SUSPENDED SOLIDS. Also called a secondary clarifier.

financial assurance   The RESOURCE CONSERVATION AND RECOVERY ACT requirement that owners and operators of hazardous waste treatment, storage, or disposal facilities demonstrate to the permitting authority that funds will be available to meet the estimated expenses of CLOSURE and POST-CLOSURE activities at the site.

finding of no significant impact (FONSI)   A document prepared by a federal agency presenting reasons why an activity or development project will not have an appreciable effect on the human environment and for which an ENVIRONMENTAL IMPACT STATEMENT will therefore not be prepared.

fines   Very small airborne particles usually less than 2 MICRONS in diameter.

finished water   Water that has completed a purification or treatment process. Compare RAW WATER.

firedamp   Methane gas found in underground coal mines. See also DAMP.

fire point   The lowest temperature at which a liquid will vaporize at a sufficient rate to support continuous combustion. Compare FLASH POINT.

firestorm   An intense fire characterized by rapid burning and strong winds generated by convection currents produced by the fire.

firm capacity   For public drinking water supplies, the system delivery capacity with the largest single water well or production unit out of service.

first law of thermodynamics   A law stating that during any chemical or physical change in a closed system, energy is not destroyed or created but is changed from one form to another. Its expression for a closed system is $Q = \Delta U + W$, where Q is the net heat absorbed by the system, W is the work performed, and ΔU is the change in internal energy. For example, chemical energy may be changed to heat energy during the burning of fossil fuels. The law is also referred to as the law of conservation of energy.

first third   See LAND DISPOSAL BAN.

first-order reaction   A chemical reaction in which the rate of reaction is directly proportional to the concentration of one of the reactants, and not to any other chemical within the reaction mixture. Compare ZERO-ORDER REACTION.

Fish and Wildlife Service   An agency, created within the United States Department of the Interior in 1940, responsible for fish and wildlife management on federal lands and for protecting these resources from possibly harmful activities of other government agencies. The Fish and Wildlife Service compiles the list of endangered and threatened species authorized by the ENDANGERED SPECIES ACT.

fish ladder   A series of small pools arranged in an ascending fashion to allow the migration of fish upstream past constructed obstacles, such as dams.

fish protein concentrate (FPC)   A dry flour or paste derived from processing fish, often fish considered undesirable for direct consumption. Rich in protein, the product has been proposed as a dietary supplement for populations in less-developed countries, but cost and residual fishy taste have hindered its acceptance.

fission   The splitting of the nucleus of an atom into two or more nuclei with the concurrent release of neutrons and a large amount of energy. The process is induced in a nuclear reactor to produce energy and in the detonation of nuclear weapons.

fission products   The nuclei produced when elements, such as uranium-235, undergo FISSION. The fragments are ISOTOPES of various elements and are frequently RADIOACTIVE. In a nuclear reactor, as these fragments accumulate, the efficiency of the reactor declines. Fission products constitute the nuclear waste in the spent fuel rods that are removed from the reactor.

fissionable material   A HEAVY ATOM that can be split by the absorption of SLOW NEUTRONS. The three most common fissionable materials are uranium-233, uranium-235, and plutonium-239. Fissionable materials are used as the fuel in a NUCLEAR REACTOR and in nuclear weapons. See also FERTILE MATERIAL.

fixation   Increasing the stability of a waste material by involving it in the formation of a stable solid derivative. Stabilization lessens the potential harmful effect of a pollutant.

fixed carbon   1. The solid nonvolatile portion of organic waste material left after combustion, excluding ash and moisture. 2. The amount of carbon dioxide converted to plant biomass by the process of photosynthesis.

fixed solids   See SUSPENDED PARTICULATE MATTER.

flame ionization detector (FID)   An analytical device used in GAS CHROMATOGRAPHY. The detector burns the gas containing the chemicals extracted from an environmental sample, which produces a current of ions or electrons proportional to the amount of specific organic materials present.

flame retardant   A chemical added to cloth to prevent or retard rapid burning.

flameless furnace   See GRAPHITE FURNACE.

flammable limits   See LOWER EXPLOSIVE LIMIT, UPPER EXPLOSIVE LIMIT.

flammable liquid   Under United States Department of Transportation regulations, any liquid with a FLASH POINT less than 100°F (38°C).

flammable material   Any solid, liquid, or gas that will burn rapidly when ignited.

Flannery decree   The 1976 consent decree subsequently codified by the 1977 amendments to the Clean Water Act that required the United States Environmental Protection Agency to develop BEST AVAILABLE TECHNOLOGY standards for over 20 industrial categories involving 65 classes of toxic water pollutants, called the PRIORITY POLLUTANTS. Named for the judge in the case, Thomas Flannery, U.S. District Court judge for the District of Columbia.

flare   A tall stack used for the routine combustion of waste gases or for the burning of materials that must be routed from a chemical reaction vessel or refining process during upset conditions.

flash point   The lowest temperature at which a flammable liquid produces a sufficient amount of vapor to ignite with a spark.

Fletcher-Munson contours   The results of sound loudness observations on young male subjects by Fletcher and Munson in 1933, expressed as plots of the SOUND PRESSURE LEVELS (in decibels) at particular frequencies that are perceived as equal in loudness. The subjects were given a 1000-Hertz tone (for example, at 40 decibels) and were asked to adjust the decibel level of another tone of different frequency until that frequency sounded as loud as the reference tone. The test group consistently adjusted lower-frequency sounds (less than 1000 Hertz) upward to make them equal the reference sound in loudness; for example, a 60-decibel 200-Hertz sound was equated to a 40-decibel 1000-Hertz sound. At 1000 to 5000 Hertz little adjustment occurred, with some upward adjustment of tones greater than 5000 Hertz. The contours illustrate the relative insensitivity of the human ear to lower frequencies. The DECIBELS, A-WEIGHTING NETWORK (dBA) scale is derived from the 40-decibel Fletcher-Munson contour. See PHONS.

floating roof   A roof that floats on the hydrocarbon liquid in an oil refinery storage tank, moving up and down with the level of crude oil or fuel in the vessel. The absence of vapor space above the liquid and the maintenance of a tight seal between the tank wall and the roof greatly reduce hydrocarbon evaporation losses to the atmosphere.

floc   In wastewater treatment, the particles formed by the COAGULATION of even smaller particles, or colloids.

flocculation   In wastewater treatment, the rapid mixing of chemicals into the wastewater to enhance the formation of FLOC. Particles must be of sufficient size for removal, filtration, or sedimentation.

flood plain   The land area bordering a river that floods when the river overflows its banks.

flora   A general term for the plant life in an area or region. Compare FAUNA.

flow system   See CONTINUOUS-FLOW SYSTEM.

flow-through test   A test of the toxicity of a chemical to aquatic organisms. The test solution containing the chemical passes through the test chamber and is not recycled. See STATIC TEST, SEMISTATIC TEST.

flue   A conduit for the passage of smoke or exhaust gases.

flue dust   Very small particles carried along in the gases that exit a furnace during combustion.

flue gas   Hot gases produced during combustion within a furnace. The gas is exhausted through the flue, carrying with it flue ash, or noncombustible particles.

flue gas conditioning   The addition of chemicals, such as ammonium salts, to an exhaust gas to raise the resistivity of the fly ash in the airstream before it enters an ELECTROSTATIC PRECIPITATOR. Greater resistivity allows the particles to maintain an electrostatic charge and promotes a higher collection efficiency in the electrostatic precipitator. Used particularly with low-sulfur coals.

flue gas desulfurization (FGD)   The removal of sulfur-containing compounds, such as sulfur dioxide, from exhaust gas. The most widely used of many sulfur dioxide removal processes is limestone scrubbing, in which powdered limestone (calcium carbonate) is injected into an exhaust gas and sulfur dioxide is absorbed and neutralized, creating calcium sulfite and calcium sulfate as byproducts.

fluidized bed combustion   A burning process designed to promote the efficient combustion of coal. The coal is first powdered and then made to flow like a liquid by the injection of a rapidly moving stream of gas. Burning is carried out while the solid powder is in this fluidlike state. Limestone can be added to absorb sulfur dioxide created from sulfur contaminants in the coal.

fluidizing air   A rapidly moving stream of air that is injected into a bed of powdered material to impart a fluidlike state to the powder. See also FLUIDIZED BED COMBUSTION.

flume   A narrow trough or chute for carrying water.

fluorescence   The emission of light or other electromagnetic energy caused by the excitation of an atom.

fluorescence detector   A detector used with LIQUID CHROMATOGRAPHY. A sample stream is subjected to shortwave radiation, wavelength-specific for a certain substance, which causes any of that substance in the sample to reemit (fluoresce) longer-wavelength energy that can be measured by the detector to indicate the amount of substance in the sample.

fluoridation   The addition of FLUORIDE to drinking water in an effort to reduce tooth decay among the human population.

fluoride   A negative ion formed from the element fluorine, or a compound containing fluorine. Fluoride combines with tooth enamel to render it less soluble in acid environments and fluoride compounds are added to public water supplies to prevent tooth decay. Fluorine is a halogen with the chemical symbol F.

fluorocarbons   See CHLOROFLUORO-CARBONS.

fluorosis   A disorder of the teeth or bones associated with excessive consumption of FLUORIDES; characterized by mottling and brittleness.

flux   1. The flow rate of mass, volume, or energy per unit of cross-sectional area that is perpendicular to the flow.   2. The movement of dissolved and suspended matter into and out of a marsh as the tides flood and ebb.

flux density   In a nuclear reactor, the number of neutrons passing through a given unit of area per unit of time, usually through one square centimeter per second. For a plane that rotates inside a sphere to meet the neutrons at right angles, $\phi = \Delta N / \Delta a \; \Delta t$, where ϕ is the flux density, Δa is the area of the circular plane, ΔN is the number of neutrons, and Δt is the time of measurement.

fly ash   Small solid particles of non-combustible residue produced by the burning of a fuel such as coal. The particles are carried from the combustion process by the flue gas.

fomite   Any object contaminated by pathogens from a diseased person.

food additive   Any chemical added to a food product during processing to enhance shelf life, appearance, flavor, or nutritional content. The list of common additives includes agents that prevent caking of the food (aluminum calcium silicate), act as chemical preservatives (ascorbic acid, propionic acid and sodium sulfite), emulsify (desoxycholic acid and propylene glycol), provide dietary supplement (biotin, leucine and zinc sulfate), bind other materials (calcium acetate, potassium citrate and tartaric acid), and stabilize (agar-agar and guar gum). Food additives are regulated by the FOOD AND DRUG ADMINISTRATION under authority of the FOOD, DRUG, AND COSMETIC ACT. See DELANEY CLAUSE, GENERALLY RECOGNIZED AS SAFE.

Food and Drug Administration (FDA)   A federal agency within the United States Department of Health and Human Services that regulates the quality and safety of foods, food colors and additives, drugs, and cosmetics. Established by the FOOD, DRUG, AND COSMETIC ACT of 1938, its role has been further defined in subsequent amendments.

food chain   The flow of CARBON and ENERGY within a specified area as a result of the feeding sequence of organisms within a COMMUNITY. The organisms are divided into TROPHIC LEVELS, which depend on how an individual organism obtains its food. The first level in the chain is occupied by the green plants, termed PRIMARY PRODUCERS; those animals that consume the plants are termed HERBIVORES or consumers and are placed in the second trophic level; and the animals that eat other animals are termed CARNIVORES or secondary consumers and are placed in the highest feeding level. The transfer of materials or mass from one trophic level to the next is approximately 10% efficient. In a simple food chain consisting of grass (primary producer) to rabbit (herbivore) to owl (carnivore), 1000 pounds of grass would be needed to support 100 pounds of rabbit which, in turn, would support 10 pounds of owl. The food chain thus described is referred to as a grazing food chain since it is based on the consumption of live, standing biomass (the grass). Other food chains are based on the consumption of the remains of dead organisms. The primary consumers in such food chains, referred to as DETRITUS based, consist of bacteria, fungi, and various worms and insects (termed the DECOMPOSER community). The next highest trophic level consists of those carnivores that consume these decomposers. See ECOLOGICAL PYRAMID.

food chain crops   Plants that, if contaminated by environmental pollutants, will directly or indirectly result in human exposure to the contaminants. These include plants grown for human consumption, plants ingested by animals that are ultimately part of the human diet, and tobacco.

food chain efficiency   An expression of the net transfer of useful energy from a food to the animal consuming it. See ECOLOGICAL EFFICIENCY.

Food, Drug, and Cosmetic Act   First passed as the 1906 Food and Drug Act, with major amendments in 1938, 1958, 1962, and 1976, the basic federal law concerning the sanitary condition and safety of food, and the efficacy and safety of drugs and cosmetics, sold in the United States. The Act regulates food additives. See DELANEY CLAUSE, GENERALLY RECOGNIZED AS SAFE.

food intoxication   A pathological condition in humans or animals that is caused by the consumption of food that contains a toxin. For example, botulism and common food poisoning are caused by the production of toxic compounds during the growth of specific types of bacteria in food prior to its consumption.

food-to-microorganism ratio (F/M ratio)   The ratio of organic material load to the microorganism mass in the aeration tank of a wastewater treatment facility. The ratio is calculated as

$$\frac{F}{M} = \frac{(Q)(BOD)}{(MLSS)(V)}$$

where Q is the flow rate of RAW SEWAGE, BOD is the BIOCHEMICAL OXYGEN DEMAND of that sewage, MLSS is the MIXED LIQUOR SUSPENDED SOLIDS concentration, and V is the aeration tank volume. Commonly expressed as kilograms of BOD per kilogram MLSS per day.

food web   The interrelationship among the biological organisms in a COMMUNITY according to the transfer of useful energy from food resources to organisms eating those resources. While a FOOD CHAIN depicts a simple linear transfer from one organism to another, most animals eat more than one type of food and this model illustrates the complexity of feeding patterns within the natural environment. For example, a single resource such as grass may serve as food for insects, mice, rabbits, and deer, while the mice may in turn be eaten by snakes, owls, and foxes.

foot-pound   A unit of WORK equivalent to 1.356 JOULES.

force mains   Pipes in which wastewater is transported under pressure; the system is used in some areas having small elevation changes with distance and therefore needing to augment the gravity flow.

forced draft (FD)   The pushing or forcing of gases through an enclosed area (such as a combustion chamber) by the use of a fan or blower. See also INDUCED DRAFT, NATURAL DRAFT.

forced expiratory volume (FEV₁) The volume of air a person can exhale in one second; a lung function test.

forced oxidation A chemical process in which pollutants in an exhaust or discharge are forced into contact with air or pure oxygen to convert the pollutants to a stable form.

forced vital capacity (FVC) The maximum volume of air a person can exhale after a maximum inhalation; a lung function test.

forcing functions Significant factors that determine the composition of natural ecosystems, such as temperature, rainfall, relative humidity, and solar radiation.

Form R The annual report of routine and accidental chemical releases from certain facilities required by Section 313 of the Emergency Planning and Community Right-to-Know Act (TITLE III).

fossil fuels Crude oil, natural gas, peat, coal, or other hydrocarbons that are derived from the remains of plants and/or animals that were converted to other forms by biological, chemical, and physical forces of nature.

free available chlorine The sum of the hypochlorous acid (HOCL) and hypochlorite ion (OCL⁻) concentrations in water.

free field Conditions in which sound can be measured without interference by echoes from barriers in the test area. Actual free-field conditions rarely exist outside of special sound chambers, but nearly free-field conditions are present if the distance to barriers is great enough to have no significant influence on the sound measurements.

free liquids Liquids capable of migrating from waste and contaminating ground water. Hazardous waste containing free liquids may not be disposed of in landfills. See PAINT FILTER LIQUIDS TEST.

free moisture Liquid that will drain freely from solid waste by the action of gravity only.

free radicals Unstable atoms or molecules with at least one unpaired electron; these materials are highly reactive and thus are short-lived. They act as important intermediates in photochemical air pollution, photolysis, chlorofluorocarbon depletion of the ozone layer, combustion processes, and polymerization reactions.

free residual chlorine The FREE AVAILABLE CHLORINE level present after the destruction of ammonia and the reduction in CHLORAMINE residuals by the progressive addition of chlorine (BREAKPOINT CHLORINATION). The residual available for disinfection in a water distribution system practicing free residual chlorination. Compare COMBINED AVAILABLE CHLORINE.

free silica See CRYSTALLINE SILICA.

freeboard The vertical distance between the waste contained in a tank or surface IMPOUNDMENT and the top of the enclosure or container.

free-living Describing a species of plant or animal that is capable of living, reproducing, or carrying out a specific function without the direct assistance of a plant or animal of a different species. Compare SYMBIOTIC.

freons A family of compounds containing carbon, chlorine, and fluorine with a typical chemical formula of CF_2Cl_2 (Freon 12). The compounds are gases at room temperature and are used as the carrier gas in aerosol cans (in countries outside of the United States), as foaming agents and solvents, and as the heat transfer gas in refrigerators and air conditioners. See CHLOROFLUOROCARBONS.

frequency For electromagnetic radiation or sound waves, the number of wave cycles passing a point in one second.

Formerly expressed as cycles per second; now expressed as HERTZ. One hertz equals one wave cycle per second.

friability The degree to which a solid can be crushed and powderized.

friable Describing a solid that is easily crushed and powdered.

Friends of the Earth (FOE) A conservation and environmental organization, founded in 1969, dedicated to preservation, restoration, and wise use of natural resources. United States headquarters in Washington, D.C., affiliates in 37 countries. Through the Friends of the Earth Foundation, the organization promotes public education and monitors enforcement of environmental policies. In 1989 membership in the United States numbered about 15,000.

front end recovery A centralized solid waste treatment process in which mechanical or manual separation of paper, glass, metals, and/or organic matter suitable for composting is performed on the collected waste before further processing.

frost heave The uneven rise in a ground surface caused by the accumulation of ice in the subsurface soil.

fuel Coal, oil, natural gas, or products thereof that are used to power equipment or provide heat. ENRICHED URANIUM used to provide energy in nuclear reactors.

fuel assembly A bundle of about 200 fuel rods, each of which contains pellets of enriched uranium. These clusters are placed in the core of a nuclear reactor to provide the fissionable material needed to power the reactor. Also called fuel element.

fuel cell A device in which hydrogen and oxygen combine to produce an electric current.

fuel cycle The steps in the production of enriched uranium for use in a nuclear reactor and the handling of fuel elements after they are removed from the reactor. The complete cycle includes the mining of uranium ore, purification of the uranium, enrichment of purified uranium in the fissionable isotope, the manufacture of pellets of enriched uranium, the fabrication of fuel rods and fuel assemblies, use of the fuel in reactors, the recovery of reusable uranium and other elements from used rods, and the disposal of waste material generated in the process. See FUEL ROD, FUEL ASSEMBLY, FUEL ENRICHMENT, FUEL REPROCESSING, FISSIONABLE MATERIAL, and ISOTOPE.

fuel element See FUEL ASSEMBLY.

fuel enrichment Increasing the abundance of FISSIONABLE MATERIAL. When uranium is purified from geological deposits, it consists of a mixture of about 0.7 percent uranium-235 and 99.3 percent uranium-238. In such mixtures, the fissionable ISOTOPE (uranium-235) is not sufficiently concentrated to support a sustained fission reaction. The amount of fissionable material can be increased by purifying the uranium-235 or by adding plutonium-239.

fuel NO$_x$ Nitrogen oxides formed by the oxidation (combustion) of organic nitrogen present in coal or oil. Compare THERMAL NO$_x$.

fuel reprocessing The recovery of usable uranium and other elements from fuel rods that have been removed from a nuclear reactor. The amount of fissionable material decreases and the amount of fission products increases when fuel rods have been employed in a nuclear reactor for about three years. As a consequence, the efficiency of the fission reaction diminishes to the point that the used rods must be removed. These used rods can be processed so that the substantial amount of usable uranium that remains can be reclaimed and other useful and waste elements are removed.

fuel rod   A long tube that contains pellets of enriched uranium dioxide used to fuel a nuclear reactor. These rods are commonly constructed of a zirconium alloy or stainless steel and are bundled into a FUEL ASSEMBLY of about 200 rods.

fugitive emissions   Any gas, liquid, solid, mist, dust, or other material that escapes from a product or process and is not routed to a pollution control device.

fume   Finely divided airborne solids formed by the condensation and solidification of material emitted as a vapor or gas; usually irritating and offensive at high concentrations.

fumigant   Any substance that is used as a gas, particulate, vapor, or smoke to kill pests (insects or rodents) in foodstuffs or structures.

fumigation   The application of a FUMIGANT to a material or area in order to kill pests or dangerous organisms.

fundamentally different factors (FDFs)   A type of variance from a water pollutant EFFLUENT LIMITATION that may be granted under provisions of the Clean Water Act if a permit applicant can show that the plant or facility is fundamentally different from the facilities, equipment, and so forth, used by the United States Environmental Protection Agency when setting the TECHNOLOGY-BASED standard for the industry category or subcategory as a whole.

fungicide   Any substance that kills fungi or molds.

furans   See DIBENZOFURANS.

fusion   A reaction that results from combining nuclei of small atoms to form larger atoms. Joining the nuclei of two atoms requires atomic collisions at very high temperatures and pressures, with a significant release of heat energy. The process is the underlying force for the release of energy from stars and from the detonation of a hydrogen bomb. Also referred to as a thermonuclear reaction.

G

gabion   A wire mesh container filled with rocks used as a barrier to retard soil erosion.

gage pressure   See GAUGE PRESSURE.

Gaia hypothesis   The proposition that the composition and temperature of the atmosphere is a product of interrelated activities in the BIOSPHERE, especially those of microorganisms, and that the biosphere behaves as a single self-regulating organism. Gaia was the ancient Greek goddess of the earth. The hypothesis was developed by the British scientist James Lovelock and American biologist Lynn Margulis.

gallons per capita per day (GPCD)   An expression of the average rate of domestic and commercial water demand, usually computed for public water supply systems. Depending on the size of the system, the climate, whether the system is metered, the cost of water, and other factors, public systems in the United States experience a demand rate of 60 to 150 GPCD.

gamete   A reproductive cell containing one copy of each gene (haploid) needed to provide the genetic information required for the development of an offspring. When gametes from the female (egg) are combined with or fertilized by those of the male (sperm), a zygote containing two copies of each gene (diploid) is produced. The zygote can develop into an offspring.

gamma radiation   See GAMMA RAY.

gamma ray A type of ELECTROMAG-NETIC RADIATION, produced by some radioactive substances, with a very short wavelength and high energy level. This type of radiation is very penetrating, potentially harmful to living things and is more energetic than x-rays. Also referred to as gamma radiation.

gammarids Any of three species of the amphipod *Gammarus (G. fasciatus, G. pseudolimnaeus, G. lacustris)* used in the laboratory analysis of the toxicity of pollutants to aquatic animals. The organism is used to determine the toxicity of liquid effluents without chemical analysis to determine the precise chemical composition of the toxicant.

garbage Waste material, typically from domestic and commercial sources, that is BIODEGRADABLE.

gas One of the three states of matter. In a gaseous state, there is little attraction between the particles, which have continual, random motion. The gas has no fixed shape or volume, can expand indefinitely, and assume the shape of the space in which it is held. It is also easily compressed, with the random collisions between particles exerting pressure on the walls of the container.

gas barrier A layer of material placed on the top and/or sides of a landfill to prevent the off-site migration of gas (and odors) produced by microbial degradation of the buried waste. Typical barrier materials are compacted clay and synthetic membranes.

gas constant See UNIVERSAL GAS CONSTANT.

gas chromatogram A graphic representation produced by the detector used in GAS CHROMATOGRAPHY. The identity of contaminants can be determined from such an output.

gas chromatograph The analytical instrument used to perform GAS CHROMATOGRAPHY.

gas chromatography (GC) An analytical technique that can yield both qualitative and quantitative evaluations of sample mixtures of volatile substances. The compounds of interest are separated by using an inert gas to flush a sample preparation through a column packed with a substance that selectively absorbs and releases the volatile constituents. A device is used to detect the level of each compound as it exits the column.

gas chromatography/mass spectrometry (GC/MS) A sensitive and accurate analytical technique, used mainly for organic compounds, in which the gas effluent from a GAS CHROMATOGRAPH is piped to a MASS SPECTROMETER for additional analysis.

gas contacting device Equipment designed to encourage substantial mixing of liquid and gas streams.

gas-cooled reactor (GCR) A nuclear reactor in which the heat produced by fission reactions is removed from the reactor CORE by a gas, as opposed to air, water, or liquid sodium. The gases most commonly used are helium and argon. See also HIGH-TEMPERATURE GAS REACTOR, WATER COOLED REACTOR.

gaseous diffusion A method used to separate isotopes of uranium. Uranium oxide that is prepared from uranium ore is first converted to uranium hexafluoride, a gaseous mixture containing both of the naturally occurring isotopes of uranium, uranium-235 and uranium-238. The isotopes are then separated on the basis of differences in their diffusion properties.

gasification See COAL GASIFICATION.

gas laws The relationships between gas temperature, pressure, and volume.

See BOYLE'S LAW, CHARLES' LAW, and IDEAL GAS LAW.

gas-liquid chromatography See GAS CHROMATOGRAPHY.

gasohol An automobile fuel that is blended as a mixture of alcohol (usually ethyl alcohol) and gasoline. The typical mixture contains 10 percent alcohol. The blend was developed to conserve supplies of gasoline; however, the mixture did not gain wide public acceptance in the 1980s, and its use was largely discontinued.

gas-phase absorption spectrum The pattern of radiation absorbance by chemical compounds in a gaseous mixture. Particular gases will absorb radiation in certain frequency ranges. The spectrum is used to monitor some air pollutants.

gas-to-cloth ratio (G/C) See AIR-TO-CLOTH RATIO.

gas volumetric flow rate See VOLUMETRIC FLOW RATE.

gastroenteritis A disorder of the stomach, small intestine, or colon that is characterized by nausea, vomiting, and/or diarrhea. The condition is caused by a variety of bacteria and viruses.

gastrointestinal tract The organs of the alimentary system through which food passes that extends from the mouth to the anus. The term is also applied specifically to the stomach, small intestine, and colon.

gauge See GAUGE PRESSURE.

gauge pressure The pressure exerted by gases relative to ATMOSPHERIC PRESSURE, commonly expressed in millimeters of mercury (mm Hg), pounds per square inch (gauge), or inches of water. If atmospheric pressure is 760 mm Hg and gauge pressure is 20 mm Hg, then the total (or absolute) pressure is 780 mm Hg.

Gaussian plume model A basic air quality dispersion model based on an assumed normal distribution of vertical and horizontal downwind concentrations from a pollutant source. The normal distribution approximates measured values. See DISPERSION COEFFICIENTS.

Gay-Lussac's law See CHARLES' LAW.

gavage In toxicological testing, the introduction of the test chemical to an animal via a stomach tube.

gene A specific segment of deoxyribonucleic acid (DNA) located on a chromosome. The individual segments can code for a specific protein, a specific type of ribonucleic acid (RNA), or a recognizable trait of an organism. Hundreds of genes can be located on a single chromosome. GAMETES have one copy of each gene (haploid), while all other cells of an organism have two copies of each gene (diploid). Mutagens to which an organism is exposed cause permanent changes in the gene structure.

gene pool The total genetic information available within a population of plants or animals that is capable of interbreeding.

General Accounting Office (GAO) An agency operating as an arm of the United States Congress with responsibility for auditing and reviewing government expenditures and programs; the office determines if federal agencies are spending public funds as intended by the legislation that authorizes the expenditures.

General Duty Clause A section of the OCCUPATIONAL SAFETY AND HEALTH ACT providing that "each employer shall furnish . . . a place of employment . . . free from recognized hazards that are causing or are likely to cause death or serious physical harm to his employees."

The clause is used by the Occupational Safety and Health Administration to force correction of workplace conditions not specifically covered by codified health and safety regulations.

generally recognized as safe (GRAS) A classification of food additives. The Food, Drug, and Cosmetic Act of 1958 provided "grandfather" approval for most additives in common use prior to the passage of that legislation. These additives cannot be removed from use until they are proven to be harmful to the public by the Food and Drug Administration. Over 600 substances are on the FDA list of GRAS substances. They were placed on this list on the basis of past experience in food use, scientific determination of their safety, or testimony of individuals knowledgeable in the area of food additives.

generator 1. A business or industrial facility that produces HAZARDOUS WASTE. 2. A device used to produce electricity.

genetic variability Natural differences within the GENE POOL of a population of organisms belonging to the same species, as indicated by observable differences among the organisms.

genetics The study of the storage, replication, transfer, and expression of information that governs the transmission of traits from parents to offspring. Some dangerous materials added to the environment disrupt these processes.

genotoxic Describing a chemical or physical agent capable of damaging the genes in a cell.

genotype The total genetic makeup of a plant or animal. All of the genetic traits contained by a specific organism are not usually expressed. Some genes are recessive, and their presence is masked by the overriding influence of dominant traits, determining the organism's actual appearance, or PHENOTYPE.

geometric growth A doubling of the number of individuals constituting a population with each time interval corresponding to a generation. Also called exponential growth, this is the characteristic growth pattern of bacteria. See EXPONENTIAL GROWTH.

geometric mean The mean of a LOGNORMAL DISTRIBUTION of values. Calculated as the nth root of the product of n values.

geometric mean diameter The mean diameter of particles of an aerosol; consequently, one-half of the particles in the aerosol will be smaller than this value, and one-half will be larger.

geometric standard deviation An expression of the dispersion of a set of measurements about a GEOMETRIC MEAN. Normally distributed measurements have a mean (average), with the dispersion indicated by standard deviations added to or subtracted from the mean. For the LOG-NORMAL DISTRIBUTION, however, the geometric mean is multiplied and divided by the geometric standard deviation to calculate the dispersion of the data set.

geomorphology The study of the nature and origin of the land features of the earth.

geostrophic wind Wind blowing parallel to pressure ISOBARS as a result of a balance between the PRESSURE GRADIENT FORCE and the CORIOLIS FORCE; this type of wind occurs at high elevations, where the effect of surface friction on the balance of forces is negligible.

geothermal Describing hot water, steam, or energy that is produced by the transfer of heat from the interior of the earth to geological deposits close to the surface. Hot springs are an example of geothermal activity.

germ cell A reproductive cell: sperm or egg.

Giardia Genus name of *Giardia intestinalis,* a protozoan parasite. See GIARDIASIS.

giardiasis A disease that results from an infection by the protozoan parasite *Giardia intestinalis,* caused by drinking water that is either not filtered or not chlorinated. The disorder is more prevalent in children than in adults and is characterized by abdominal discomfort, nausea, and alternating constipation and diarrhea.

gigawatt (GW) A unit of power commonly used when referring to nationwide energy production or consumption, especially electric energy. Equal to one billion watts, or 1000 MEGAWATTS.

gill net A net placed vertically in estuaries, streams, or rivers to capture fish of a certain size by allowing their heads through the openings but catching their gills in the mesh. The nets are so efficient in catching fish that their use can seriously reduce the breeding stock of some fish species.

glass frit The calcined or partly fused raw material for making glass.

glassification See VITRIFICATION.

glazing 1. Injury to the lower leaves of certain broadleaf plants from excessive exposure to the air pollutant PEROXYACETYL NITRATE, causing the leaves to appear silver. 2. A colorless glass coating applied to the surface of a container or product to reduce its porosity. Glazing containing lead has been a source of lead poisoning when used on earthenware in which acidic foods or beverages are stored.

Global Environment Monitoring System (GEMS) A part of the United Nations Environment Programme's EARTHWATCH, based in Nairobi, Kenya. The purpose of GEMS is to coordinate environmental monitoring and assessment activities worldwide and to provide an environmental data exchange service.

global warming See GREENHOUSE EFFECT.

globe temperature The temperature reading that indicates the level of radiant heat or heat transferred by INFRARED RADIATION. It is read from a thermometer with its bulb set in the center of a black 6-inch sphere, which acts to absorb radiant heat in its surroundings.

glove box An airtight container with flexible gloves attached that extend into the box, allowing a worker to manipulate the box contents without exposure to the material(s). Such boxes are used for handling highly radioactive substances and highly infectious materials.

Gold Book The United States Environmental Protection Agency document *Water Quality Criteria 1986,* which contains information on ambient water quality toxicity for acute and chronic exposures to aquatic life and health risk for human exposures to pollutants from water or fish consumption. These criteria are not legally enforceable standards.

good engineering practice stack height (GEP stack height) A regulatory requirement of the U.S. Environmental Protection Agency limiting the height of stacks emitting air pollutants; the prescribed height depends on when the stack was constructed, the height or width of nearby structures, and/or the ground level impact.

Good Laboratory Practice Standards (GLP standards) Standards that must be followed by individuals conducting studies relating to health effects, environmental effects, and chemical fate testing with the support of the Environmental Protection Agency or the Food and Drug Administration. The regulations address the areas of personnel, management, quality assurance, care of laboratory animals, and substance handling.

Government Printing Office (GPO) The official publisher of congressional documents and publications of agencies and departments of the executive branch. The office operates bookstores in over 20 cities and distributes government documents to depository libraries across the United States.

grab sample   Typically, a single air or water sample drawn over a short time period. As a result, the sample is not representative of long-term conditions at the sampling site. This type of sampling yields data that provides a snapshot of conditions or chemical concentration at a particular point in time.

gradient   See CONCENTRATION GRADIENT, PRESSURE GRADIENT.

gradient wind   Wind that is assumed to move parallel to the curved path of atmospheric isobars with no deviations due to friction between the wind and the ground.

Graham's law   Law stating that gases diffuse at a rate proportional to the square root of their density, with lighter molecules diffusing faster than heavy molecules. This property of gases can be used to separate uranium-235 hexafluoride (which is lighter) from uranium-238 hexafluoride (which is heavier).

grain   A mass unit used in expressions of air concentrations of particulate matter, such as grains per cubic foot. One grain equals 65 milligrams or 0.00014 pound.

grain loading   An expression of the air concentration of pollutants that are flowing into an emission control device or are being emitted by a smokestack; usually expressed in GRAINS per cubic foot.

gram (g)   A unit of mass equal to 10^{-3} kilograms or 1/28th of an ounce. One milliliter (cubic centimeter) of water contains one gram.

gram molecular weight (GMW)   The mass, in grams, of a substance equal to its MOLECULAR WEIGHT. For example, the molecular weight of water (H_2O) is 18 (the sum of the atomic weights of two hydrogen atoms and one oxygen atom), so its gram molecular weight is 18 grams. The amount of a material equal to its gram molecular weight comprises one gram-MOLE of the substance.

Gram negative/Gram positive   The response of bacteria to a procedure called the Gram stain. When treated with crystal violet, Gram's iodine, 95% ethanol, and safranin, bacteria usually retain either a red color (Gram negative) or a purple/blue color (Gram positive). In addition to the differences in color upon staining, these two types of bacteria represent bacteria that are fundamentally different in terms of structure, physiology, ecology, and pathogenicity. Common Gram-negative species are *Escherichia, Salmonella,* and *Pseudomonas;* common Gram-positive species are *Bacillus, Staphylococcus,* and *Streptococcus.*

Gram stain   See GRAM NEGATIVE/GRAM POSITIVE.

grandfather clause   In environmental or occupational safety and health laws or regulations, a statement that new rules or strictures apply only to persons or industries beginning business or building new emission sources after a certain date. Operations in existence before that date are "grandfathered" in and do not have to comply with the new rules.

granitic crust   That part of the earth's crust containing the continents.

granular activated carbon (GAC)   A carbon used in the treatment of drinking water. The extremely large surface area ADSORBS organic compounds as the water flows through GAC filter beds one to several meters deep.

Graphical Exposure Modeling System (GEMS)   A set of interactive computer programs developed by the United States Environmental Protection Agency that are used for statistical analysis and modeling of environmental data. The programs support policy and planning studies of the release of toxic materials into the air, surface water, groundwater, or soil. Census data for population exposure analysis are incorporated into the package.

graphite furnace   An ATOMIC ABSORPTION SPECTROPHOTOMETER that uses an electric current, instead of a flame, to heat and atomize the sample. Also called a flameless furnace.

grassland   A geographical region dominated by shrubs and grasses, receiving 10 to 30 inches of rain annually. Alpine grasslands are in cool, high-elevation areas. Temperate grasslands, called prairie (North America), pampas (South America), steppe (Asia), or veldt (South Africa), are found in regions with moderate temperatures. Tropical grasslands, also called savannas, are found in warmer climates.

grate siftings   Material that falls through the openings in the fuel bed of an incinerator that burns solid waste.

gravimetric   Pertaining to measurements of the weight (mass) of samples or materials.

gravitational acceleration   The change in velocity per unit time of a falling body; for practical purposes, equal to 9.8 meters (32 feet) per second per second.

gravitational constant   The constant of proportionality *(G)* in the equation describing the gravitational attraction between two bodies with masses m_1 and m_2:

$$F_G = G \frac{m_1 m_2}{r^2}$$

where F_G is the gravitational force on either of the two bodies and r is the

distance between them. G is equal to 6.67×10^{-11} newton-meter^2 kilogram^{-2}.

gravitational force   The mutual attraction between two physical bodies, as expressed in NEWTON'S LAW OF UNIVERSAL GRAVITY. The force of gravity is generally considered to be identical to the weight.

gravity flow   The downhill flow of water or other liquid through a system of pipes, generated by the force of gravity.

gravity transport   See GRAVITY FLOW.

gray (Gy)   The basic SI UNIT of radiation dose absorbed per unit mass of tissue. One gray represents an absorbed dose of one joule of energy per kilogram of tissue. The unit can be used to express the absorption of any type of ionizing radiation, and is based on the physical properties of the particular radiation.

grazing   The consumption of live plant biomass by a HERBIVORE.

grazing food chain   The feeding pattern of an animal community that is based on the consumption of live plant biomass by the primary consumers. This is contrasted with a DETRITUS FOOD CHAIN, which is based on the consumption of dead plant biomass by the primary consumers.

Green parties   Political parties whose primary interests revolve around environmental issues. They offer candidates for election to public office. Such political parties are more active in Europe than in the United States; however, they are not major political factors even in Europe.

green revolution   The advances in farming techniques, including increased irrigation and fertilizer use, and crop varieties that have allowed greatly increased crop yields in many areas of the world. See HIGH-YIELDING VARIETIES.

greenhouse effect   The predicted excessive warming of the atmosphere resulting from the accumulation of atmospheric carbon dioxide. The atmosphere is normally warmed when infrared radiation emitted by the earth is absorbed by carbon dioxide gas and water vapor in the air. As the amount of carbon dioxide increases due to the combustion of fossil fuels and deforestation, especially of tropical rain forests, it is proposed that more heat energy will be retained by the earth's atmosphere, resulting in a change in rainfall and wind patterns and melting of polar ice, thus raising the global sea level. The change in weather patterns could have devastating consequences to the world's present prime agricultural areas. A significant rise in sea level could flood many coastal cities and damage ecologically important coastal wetlands. Other heat-absorbing gases that are increasing in the atmosphere as a result of human activities are methane, nitrous oxide, and the chlorofluorocarbons.

greenhouse gases   Atmospheric gases or vapors that absorb outgoing infrared energy emitted from the earth, contributing to the GREENHOUSE EFFECT. The more important ones are carbon dioxide, water vapor, methane, nitrous oxide, and the chlorofluorocarbons.

Greenpeace   An international environmental organization noted for its aggressive and highly physical protest activities, such as members steering small boats in the way of whaling vessels and into waters used for nuclear testing. In 1985, two bombs sunk the Greenpeace ship *Rainbow Warrior* in harbor at Auckland, New Zealand, before it was to lead a group of vessels into the French nuclear test site near Tahiti. The sinking, which killed one Greenpeace member, drew worldwide attention. In 1989, United States membership was about 500,000; worldwide membership was about 3 million. International headquarters in Lewes, England; U.S. headquarters in Washington, D.C.

grit   Sand or fine gravel carried in wastewater.

grit chamber   The initial treatment device at a sewage treatment plant; dense material from the incoming wastewater settles to the bottom of the chamber and is removed.

grit removal   The process of removing sand and fine gravel from a stream of domestic waste in a GRIT CHAMBER.

grit tank   See GRIT CHAMBER.

groundwater plume   A volume of contaminated groundwater that extends downward and outward from a specific source; the shape and movement of the mass of the contaminated water is affected by the local geology, materials present in the plume, and the flow characteristics of the area groundwater.

groundwater velocity   The rate of water movement through openings in rock or sediment. Estimated using DARCY'S LAW.

ground zero   The point on the ground at which a nuclear weapon detonates, or that point on the ground directly under an atmospheric detonation of a nuclear weapon. The center of the area of greatest damage caused by a nuclear weapon.

guideline model   Any of the AIR QUALITY DISPERSION MODELS approved by the United States Environmental Protection Agency for use in the permit review process for new or modified facilities, or in the evaluation of a control strategy to solve air quality problems in an area. Each of the models has an appropriate application(s).

guillotine damper   A flat grate or plate that can be inserted perpendicular to the flow of a gas within a confined passage to regulate the rate of flow.

gully reclamation   Projects designed to prevent erosion in gullies by either filling them in or planting vegetation to stabilize the banks.

gypsum   Calcium sulfate; a by-product of the reaction between limestone (calcium carbonate) and sulfur dioxide in control devices that reduce sulfur dioxide emissions from stacks. See FLUE GAS DE-SULFURIZATION.

H

Haber process   The industrial method used for the fixation of atmospheric nitrogen for use as fertilizer. The process forms ammonia by the direct combination of atmospheric nitrogen and molecular hydrogen from natural gas. The reaction requires temperatures of 500–1000°C, pressures of 100–1000 atmospheres, and the presence of a catalyst. The ammonia can be chemically combined with carbon dioxide to produce urea, or further reacted with oxygen to make nitric acid. Nitric acid added to ammonia will form ammonium nitrate, which, along with urea, is a widely used nitrogen fertilizer.

habitat   The specific surroundings within which an organism, species, or COMMUNITY lives. The surroundings include physical factors such as temperature, moisture, and light together with biological factors such as the presence of food or predator organisms. The term can be employed to define surroundings on almost any scale from marine habitat, which encompasses the oceans, to microhabitat in a hair follicle of the skin.

haematophagous   Describing arthropods such as ticks, fleas, and mosquitoes that feed on blood. Some members of this group have the capacity to transmit infectious diseases, such as, Rocky Mountain spotted fever and yellow fever.

half-life   The time required for one-half of a radioactive substance to degrade to another nuclear form or to lose one-half of its activity, or for one-half of a chemical material to be degraded, transformed, or eliminated in the environment or in the body. Each radioactive substance has a predictable rate of decay, from millionths of a second to millions of years. Half-lives of chemical materials in the environment vary with the type of chemical. (See diagram on page 116.)

half-value layer (HVL)   The shielding material thickness that will reduce the quantity of a beam of radiation to one-half of the strength it possessed before it traversed the material. Two half-value layers reduce the radiation to one-fourth of its value. Half-value layers vary with the type of material (lead half-value layers are thinner than those for concrete), and the HVL for a particular material is specific to the type and source of radiation. For example, the HVL for steel against the gamma radiation emitted by cobalt-60 is 0.82 inch; the HVL for steel against the gamma radiation emitted by cesium-137 is 0.64 inch.

halocarbons   See   CHLOROFLUORO-CARBONS

halocline   The boundary between surface fresh water and underlying saltwater in a stratified coastal environment. A location where there is a marked change in salinity.

halogen   One of the reactive nonmetals: fluorine, chlorine, bromine, iodine, and astatine. These elements are known for their reactivity and are only found combined with other chemicals in nature. At room temperature, fluorine and chlorine are gases, bromine is a liquid, and iodine and astatine are solids. Chlorine is used extensively as an industrial chemical and is a component in many compounds of environmental interest. See CHLORINE and CHLORINATED HYDROCARBONS.

halogenated   Describes a chemical compound containing one or more of

HALF-LIFE

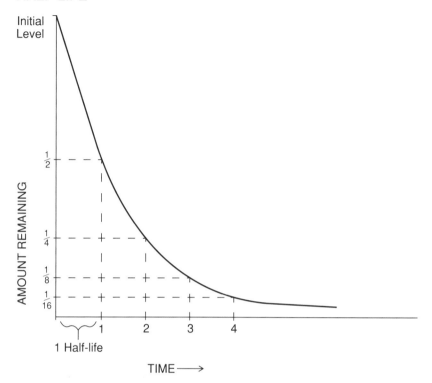

1 Half-life

TIME ⟶

the halogens, usually chlorine, bromine, or fluorine.

halogenated dibenzofuran (HDF) See DIBENZOFURANS.

halogenated dibenzo-*p*-dioxin (HDD) See TETRACHLORO-DIBENZO-PARA-DIOXIN.

halogenated organic compounds (HOCs) Chemical materials containing carbon, hydrogen, and one or more of the halogens, usually chlorine, bromine, or fluorine.

halomethane See TRIHALOMETH-ANES.

halons Compounds composed of carbon together with fluorine, bromine, and/or chlorine used mainly as fire-extinguishing gases, such as Halon 1211

(CF$_2$BrCl) and Halon 1301 (CF$_3$Br). The chemicals are long-lived in the atmosphere and are implicated in OZONE LAYER DEPLETION.

halophytes Plants that require the presence of saltwater or salty soil for growth.

hammer mill A machine in which solid material, such as community trash or other solid waste, is pulverized by hammers into smaller pieces.

hammer provision A provision in an environmental statute that makes effective certain requirements on a specified date if the United States Environmental Protection Agency has not yet issued regulations concerning the environmental issue addressed by the statute. Such provisions are used by Congress in response to common delays by the agency in issuing regulations implementing environmental laws. The HAZARDOUS AND SOLID WASTE AMENDMENTS of 1984 contained several hammer provisions concerning the prohibition of certain wastes from land disposal. See LAND DISPOSAL BAN, SOFT HAMMER.

hard pesticide A chemical agent used to kill insects that does not undergo ready biological decomposition. MINERALIZATION is very slow if it takes place at all. Compare SOFT PESTICIDE.

hard water Drinking water that contains sufficient concentrations of metal ions, primarily those of calcium and magnesium, to result in the formation of precipitates with soaps and detergents. Compare SOFT WATER.

hardness 1. A measure of the amount of calcium and magnesium salts dissolved in water. 2. The relative penetrating ability of x-rays.

hardpan A layer on or beneath the soil surface, usually composed of clay particles, sand, gravel, or calcium carbonate, that is compacted and relatively impermeable.

hardwoods Deciduous trees with broader leaves and, usually, slower growth rates than the conifers, or SOFTWOODS. Common temperate-region hardwoods include oak, maple, cherry, walnut, beech, birch, cypress, elm, and hickory.

hazard A physical or chemical agent capable of causing harm to human health or the environment. See also HAZARDOUS SUBSTANCE.

hazard and operability study (HAZOP) A technical, detailed review of the potential for failure of a system or system components. The review is usually performed for industrial facilities and seeks to identify all credible scenarios that may lead to a fire, explosion, or chemical release. Steps are then taken to reduce the likelihood of those events.

Hazard Assessment Computer System (HACS) See CHEMICAL HAZARD RESPONSE INFORMATION SYSTEM (CHRIS).

hazard identification The determination of the possible adverse health effects produced by a chemical or physical agent. The analysis includes an estimation of the amount, frequency, and duration of exposure that may lead to the adverse effects and the identification of any susceptible population subgroups.

Hazard Information Transmission (HIT) A service of the CHEMICAL MANUFACTURERS ASSOCIATION that provides a computer printout of emergency information to first responders to a hazardous material incident. When given the name or identifying number of the chemical that has been released or is in danger of being released, the system lists hazards of the material, appropriate personal protective equipment, first aid directions, and methods for controlling any release or fire. The service is free of charge to organizations that register with the Chemical Manufacturers Association. See also CHEMICAL TRANSPORTATION EMERGENCY CENTER.

Hazard Ranking System (HRS) A method for ranking hazardous waste disposal sites for possible placement on the NATIONAL PRIORITIES LIST, as provided for by the COMPREHENSIVE ENVIRONMENTAL RESPONSE, COMPENSATION, AND LIABILITY ACT. The ranking uses information gathered by the PRELIMINARY ASSESSMENT AND SITE INSPECTION and the LISTING SITE INSPECTION. The need for remedial action is scored on the basis of potential harm to human health resulting from (1) releases into groundwater, surface water, or the atmosphere; (2) fire and explosion; and/or (3) direct contact with hazardous materials. The HRS evaluation assigns an overall numerical value to each site, which determines its priority for cleanup.

Hazardous and Solid Waste Amendments (HSWA) A 1984 fed-

eral statute amending the RESOURCE CON-
SERVATION AND RECOVERY ACT. The
amendments expanded and strength-
ened the regulation of hazardous and
solid wastes.

**Hazardous Materials Transporta-
tion Act (HMTA)**   The 1975 federal
law governing the transportation of hazardous materials and hazardous wastes.
Regulations implementing the HMTA are
issued by the United States Department
of Transportation (DOT). (The hazardous
waste MANIFEST SYSTEM is regulated by
the United States Environmental Protection Agency under the RESOURCE CON-
SERVATION AND RECOVERY ACT.) The DOT
regulations cover container, labeling, and
marking standards as well as the use of
PLACARDS and routing.

hazardous substance   Any chemical
that poses a threat to human health or
the environment if released in significant
amounts. Federal statutes and regulations include descriptions and lists of hazardous substances. Materials are regulated as hazardous under Section 311 (b)
(2) (A) or Section 307(a) of the CLEAN
WATER ACT, Section 112 of the CLEAN AIR
ACT, Section 3001 of the RESOURCE CON-
SERVATION AND RECOVERY ACT, Section
102 of the COMPREHENSIVE ENVIRONMEN-
TAL RESPONSE, COMPENSATION, AND LIA-
BILITY ACT, or Section 7 of the TOXIC
SUBSTANCES CONTROL ACT. Title 40, Section 302.4 of the CODE OF FEDERAL REG-
ULATIONS provides a list of designated
hazardous substances.

hazardous substance list (HSL)   A
listing of designated HAZARDOUS SUB-
STANCES appearing at Title 40, Section
302.4 of the CODE OF FEDERAL REGULA-
TIONS.

**Hazardous Substances Response
Trust Fund**   The original official name
of the Superfund; changed in 1986 to
the HAZARDOUS SUBSTANCES SUPERFUND.

**Hazardous Substances Super-
fund**   A federal trust fund for use in the
cleanup of spills or sites containing hazardous waste that pose a significant threat
to the public health or the environment,
also known as the Superfund. The fund,
originally called the Hazardous Substances Response Trust Fund, was established by the COMPREHENSIVE ENVIRON-
MENTAL RESPONSE, COMPENSATION, AND
LIABILITY ACT in 1980. Beginning that
year $1.5 billion was to be collected over
five years, mainly from taxes on crude
oil, petroleum products, petrochemicals,
and certain inorganic chemicals. The 1986
reauthorization of the law, which changed
the fund's name to the Hazardous Substances Superfund, increased the fund to
$8.5 billion and broadened the tax base
to include a general corporate Superfund
tax. Another one-half billion dollars was
included to clean up leaks from underground storage tanks.

hazardous waste   Any solid waste
listed as hazardous under the RESOURCE
CONSERVATION AND RECOVERY ACT regulations, Title 40, Part 261 of the CODE OF
FEDERAL REGULATIONS, or that poses a
significant threat to human health or safety
because it is toxic, ignitable, corrosive, or
reactive, as determined by specified tests.
See LISTED HAZARDOUS WASTE, CHARAC-
TERISTIC HAZARDOUS WASTE.

**hazardous waste management fa-
cility (HWM facility)**   All contiguous
land and structures used in treating, storing, or disposing of HAZARDOUS WASTE.

hazardous waste site   A location for
the treatment, storage, or disposal of
HAZARDOUS WASTE.

head   1. The removable top section of
a pressure vessel containing the CORE of
a nuclear reactor. It is removed during
the refueling of the reactor to provide
access to the fuel rods in the reactor core.
2. The pressure exerted by a fluid. See
PRESSURE HEAD; HEAD, TOTAL.

head loss   The loss of energy in a
hydraulic system caused by friction between the moving fluid (water) and the

pipe as well as by other, smaller factors, such as changes in pipe diameter, bends, and valves.

headspace   The zone above the contents in a closed sample container, possibly containing vapors or gases evaporated from the sample material.

head, total   In hydraulics, a term for the energy at any point in a hydraulics system. The total head is the sum of the ELEVATION HEAD, PRESSURE HEAD, and VELOCITY HEAD. Head is expressed in length units, such as feet, and is another way of describing water pressure. Feet of head refers to the height of a column of water; one foot of water head equals 0.433 pound per square inch.

health effects   Adverse effects of substances on the normal functioning of the human body through contact, exposure, and so on.

hearing threshold level   At a specific frequency, the SOUND PRESSURE LEVEL a person can just detect, for each ear.

heat   The kinetic energy of a material arising from the molecular motion of the material. Heat is often expressed in CALORIES or JOULES. The heat contained in a body depends on its chemical makeup, mass, and temperature. Heat is transferred from materials at higher temperatures to those at lower temperatures when they come in contact, by CONVECTION or as RADIANT HEAT.

heat exchangers   Any mechanical device designed to transfer heat energy from one medium to another. The cooling system of an automobile consists of a heat exchange unit consisting of a liquid coolant, pumps, hose, and radiator, all of which are in place to transfer heat energy from the engine block through the radiator to the atmosphere. A nuclear reactor has a heat exchange unit designed to transfer heat from the primary coolant in which the core is immersed to

the water that is boiled to produce the steam.

heat island   See URBAN HEAT ISLAND.

heat of combustion   The heat energy liberated by the chemical reaction between an organic fuel and oxygen to form carbon dioxide and water.

heat of condensation   The heat released when a vapor changes state to a liquid. See HEAT OF VAPORIZATION.

heat of fusion   For a solid at its melting point, the heat required to change it to the liquid state; the value is expressed in calories of heat energy per gram of solid melted. For solid water (ice) at 0°C, the heat of fusion is about 80 calories per gram. This is equal to the heat energy per gram that must be removed to convert water at its freezing point (0°C) to a solid.

heat of vaporization   For a liquid at its boiling point, the amount of heat required to cause it to change to the vapor state; expressed in calories of heat energy per gram of liquid vaporized. For water at 100°C and standard atmospheric pressure, the heat of vaporization is 540 calories per gram. Conversely, when a liquid condenses, it loses the heat absorbed upon vaporization, giving off HEAT OF CONDENSATION.

heat rate   The net efficiency of an electric power generating facility. The measure is calculated by dividing the total heat content of the fuel burned (in BTUs) by the amount of electricity generated (in kilowatt-hours).

heat sink   Any material used to absorb heat. In the environment, this is usually air or water that absorbs waste heat produced in the operation of electric power plants or other industrial facilities.

heat stress index (HSI)   An expression of the evaporative capacity of an individual to maintain a normal body temperature compared to the maximum possible evaporative capacity under particular environmental conditions. The index is calculated by using the DRY-BULB TEMPERATURE, WET-BULB TEMPERATURE, GLOBE TEMPERATURE, air velocity, and the level of work the individual is performing (metabolic rate).

heat transfer agent   A liquid or gas that functions in a HEAT EXCHANGER to facilitate the movement of heat from one location to another. For example, the engine coolant in an automobile serves to transfer heat from the engine block to the atmosphere. Likewise, water facilitates the movement of heat from the reactor core to the outside of a nuclear reactor.

heating degree-day   See DEGREE-DAY.

heating value   The HEAT OF COMBUSTION per unit mass of a material; typically expressed as joules/kilogram, calories/gram, or BTU/pound.

heatstroke   An acute condition resulting from excessive heat exposure, characterized by hot, dry skin and mental confusion. If not treated, the condition can lead to convulsions and death. The disorder stems from the failure of the body's sweating mechanism, which allows a sustained rise in internal body temperature.

heavy atom   An ISOTOPE of an element that contains more neutrons than in the most frequently occurring form of that element. For example, the most common isotope of hydrogen has a single proton in the nucleus (atomic weight of 1), whereas the heavy form of hydrogen contains a proton and a neutron in the nucleus (atomic weight of 2).

heavy hydrogen   The ISOTOPE of hydrogen that contains one proton and one neutron in the nucleus rather than the more common form having one proton and no neutron. Also called deuterium. See HEAVY WATER.

heavy metals   The metallic elements of relatively high molecular weight, such as lead, mercury, arsenic, cadmium, chromium, and zinc. Chronic exposure to excessive concentrations of the ionic forms of these elements is associated with a variety of adverse health effects.

heavy water   Water in which a significant portion of the normal hydrogen atoms, hydrogen with a mass number of 1, have been replaced with the heavy isotope of hydrogen (HEAVY HYDROGEN or deuterium), hydrogen with a mass number of 2. The chemical formula for water is H_2O, and that for heavy water is D_2O. Used in a HEAVY-WATER REACTOR.

heavy-water reactor (HWR)   A nuclear reactor in which HEAVY WATER is utilized as both the coolant and the moderator. Heavy water is a much more efficient moderator than normal water, and therefore allows the use of naturally occurring, unenriched uranium as the nuclear fuel for the fission process. Compare LIGHT-WATER REACTOR.

hectare   A unit of area equal to 10,000 square meters; one hectare equals 2.47106 acres.

heliostat   A device designed to rotate slowly in order to continuously reflect the rays of the sun in a fixed direction. Used in some applications of solar energy.

heme   One member of a class of compounds derived from porphyrins. These compounds complex with metal ions (such as iron) and proteins to form the central structure in molecules such as hemoglobin and chlorophyll. The iron-porphyrin complex is responsible for the oxygen-carrying ability of hemoglobin. The word porphyrin is derived from the Greek word meaning purple *(porphyra)*, because porphyrin-containing molecules are highly colored. All porphyrin derivatives have the same central structure but differ in the metal ion included and the structures along the periphery.

hemoglobin   The protein in red blood cells that carries oxygen absorbed in the lungs to the body tissues. See HEME.

Henry's law   The expression stating that, for a gas in contact with water at equilibrium, the ratio of the atmospheric concentration of the gas and its aqueous concentration is equal to a constant (HENRY'S LAW CONSTANT), which varies only with temperature. One form of Henry's law is

$$K_H = \frac{P}{C_w}$$

where K_H is Henry's law constant, P is the equilibrium partial pressure of a gas in the air above the water, and C_w is the aqueous concentration of the gas dissolved in the water.

Henry's law constant   The ratio, at equilibrium and for a given temperature, of the atmospheric concentration of a gas to its aqueous concentration; typical units are ATMOSPHERES per MOLE FRACTION. Henry's law constant is sometimes given as the equilibrium ratio of the aqueous concentration of a gas to its atmospheric concentration; in this case the units will be inverted, in mole fraction/atmosphere.

hepatitis   Inflammation of the liver. A virus-caused disorder transmitted to humans by the consumption of raw oysters taken from water contaminated with sewage.

hepatotoxicity   The relative ability of a chemical agent to cause selective damage to the liver.

heptachlor   An organochlorine insecticide belonging to the family of chemicals known as the cyclodienes (a family that also includes ALDRIN and DIELDRIN). Heptachlor is persistent in the environment; exposure to the agent has been shown to cause cancer in at least one species of mammals and to result in a reduced immune response in mammals.

herbaceous   Describing a nonwoody vascular plant. No parts persist above ground during the winter. Compare EVERGREEN.

herbicide   A chemical agent (often synthetic) capable of killing or causing damage to certain plants (usually weeds) without significant disruption of other plant or animal communities.

herbivore   An animal that eats only plants, such as a grasshopper, other insects, and cattle. Compare CARNIVORE, OMNIVORE.

herding agent   A chemical applied to the surface of water to control the spread of a floating oil spill. Piston film (federal supply system chemical FSN 9G 6810-172-9110) is commonly used.

heritable   A property of an organism that is passed on to offspring through GAMETES.

herpetofauna   Animals that belong to the amphibian or reptile groups, such as snakes, frogs, and lizards.

hertz (Hz)   A unit describing the frequency of electromagnetic radiation, wave motion, or sound. One hertz is equal to one cycle per second.

heterotroph   An organism that cannot satisfy its carbon nutrition requirements by converting inorganic materials to organic materials, but must derive its carbon nutrition from other organic forms of carbon, such as sugars and amino acids. Compare AUTOTROPH.

heuristic routing   An approach to the design of routes for solid waste collection vehicles, which uses such basic rules as the minimization of left turns and dead ends and the use of clockwise loops.

hexachlorobenzene (HCB)   A fungicide that has been shown to cause severe skin reactions in humans. Chemically, the agent is an AROMATIC benzene ring with a chlorine atom substituted for each of the six hydrogens normally present on benzene. The agent is persistent in the environment.

high-density polyethylene (HDPE) A low-permeability synthetic organic material sometimes used as a landfill LINER.

high-efficiency particulate air filter (HEPA filter) A filter having at least a 99.97% removal efficiency for particles with a diameter of 0.3 micron. Required for the control of certain high-hazard dusts, such as asbestos and radioactive materials.

highwall A cliff cut into a mountain or hillside during the process of removing coal deposits by surface or strip mining.

high-level liquid waste An aqueous solution or suspension containing mixed fission products derived from the recycling of nuclear fuel rods that have been removed from reactors. The waste normally contains more than 100 microcuries per milliliter of fission products. See HIGH-LEVEL WASTE.

high-level radioactive waste See HIGH-LEVEL WASTE.

high-level waste (HLW) Highly radioactive material requiring perpetual isolation; waste that includes untreated SPENT FUEL from a nuclear power plant, the residue from the chemical processing of spent fuel, and much of the waste from nuclear weapons production.

high-temperature gas reactor (HTGR) A nuclear reactor that uses enriched uranium as the fission fuel and graphite as the moderator. Helium gas is employed as the coolant. The reactor operates at a CORE temperature in excess of 1000°C, which is about three times the core temperature of water-cooled reactors.

high-volume air sampler (hi-vol) A device for sampling airborne particu-late matter. A motor pulls air through a preweighed filter for 24 hours, capturing any airborne particles. The filter is reweighed to determine the mass of particulate in the sampled air. The mass is divided by the volume of air sampled to obtain a mass/volume concentration.

high-yielding varieties (HYVs) Strains of food-crop plants, such as wheat and rice, that have been developed to allow a greater production of grain per acre of land cultivated than traditional varieties. The new varieties often require a greater input of fertilizer and water to reach full maturity. These strains have contributed to the so-called GREEN REVOLUTION.

holding medium A special fluid employed for maintaining fecal bacteria in a viable state between the time that water samples are processed by filtration and the time that the filters used to remove the bacteria from water can be INCUBATED properly. The medium protects viability between sampling and analysis.

holding time The time allowed between removal of samples from water sources for bacteriological analysis and the processing of those samples.

holistic Of or related to a view of the natural environment that encompasses an understanding of the functioning of the complete array of organisms and chemical-physical factors acting in concert rather than the properties of the individual parts. Compare REDUCTIONISTIC.

holothurian A group of marine, bottom-dwelling animals related to the sea stars and sand dollars (echinoderms). Unlike these relatives, the holothurians have soft bodies and are long and slender in shape, such as the sea cucumber.

homeostasis A condition in which the systems of the body act together to maintain a relatively constant internal environment even when external conditions may vary.

homeotherm   See ENDOTHERM.

hood   A device that captures or encloses air contaminants. It is usually ventilated to the outside with the aid of a strong exhaust fan. Volatile or irritating chemicals are handled within an enclosure hood to prevent exposure of personnel to the agent.

hopper   A container for dusts collected by ELECTROSTATIC PRECIPITATORS or CYCLONES.

horizon   A specific layer of soil that is different from adjacent layers in texture, color, mineral content, and other qualities. Also called soil horizon. See also SOIL PROFILE.

horizontal   dispersion   coefficient   See DISPERSION COEFFICIENTS.

horsepower   An engineering unit of power describing the rate at which WORK is performed. The unit is based on the English system of measurements and is equal to 550 foot-pounds per second or 33,000 foot-pounds per minute.

host   The larger organism within which a parasite lives. A tick that carries the bacterium that causes Rocky Mountain spotted fever, for example, is said to be the natural host of that pathogen.

hot   An informal or colloquial expression meaning highly radioactive.

hot soak   The gasoline that evaporates from an automobile carburetor after the engine is shut off. Emission control devices are now installed to collect this evaporating gasoline in a canister filled with ACTIVATED CHARCOAL. When the engine is restarted, the gasoline in the canister returns to the engine to be combusted.

hot spot   An informal expression designating a specific area as being contaminated with radioactive substances, having a relatively high concentration of air pollutant(s), or experiencing an abnormal disease or death rate.

hot-side ESP   An ELECTROSTATIC PRECIPATOR located on the upstream side of the air preheater, where heat from the exhaust gases has not been partially transferred to incoming boiler air. Air temperatures in the hot-side ESP range from 600 to 800°F (320–420°C).

human equivalent dose   The dose to humans of a chemical or physical agent that is expected to exert the same effect that a certain dose has produced in animals.

humic   Of or containing humus.

humus   Organic matter in soil derived from the partial decomposition of plant and animal remains. Generally, the decomposition has proceeded sufficiently to make it impossible to recognize the original material.

hydraulic conductivity   An empirically derived expression of permeability, used as a coefficient in DARCY'S LAW to calculate the velocity of groundwater flow; the units are in length per time. Also called coefficient of permeability.

hydraulic fracturing   Any technique involving the pumping of fluid under high pressure into an oil or gas formation to create fissures and openings in the reservoir rock and increase the flow of oil or gas.

hydraulic gradient   The change in the elevation of the WATER TABLE per unit horizontal distance, often expressed in feet per mile; an important variable influencing the velocity of groundwater flow. See DARCY'S LAW.

hydraulic head   An expression of water pressure in length units; the height to which water will rise in a pipe if one end is inserted into the hydraulic system; Hydraulic head *(H)* is given by $H = Z + p/W$, where Z is the ELEVATION HEAD, p is pressure, and W is the SPECIFIC WEIGHT of water. p/W is the PRESSURE HEAD. Also called piezometric height.

hydraulic loading For a SAND FILTER wastewater treatment unit, the volume of wastewater applied to the surface of the filtering medium per time period. The loading is often expressed in gallons per day per square foot (gpd/ft^2), or cubic meters per square meter per day (m^3/m^2/d).

hydraulic radius For a flowing fluid (water), a measure of resistance to flow in pipes or open channels. The radius is expressed in length units, and calculated as the flow area divided by the wetted perimeter. For circular pipes with full flow, flow area equals $\pi D^2/4$, where D is the pipe diameter, and the wetted perimeter is the inside circumference πD; therefore the hydraulic radius is $D/4$.

hydraulics The science concerned with water and other fluids at rest or in motion.

hydric Containing an abundance of water.

hydrocarbons Chemical compounds containing carbon and hydrogen as the principal elements.

hydrogen bonding A type of weak attraction between molecules. The interaction between individual water molecules serves as an example. The water molecule is composed of one oxygen atom and two hydrogen atoms. These atoms are held together by strong covalent bonds which are formed by the sharing of electrons by the two atoms. The covalent bonds are very difficult to break. In addition, there is a weak attraction between the separate water molecules by way of the oxygen atom of one molecule and the hydrogen atom of an adjacent

molecule. This attraction, termed a hydrogen bond, constantly forms and breaks among the multitude of water molecules in liquid water. Evaporation of water occurs when individual water molecules at the air-water interface break free of all hydrogen bonds and enter the gas phase. Hydrogen bonds are also found between many organic molecules, for example, between the two strands of deoxyribonucleic acid that constitute individual genes.

hydrogen ions Hydrogen atoms without an electron; chemical symbol: H^+.

hydrogen sulfide A foul-smelling gas (H_2S) produced during the ANAEROBIC decomposition of organic material by bacteria and by the metabolism of sulfate-reducing bacteria in the anaerobic sediments of fresh water and marine systems. Also called sewer gas because of the distinctive odor. The gas is responsible for the "rotten egg" odor of some well water. Hydrogen sulfide is very reactive chemically, forming sulfide salts with a host of metals; these metal sulfides are dark in color and are responsible for the black color of many muds. Significant concentrations of the gas are toxic to humans; consequently, the gas represents an occupational hazard to individuals working in industries involved in drilling for, and refining, petroleum.

hydrograph A graph of a stream or river DISCHARGE at a certain point over a period of time.

hydrologic cycle The stocks and flows of water in the ecosphere including the processes of evaporation, precipitation, CONDENSATION, TRANSPIRATION, and surface and subsurface runoff.

hydrology The study of the distribution, movement, and chemical makeup of surface and underground waters.

hydrolysis The chemical reaction of a substance with water, resulting in the splitting of the larger molecule into smaller parts; an important degradation mechanism for pollutants in water or on land.

hydrophyte   A plant that lives in water. Compare MESOPHYTE, XEROPHYTE.

hydropower   The utilization of the energy available in falling water for the generation of electricity.

hydrosphere   That portion of the earth composed of liquid water, such as lakes, rivers, ponds, and oceans. Together with the atmosphere, lithosphere, and biosphere, constitutes the earth's ecosphere.

hydrostatic equation   See PRESSURE.

hydrostatic pressure   Pressure exerted by nonmoving water due to depth alone. Expressed as $P = WH$, where P is the GAUGE PRESSURE at a depth H and W is the SPECIFIC WEIGHT of water.

hygrometer   An instrument used to measure moisture in the air.

hygroscopic   Describing a chemical substance with an affinity for water, one that will absorb moisture, usually from the air. Silica gel and zinc chloride are hygroscopic materials that are used as drying agents.

hypolimnion   The lower, cooler water layer found in stratified lakes. See EPILIMNION.

hypothesis   An informed theory that best describes a set of available data. The assumption is stated is such a way that subsequent experimentation or observation can test the validity of the theory.

hypoxia   A condition in which natural waters have a low concentration of DISSOLVED OXYGEN (about 2 milligrams per liter compared with a normal level of 8 to 10 milligrams per liter). Most game and commercial species of fish avoid waters that are hypoxic.

I

ideal gas   A gas that would perfectly obey the IDEAL GAS LAW. At normal temperatures and pressures, most gases behave similarly to an ideal gas, and the ideal gas law is applied routinely in air pollution calculations.

ideal gas law   The equation that describes the relationships among volume, pressure, and temperature for an IDEAL GAS. For n moles of gas, $PV = nRT$, where P = ABSOLUTE PRESSURE, V = total volume, R = UNIVERSAL GAS CONSTANT, and T = ABSOLUTE TEMPERATURE.

igneous rock   Rock formed directly from cooled magma, such as granite, that has erupted from deeper in the earth's crust.

ignitability   A characteristic used to define a HAZARDOUS WASTE. An IGNITABLE WASTE meets the hazardous waste characteristic of ignitability.

ignitable waste   A substance or mixture that meets any of several definitions: (1) a liquid with a FLASH POINT of 60° Celsius or less; (2) a nonliquid that can catch fire through friction, absorption of moisture, or by spontaneous chemical change; or (3) ignitable compressed gases or oxidizers, as defined by United States Department of Transportation regulations.

illuvial   Describing soil material, usually minerals and colloidal particles, that is removed from the upper soil horizon to a lower soil horizon. Illuvial deposits can form a HARDPAN.

Imhoff tank   A two-chamber sewage treatment device in which SEDIMENTATION takes place in an upper chamber, after which the solids fall into a lower chamber where they are digested by microorganisms. Named for its developer, Karl Imhoff.

immediately dangerous to life and health (IDLH) The maximum air concentration of a chemical substance from which a healthy worker could escape within a 30-minute exposure without irreversible adverse health effects or experiencing escape-impairing health conditions (dizziness, unconsciousness, etc.). The level is set by the National Institute for Occupational Safety and Health. This acute air exposure standard is often used to estimate adverse effects on nearby human populations that may be caused by spills or other short-term accidental releases of toxic chemicals.

immigration The movement of individuals into an area or country to assume permanent residence. Compare EMIGRATION.

immiscible Describing two liquids, neither of which will act as a solvent for the other to any appreciable extent. As a result, the liquids will form two layers when mixed and allowed to stand. For example, a gasoline-water mixture will separate into two layers with gasoline on the top and water on the bottom (because it is denser than gasoline).

immobilization Synonym for STABILIZATION/SOLIDIFICATION.

impact noise A brief, punctuated sound lasting generally less than one-half second and not repeated more than once per second, for example the sound from a hammering device. Impact noise is difficult to measure accurately.

impaction The collision of particles with the lung surfaces as contaminated air is taken into the lungs. The velocity of the air and the aerodynamic diameter of the particles are the primary variables governing impaction.

impedance An expression of the resistance to an energy flow calculated as the magnitude of the cause of the flow (force, pressure, voltage) divided by the energy flow. For example, Ohm's law

states $Z = E/I$, where Z = impedance, E = voltage, and I = electric current. Energy transfer is most efficient if the impedance of the source is matched to that of the receiver.

impermeability A characteristic of a material that prevents the passage of a fluid through that material, as in the relative impermeability of a LINER at a solid waste landfill. Liners are rated in terms of how slowly a liquid can move through them, or how close they are to being truly impermeable.

impingement separator An air pollution control device that removes particulate matter from a gas stream by causing the particles to strike and adhere to plates as the gas direction is changed.

impinger An air-sampling device that collects gases or particulate matter by directing sample air down a glass tube running inside a collection bottle that is filled with an absorbent liquid. The sampled air is mixed with the liquid by the high-velocity air "impingement" near the bottom of the bottle, which increases the collection efficiency.

impoundment Any land area or formation that can hold liquid. Impoundments, which are open to the atmosphere, are used to treat, store, or dispose of waste, and include aeration tanks, holding ponds, and lagoons. Also called surface impoundment.

in situ In place. An in situ environmental measurement is one that is taken in the field, without removal of a sample to the laboratory.

in situ gasification A method of energy extraction that involves igniting and aerating an underground coal seam, then capturing a low-energy gas from another well sunk into the coal seam. A portion of the energy of the coal is thereby recovered without removing the coal from the ground.

in vitro In glass, outside an intact living organism. Refers to experiments that are conducted in Petri dishes, test tubes, and like apparatus. Chemical toxicity tests performed using cell or tissue cultures are in vitro tests. Compare IN VIVO.

in vivo In a living organism. Experiments performed inside living organisms, such as chemical toxicity tests that involve the introduction of the tested substance into or on the body of an animal. Compare IN VITRO.

inbreeding The mating of animals that are closely related; the process can result in an increase in the expression of recessive genes, which are often maladaptive. The adverse aspects of inbreeding can further harm endangered species, as the small numbers of remaining individuals mate, by necessity, with related individuals.

inbreeding depression The lowering of the quality and vigor of a population of organisms (either plant or animal) due to breeding between individuals that are closely related in a genetic sense. Inbreeding increases the chance that recessive genes, which result in the expression of traits that diminish the health, development, or reproduction of the offspring, will become apparent. The phenomenon is the rationale behind the use of hybrid seed in agriculture and the prohibition against marriage to close relatives among humans.

incidence The rate of occurrence of a specific disease or event within a given number of individuals over a standard time period. For example, the number of lung cancer cases per 100,000 people per year.

incineration Burning organic waste materials. This disposal technique is used to destroy dangerous organic compounds with the reduction of the material to its mineral constituents. For example, chlorinated hydrocarbons are converted to carbon dioxide (CO_2), water (H_2O),

and hydrochloric acid (HCl). In addition, bulk yard trash and similar solid waste can be reduced to noncombustible ash.

incipient LC$_{50}$ The computed concentration of some toxic substance in water that would be lethal to 50 percent of a test population of aquatic organisms. The value is extrapolated from experimentally derived data when laboratory exposure to the chemical does not result in a mortality exceeding 10 percent of the test organisms upon exposure to the test pollutant over a 24-hour test period. See also LETHAL CONCENTRATION–50 PERCENT, LETHAL DOSE–50 PERCENT.

incompatible waste Hazardous wastes that, if mixed, will chemically react to form hazardous products or excessive thermal discharge. Also, wastes that may damage or corrode a container or containment structure, leading to a release.

increment The allowable increase in ambient air concentrations of certain pollutants over the BASELINE level established under the PREVENTION OF SIGNIFICANT DETERIORATION program. Increments are set for sulfur dioxide, total suspended particulates, and nitrogen dioxide.

increment consumption Under the PREVENTION OF SIGNIFICANT DETERIORATION air pollution management program, the modeled increase in certain air pollutant concentrations over the BASELINE ambient air quality levels. The allowable degradation in air quality is called the INCREMENT. Any air emissions considered to be additional to an area after the baseline date are counted against (consume) the allowable increment.

incubate To maintain environmental conditions that are optimum for the growth of bacteria. For example, coliforms grow best when held at 37°C.

incubation period The average time between exposure to an infectious, disease-causing agent and the manifestation of the signs and/or symptoms of the disease.

independent association In the study of risk factors for a disease, a relationship between a suspected factor and disease risk that remains after adjusting or controlling for the influence of other variables.

independent variable A measurable quantity that, as it takes different values, can be used to predict the value of a DEPENDENT VARIABLE. In EPIDEMIOLOGY, an independent variable is an exposure or characteristic that may influence a particular health condition or effect.

indicator organisms The microorganisms that, if present above certain levels in water, indicate contamination by human sewage. The COLIFORM BACTERIA are used commonly as indicators because the test for them is reliable, relatively inexpensive, and produces timely results.

indicator parameters/constituents Groundwater quality measurements specified in a hazardous waste landfill or surface impoundment TREATMENT, STORAGE, OR DISPOSAL facility permit that must be made on samples taken at the POINT OF COMPLIANCE. The parameter may be an analysis of general water quality, such as SPECIFIC CONDUCTANCE, or measurements of one or more specific chemical constituents, such as arsenic, benzene, or chloroform. The measurements of indicator parameters/constituents are compared statistically to their background levels measured at an UPGRADIENT WELL to determine the possibility of adverse influence by the facility on the groundwater.

indigenous Native; naturally present in an area. Autochthonous. Compare ALLOCHTHONOUS.

indirect association See SECONDARY (INDIRECT) ASSOCIATION.

indirect hot air A system designed to reheat an exhaust gas that has been cooled by passage through a pollution-control device. Ambient air is heated with steam and then added to the flue gas to provide a buoyant density to the exhaust.

indirect source A business, shopping center, or highway, among others, that attracts mobile sources of air pollution.

indirect source review (ISR) Environmental agency review and possible controls on the siting of shopping centers, office complexes, highways, and airports, among other indirect sources of air pollution, to improve or maintain air quality. The United States Environmental Protection Agency is not permitted to require an ISR program as part of a STATE IMPLEMENTATION PLAN under current law.

individual lifetime risk The estimated increase in the lifetime risk of an adverse health effect in a person exposed to a specific amount of a physical or chemical agent for a given period of time. See UNIT RISK ESTIMATE and POPULATION RISK.

indoor air pollution The presence of excessive levels of air contaminants inside a home or building from sources such as cigarette smoking, fuel combustion for heating or cooking, certain wallboards, carpets, insulation and the geology of the area (RADON in soil or rocks beneath the structure). Emissions are more likely to accumulate in structures having limited air exchange with the outside. Many air pollutants typically have higher concentrations indoors than outdoors. See INDOOR/OUTDOOR CONCENTRATION RATIO.

indoor/outdoor concentration ratio ([I]/[O]) The measured or typical concentration of an air pollutant indoors divided by the concentration of the same substance outdoors. If the ratio for a certain pollutant is greater than 1, the indoor air concentration is greater than the outdoor concentration, and vice versa for a ratio less than 1. See INDOOR AIR POLLUTION.

Indoor Radon Abatement Act A 1980 amendment to the TOXIC SUB-

STANCES CONTROL ACT that authorizes the United States Environmental Protection Agency to issue regulations necessary to meet the stated goal of reducing RADON levels inside buildings to the levels present outside the same buildings.

induced draft (ID)   The type of fan that pulls air through a furnace, boiler, or cooling tower, aiding the flow of air out of the smokestack or other exit ducts. See also FORCED DRAFT, NATURAL DRAFT.

induced radioactivity   The creation of an unstable or radioactive atom by the bombardment of a stable isotope of an element with neutrons. For example, the stable isotope of nickel has an atomic mass of approximately 58. Bombardment of that form of nickel with neutrons produces nickel-59, which is a radioactive isotope.

industrial melanism   An increase in the relative abundance of darker strains of insects and spiders in regions around industrial facilities. There is a natural variation in the pigmentation of the various species, and the survival of the darker pigmented varieties is favored in the darkened surroundings of industrial environments. The darker pigmentation provides those strains with camouflage and, therefore, protection from predators. Conversely, the lighter pigmented varieties are easier to see and therefore more susceptible to capture by predators. The result is a gradual enrichment of the entire population for the dark variety and a diminution of the light variety.

Industrial Source Complex model (ISC model)   An air quality dispersion model, approved by the EPA, for estimating short- and long-term ambient air concentrations in an area using emission rates from smokestacks and/or area sources and a compilation of the meterological data from the area. Annual averages are estimated by the ISCLT (Long-Term) model and the ISCST (Short-Term) model concentrations for averaging times of 24 hours or less.

industry category   Under the Clean Water Act, the United States Environmental Protection Agency has divided facilities into industry categories and subcategories, for which water EFFLUENT LIMITATIONS are written, based on the type of process and the technology appropriate to the industry. For example, the battery manufacturing category has seven subcategories, each of which has specific effluent limitations.

inert   Describing a substance that does not react chemically with other materials under ordinary conditions. For example, helium does not undergo chemical reactions readily. Compare ACTIVE.

inertial separator   An air pollution control device for particulate matter removal that operates by forcing a dust-containing airstream to abruptly change direction, causing the particulates to collide with a wall, baffle, or louver by inertia. The solids fall into a collection hopper.

infectious   Describing a virus, bacterium, fungus, or protozoan that can invade a host to produce disease. The term is also applied to a disease caused by some pathogenic microbe (like typhoid fever) as opposed to a disease that is not (like coronary artery disease).

infectious waste   Discarded items from medical or related facilities containing viable organisms capable of causing INFECTIOUS diseases.

infiltration   1. The wind-induced movement of air into a building through openings in walls, windows, or doors. An important consideration in energy conservation and indoor air quality analysis. 2. The entrance of groundwater into sewer pipes.

infiltration and inflow (I & I)   The entrance of groundwater (infiltration) or of surface water (inflow) into sewer pipes. Groundwater can seep through defective pipe joints or cracked pipe sections; roof or basement drains are sources of surface water inflow. Excessive infiltration and inflow can cause sewers to back up or can overload sewage treatment plants, causing a reduction in treatment time or a complete bypass of the treatment process during periods of significant rainfall.

information standard   A type of environmental, health, or safety standard that relies on warnings on the container label or on other guidance to prevent or minimize adverse effects from the use or disposal of a material or product; for example, the dosage information on over-the-counter medicines, application and disposal information on pesticide labels.

infrared radiation   Electromagnetic radiation with wavelengths longer than visible light (750 nanometers) and shorter than radio waves (3×10^5 nanometers). Outgoing infrared radiation from the earth warms the atmosphere. See GREENHOUSE EFFECT.

ingestion   The taking in of substances by mouth (eating or drinking); one method of chemical exposure.

inhalable diameter   The diameter of particles (about 15 micrometers or less) that can be deposited in the human respiratory tract.

initiation/promotion   The two-stage theory of carcinogenesis. The first causes the development of premalignant cells, and the second completes the conversion of the cell to the malignant state. See CARCINOGENESIS.

initiator   In the two-stage model of carcinogenesis, an initiator produces a change in a cell, commonly called a MUTATION, which leaves the cell in a premalignant condition. Without exposure to a PROMOTER, the changed cell does not become a tumor cell. See CARCINOGENESIS.

injection well   A well used to inject or reinject fluids underground. See CLASS I,

II, III, IV, AND V INJECTION WELLS, UNDERGROUND INJECTION CONTROL.

inoculate   1. To add a viable culture of a virus, bacterium, or fungus to a growth medium or animal host for the purpose of cultivating the microbe. 2. To add specific bacteria to waste disposal ponds for the purpose of stimulating the decomposition of some special waste.

inoculum   The organisms with which one INOCULATES a culture medium.

inorganic   Of or related to chemicals that do not contain carbon atoms. Compare ORGANIC.

insolation   The interception rate of solar energy on a surface; insolation intensity is greatest when the surface is perpendicular to the solar radiation. Typically expressed in WATTS per square meter.

inspection and maintenance (I & M)   An air quality program requiring automobiles after a certain model year to be inspected for the presence and proper operation of emission control devices. The owner may be required to repair or replace the control equipment. Typically implemented locally, in areas with air quality problems caused or related to automobile emissions.

in-stream aeration   The addition of air to a flowing stream to maintain the DISSOLVED OXYGEN content of the water at an acceptable level.

instrument detection limit (IDL)   The minimum concentration of a substance that is detectable by an instrument. Some compounds of interest to environmental concerns occur in such small concentrations that an insufficient amount is present to allow detection with available instrumentation. Compare METHOD DETECTION LIMIT, QUANTITATION LIMIT.

integrated pest management (IPM)   The use of a combination of the following to limit pest damage to agricultural crops: (1) agricultural practices, such as field tilling to disrupt insect egg development; (2) biological control agents, such as viruses, fungi, or bacteria; (3) introduction of large numbers of sterile male insects; (4) timed application of synthetic chemical pesticides; (5) application of PHEROMONES and JUVENILE HORMONES. IPM stresses the control of pests at manageable levels, rather than their eradication.

Integrated Risk Information System (IRIS)   An electronic on-line database provided by the United States Environmental Protection Agency containing chemical toxicity data and regulatory information for use in health risk assessment and management.

Interagency Regulatory Liaison Group (IRLG)   A group formed in 1977 to coordinate federal regulation of toxic substances by sharing information, planning agency studies to avoid duplication, and developing consistent federal policy approaches. The group is composed of representatives from the Environmental Protection Agency, the Occupational Safety and Health Administration, the Food and Drug Administration, the Consumer Product Safety Commission, and the Department of Agriculture Food Safety and Quality Service.

interceptor   A relatively large drain that conveys municipal wastewater to a treatment plant.

interim status   Under the RESOURCE CONSERVATION AND RECOVERY ACT, a regulatory designation allowing TREATMENT, STORAGE, OR DISPOSAL facilities in existence on November 19, 1980, to continue to operate under a specific set of regulations (PART A permit) until a final (PART B) permit for the facility is applied for and approved.

intermediate   A compound produced during the metabolic conversion of organic compounds by biological organisms. For example, when an organism takes in glucose for use as a nutrient, there is a sequence of biochemical steps between the glucose molecule and the ultimate release of carbon dioxide by the organism. These biochemical steps produce compounds such as fructose-6-phosphate, glyceraldehyde phosphate, pyruvic acid, citric acid, and oxaloacetic acid, among many others. Each of these would be considered an intermediate in the metabolic degradation of glucose. Such a series of intermediates could be identified in the degradation or synthesis of almost any organic compound. See DICHLORODIPHENYLDICHLOROETHENE.

intermittent control system (ICS)   The deliberate variation in an air emission rate of a pollutant to maintain a ground-level concentration of the pollutant below a certain value. The emission rate may be changed to match current weather (dispersion) conditions or to respond to changes in measured ground-level concentrations. New applications of this control technique after December 31, 1970 are not allowed under the CLEAN AIR ACT.

intermittent noise   A noise that lasts for at least one second and then ceases for at least one second.

internal combustion engine   An engine in which the fuel is burnt inside a closed cylinder and converted to mechanical energy by a piston, such as in gasoline or diesel engines.

internal conversion   The process whereby an atomic nucleus releases energy to an orbiting electron, which can then be ejected from the atom.

internal energy   The sum of the KINETIC ENERGY and POTENTIAL ENERGY in the atoms and/or molecules that make up a body or system. Internal energy is affected by changes in temperature and by energy added to or removed from the system as WORK is done on or by the system.

internal radiation Radiation received from radioactive substances located within the body of an organism. For example, strontium-90, absorbed from the atmosphere or from food by an animal or human, can deposit in the bones and become a source of damaging radiation for that organism over a long period of time.

International Agency for Research on Cancer (IARC) A division of the WORLD HEALTH ORGANIZATION that assigns chemicals to specific categories based on their ability to cause cancer in humans or other animals. The categories are Group 1: The agent is carcinogenic to humans; Group 2A: The agent is probably carcinogenic to humans; Group 2B: The agent is possibly carcinogenic to humans; Group 3: The agent is not classifiable as to its carcinogenicity to humans; Group 4: The agent is probably not carcinogenic to humans. These categories are then used to formulate regulations concerning industrial and public exposures. Publications include *IARC Monographs on the Evaluation of Carcinogenic Risk of Chemicals to Humans.* The agency is headquartered in Lyon, France.

International Atomic Energy Agency (IAEA) A multinational organization that regulates the safety of nuclear power stations, the management of nuclear wastes, and nuclear fuel cycle services; issues radiation protection guidelines; and maintains safeguards agreements with nations that are parties to the Treaty on the Non-Proliferation of Nuclear Weapons. It is headquartered in Vienna, Austria.

International Commission on Radiological Protection (ICRP) A group based in Surrey, England, that recommends maximum permissible exposures to IONIZING RADIATION; the commission is financially supported by the World Health Organization, the International Atomic Energy Agency, and the United Nations Environment Program, among others.

International Environmental Information System (INFOTERRA) A part of the United Nations Environment Program's EARTHWATCH. The system is based in Nairobi, Kenya, and provides a global network of information sources on environmental subjects.

International Register of Potentially Toxic Chemicals (IRPTC) A service that provides technical and regulatory information on chemical hazards. Based in Nairobi, Kenya. Part of the United Nations Environment Program's EARTHWATCH.

international system of units See SI UNITS.

International Union for Conservation of Nature and Natural Resources (IUCN) A nongovernmental agency of international scope that promotes measures to conserve wildlife and natural resources. Based in Gland, Switzerland.

International Whaling Commission A multinational group organized in 1946 under the International Convention for the Regulation of Whaling. Purposes of the organization include conservation and development of whale populations, the support of research relating to whales, and the collection of information on current whale population levels. Quotas and restrictions adopted by the Commission are not enforceable, but rely on self-regulation by member countries. In 1981 the Commission established a quota of zero for northern Pacific sperm whales and set a five-year moratorium on commercial whaling beginning in 1986. A number of nations have not observed the moratorium, notably Japan. Based in Cambridge, England.

interspecies extrapolation The application of toxicological data obtained from one species to another species, most often from laboratory animals to humans. See SCALING FACTORS.

interstices   Pore spaces in soil or rock.

interstitial water   Water in the pore spaces in soil or rock.

intertidal zone   That area of coastal land that is covered by water at high tide and uncovered at low tide.

intoxication   A disorder caused by the presence of some material (toxin) that causes damage to the body. For example, botulism is caused by the presence of a toxin produced by certain bacteria growing in food.

inverse square law   The relationship describing the reduction in a physical quantity (such as radiation or noise) with increasing distance from a source. The emitted quantity decreases by a factor equal to one divided by the square of the distance from the source ($1/d^2$).

INVERSE SQUARE LAW

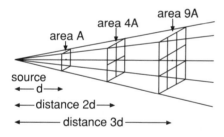

inversion   See TEMPERATURE INVERSION.

invert   The bottom of a pipe or duct.

inverted siphon   A section of a sewer line that is placed deeper in the ground than normal in order to pass under utility piping, waterways, rail lines, or other obstacles. The sewer line is raised again after passing under the obstacle. Also called a sag pipe.

ion   An atom or molecule that carries a net charge (either positive or negative) because of an imbalance between the number of protons and the number of electrons present. If the ion has more electrons than protons, it has a negative charge. If the converse is true, the ion has a positive charge. Ions with a positive charge are called cations; ions with a negative charge are called anions.

ion exchange   The substitution of one ion for another in certain substances. Either ANION exchange or CATION exchange is possible. The most common cation exchange involves the conversion of hard water to soft water. Hard water contains the divalent ions of calcium (Ca^{+2}) and magnesium (Mg^{+2}), which cause soap and detergents to form PRECIPITATES in water. A water softener consists of a resin that is saturated with sodium ions (Na^+). As hard water percolates through the resin, the ions of calcium or magnesium are removed as they attach to the resin, thus releasing (being exchanged for) sodium ions.

ionization   The process by which an atom or molecule acquires an electric charge (positive or negative) through the loss or gain of electrons in a chemical reaction, in solution, or by ionizing radiation.

ionization meter   A device used to measure dose rates for BETA RADIATION, GAMMA RAYS, and X-RAYS. Ions are formed as the radiation passes through a gas-filled chamber and are detected either by the electric current flow, if a constant voltage is applied in the chamber, or by the amount of discharge, if the chamber is charged as a condenser and the ionization drains the charge away.

ionizing radiation   Electromagnetic radiation or atomic particles capable of displacing electrons from around atoms or molecules, thereby producing charged atoms or molecules or ions. The more common types of ionizing radiation are X-RAYS, GAMMA RAYS, ALPHA PARTICLES, and BETA PARTICLES.

ionosphere The upper layer of the atmosphere above the stratosphere, from a distance of about 80 kilometers from the Earth's surface. Incoming solar radiation is sufficiently intense to cause the ionization of the sparse gas molecules present.

irrigation return flows Field runoff of agricultural irrigation water; this water is excluded by statute from the definition of solid or hazardous wastes and from the permit requirements of the CLEAN WATER ACT.

irritant Any material that causes an inflammatory reaction in skin or mucus membranes, resulting in a reddening, swelling, soreness, or other physical discomfort.

isobar A line on a map (such as a weather map) connecting points having the same barometric pressure.

isokinetic sampling In sampling air contaminants inside a stack, the maintenance of the same air velocity in the sample probe inlet as in the exhaust gas carried by the stack. Otherwise, the sample will underestimate or overestimate the pollutant concentration in the stack gas.

isomer A chemical compound (usually organic) that has the same molecular formula as another compound but a different molecular structure and therefore different chemical and physical properties. For example, butane and isobutane both have the molecular formula C_4H_{10}, but they have a different arrangement of the carbon and hydrogen atoms.

isopleth A line on a map connecting points at which a certain variable has the same value, for example, a ring around a smokestack connecting locations with equal estimated ambient air concentration.

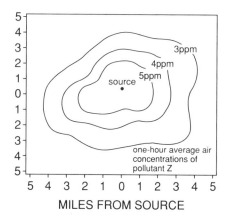

ISOPLETH

MILES FROM SOURCE

isotherm 1. A line on a map connecting points having the same measured or average temperature. 2. A curve showing the changes in a variable while held at a given temperature. For example, adsorption isotherms show the amount of pollutant adsorbed per mass of adsorbing material compared to the amount (concentration) of the pollutant left in water or air at equilibrium.

isotopes Atoms of a single element that have the same number of protons (hence they have the same atomic number) but a different number of neutrons. Consequently, they have different atomic masses and different nuclear properties. For example, carbon-12, carbon-13, and carbon-14 each have six protons (therefore an atomic number of 6), but they have six, seven, and eight neutrons, respectively. Each form behaves as carbon in chemical reactions; however, carbon-14 is radioactive (unstable), whereas the other two isotopes are not.

isotropic Displaying equal properties in all directions. For example, isotropic atmospheric pressure at a point 6 feet above the ground is equal in all directions: up, down, left, right, and all points in between.

Itai-Itai disease A disease characterized by kidney dysfunction, bone deformities, and bone pain that is caused by long-term excessive CADMIUM exposure. The name is Japanese for "ouch-ouch," because the symptoms were noted in a group of Japanese with high cadmium doses resulting from the consumption of rice and soybeans grown in soil contaminated by cadmium emissions from mining operations.

J

Jackson turbidity unit (JTU)   A unit that expresses the cloudiness (turbidity) of water; the measure is related to the distance through the water that light can be seen by the unaided eye. The measurement apparatus, a candle turbidimeter, consists of a glass tube with a candle aligned under it. Water is added to the glass tube until the light of the candle cannot be seen; a light path of 2.3 centimeters (cm) equals 1000 JTUs, one of 21.5 cm (clearer water) equals 100 JTUs, etc. (the scale is nonlinear). Named for D. D. Jackson, who, together with G. C. Whipple, published the method in 1900.

jaundice   A condition caused by the presence of abnormal amounts of bilirubin, an end product of hemoglobin metabolism, in the blood, and characterized by a yellowing of the skin and eyes. Exposure of humans to toxic substances that damage the liver can result in this condition.

joint and several liability   The legal principle, held by the courts to apply to hazardous waste site cleanups under the COMPREHENSIVE ENVIRONMENTAL RESPONSE, COMPENSATION, AND LIABILITY ACT (CERCLA), that allows one party to be held responsible for an entire liability if that party contributed to the liability in part. Thus a hazardous waste generator with waste at a CERCLA site, a former owner or operator of the disposal facility, or a transporter that chose the facility can be sued by the United States Environmental Protection Agency for the full cost of cleanup at a hazardous waste site used by many disposers. This principle does not prevent the party held to be jointly and severally liable from legally pursuing other parties for their share of the cleanup.

joule (J)   A unit of energy. In the SI system, the unit equal to the work done by a force of one newton over a distance of one meter in the direction of the force (1 J = 1 Nm). Also equal to 0.239 CALORIES, or 9.484×10^{-4} BRITISH THERMAL UNIT (BTU).

juvenile hormones   Organic compounds produced by insects undergoing metamorphosis that maintain the insect in a certain stage in its life cycle; once the hormone is no longer secreted, maturation occurs. Synthesized versions of species-specific juvenile hormones are applied to growing crops to prevent a particular pest species from maturing and reproducing, thus controlling the pest.

K

Kelvin (K)   The SI unit of TEMPERATURE. Zero Kelvin is ABSOLUTE ZERO, and an interval of 1 K is equal to 1° on the Celsius scale and 1.8° on the Fahrenheit scale. 0°C = 273.15 K.

kepone   A chlorinated hydrocarbon insecticide that is persistent in the environment. The agent was banned in 1976 after several workers in a kepone manufacturing unit in Virginia fell ill, and fish in the nearby James River were found to contain unsafe levels. Also referred to as chlordecone.

kerogen   A solid hydrocarbon embedded in subsurface rock formations.

(Sometimes erroneously called oil shale, it is a solid, not liquid oil, and the rock is not shale.) The largest known deposits are in parts of Colorado, Utah, and Wyoming. Although kerogen deposits are estimated to exceed proven oil reserves in the United States by 25–50 times and it is a potential source of fuel, the high extraction and environmental costs have prevented exploitation.

kilocalorie (kcal) 1000 CALORIES. 1 kcal = 1 dieter's calorie, the unit used to indicate the caloric content of food.

kilogram (kg) The SI unit of mass. One kilogram, or 1000 grams, equals about 2.205 pounds.

kilopascal 1000 PASCALS; a common unit for pressure measurements. One kilopascal equals 4.02 inches of water, 0.145 pound per square inch, or 7.5 millimeters of mercury.

kiloton A unit of explosive energy applied to nuclear weapons; equal to 1000 tons of TNT.

kilowatt (kW) A unit of power, which is energy used per time, equal to 1000 WATTS. Since one watt equals one JOULE per second, a kilowatt equals 1000 joules per second. See KILOWATT-HOUR.

kilowatt-hour (kWh) A unit of electrical energy equal to 1000 WATT-hours or a power demand of 1000 watts (e.g., ten 100-watt incandescent light bulbs) for one hour. The equivalent of 3,600,000 joules. Utility rates are typically expressed in cents per kilowatt-hour.

kinematic viscosity The VISCOSITY of a fluid divided by its mass density (mass per unit volume). Units are length squared divided by time (e.g., square meters per second). The values are also expressed in centistokes, or 0.001 square meter per second.

kinetic energy (k) The energy inherent in a substance because of its motion,

expressed as a function of its velocity and mass, or $MV^2/2$.

Kirchhoff's law The law stating that, for a given temperature, the ABSORPTIVITY and EMISSIVITY of a substance for a certain wavelength of radiation are equal. Named for the German physicist Gustav Robert Kirchhoff (1824–1887).

Kjeldahl nitrogen The amount of nitrogen contained in ORGANIC material as determined by a method based on the digestion of the sample in a sulfuric acid–based reagent, which converts the nitrogenous organic material to carbon dioxide, water, and ammonia. Subsequently, the ammonia is quantified. Named for Johann Kjeldahl of Denmark (1849–1900).

knockout tray A device designed to capture the majority of solids, droplets, and mists carried over from a SCRUBBER.

Koschmieder's relationship An equation that gives the visual range (L_v) in the atmosphere, using estimates for the scattering and absorption of light by naturally occurring and pollutant gases and particles; expressed as

$$L_v = \frac{3.92}{b_{ext}}$$

where b_{ext} is the EXTINCTION COEFFICIENT. If the extinction coefficient has units of 1/kilometer, then the visual range is in kilometers.

krill Small crustaceans that are abundant and form an important part of the food chain in Antarctic waters; a possible source of human dietary protein, either directly or via animal feed. The effect of large-scale harvesting of krill on ocean food chains is unknown.

K-selected A type of reproductive strategy of a species that allocates a relatively high proportion of available energy on the maintenance and survival of individual members of the species rather than on reproduction. The number of such organisms in a defined area does not tend to increase beyond the capacity of the environment to support the species and is typical of a stable environment. Compare R-SELECTED.

kwashiorkor Malnutrition caused by protein deficiency, although caloric intake is adequate. In the absence of famine, the disorder is usually seen in one-to three-year-old children in very poor areas as a result of premature weaning. Characterized by retarded growth, hair loss, and accumulation of fluids in the abdominal area. Compare MARASMUS.

L

laboratory blank An artificial sample, usually distilled water, introduced to a chemical analyzer to observe the response of the instrument to a sample that does not contain the material being measured. The blank can also detect any contamination occurring during laboratory processing of the sample.

lacrimator An air contaminant that causes eye tearing upon excessive exposure; this type of contaminant is often present as an ingredient in photochemical smog, such as peroxyacetylnitrate (PAN).

lacustrine Pertaining to lakes, as in lacustrine environment.

lag phase The period following the inoculation of a growth medium with a culture of bacteria. Some time is required for the organism to adjust to the new environment; during this interval there is no increase in the number of bacteria per unit volume of culture fluid.

lagoon A pond used for the stabilization and decomposition of organic ma-

terials in wastewater. These ponds allow for the settling and subsequent ANAEROBIC mineralization of particulate material and for the AEROBIC DECOMPOSITION of dissolved constituents. Often, aeration devices that agitate the contents of the pond are added to enhance the activities of aerobic bacteria.

laminar flow Smooth, nonturbulent flow of a fluid (water, air) in a pipe or around obstacles. Compare TURBULENT FLOW. See also REYNOLDS NUMBER.

land application A method for the disposal of treated domestic wastewater. The wastewater, which has been subjected to PRIMARY and SECONDARY TREATMENTS as well as disinfection to kill or inactivate dangerous microorganisms, is sprayed over the ground to remove plant nutrients and promote the growth of vegetation. A type of TERTIARY TREATMENT.

land ban See LAND DISPOSAL BAN.

land breeze The land-to-sea surface wind that occurs in coastal areas at night. It is caused by the rising of the air above the ocean, which is warmer than the land due to the rapid cooling of the land after sunset. Compare SEA BREEZE.

land disposal ban A process initiated by the United States Congress under the HAZARDOUS AND SOLID WASTE AMENDMENTS of 1984. The land disposal of specific hazardous wastes was prohibited automatically unless the United States Environmental Protection Agency ruled that the disposal ban was unnecessary to protect human health and the environment, and issued treatment standards for the wastes. The process included a schedule of phased restrictions on the following types of wastes, in this order: BULK LIQUID hazardous wastes or hazardous wastes containing FREE LIQUIDS, dioxin-containing wastes and spent solvents, CALIFORNIA LIST WASTES, and all remaining listed or known hazardous wastes in three separate groups (known as the first third, second third, and third third). The

Environmental Protection Agency has published treatment standards that allow many of these named wastes to be disposed of in landfills. See HAMMER PROVISION, SOFT HAMMER.

land farming   A technique for the controlled biodegradation of organic waste that involves the mixture of waste sludges with soil. Microorganisms in the soil degrade the organic wastes. The biodegradation is enhanced by tilling the soil-waste mixture to ensure adequate oxygen and the control of moisture content, nutrient levels, and soil pH.

landfill   An area where solid or solidified waste materials from municipal or industrial sources are buried.

Langelier index   An expression of the ability of water to dissolve or deposit calcium CARBONATE scale in pipes. The index is important in industrial water systems where the formation of scale or sludge can cause equipment or process failure. The index is calculated from direct measurements of the following in the water system: pH, ALKALINITY, calcium concentrations, TOTAL DISSOLVED SOLIDS, and temperature. A positive value indicates a tendency to form scale, and a negative value means the water will dissolve scale and may be corrosive. Named for W. F. Langelier, who devised the index in 1949. Also called the stability index.

langley (ly)   A unit expressing the quantity of electromagnetic radiation received or emitted by a unit area of surface. One langley equals one CALORIE per square centimeter. The rate of solar INSOLATION for a point on the surface of the earth perpendicular to the rays of the sun is about 1.3 langleys per minute; at the upper reaches of the atmosphere, this perpendicular insolation is 2.0 langleys per minute, also called the SOLAR CONSTANT.

lapse rate   See ADIABATIC LAPSE RATE and ENVIRONMENTAL LAPSE RATE.

latency   See LATENT PERIOD.

latent heat transfer   The removal or addition of heat when a substance changes state. In the environment, this almost always refers to the release of heat from water upon condensation and the absorption of heat by water upon evaporation. See HEAT OF VAPORIZATION.

latent period   The time between exposure to a dangerous substance or radiation and the development of a disease or pathological condition resulting from that exposure.

lateral   A municipal wastewater drain pipe that connects a home or business to a branch or MAIN.

lateral expansion   Any horizontal expansion of the area that will receive MUNICIPAL SOLID WASTE at an existing landfill; such an expansion is considered a new unit by federal regulations and must meet the (usually more stringent) standards that apply to new facilities.

lateritic soil   Land that consists of minerals that are rich in iron and aluminum compounds, other minerals having been removed by LEACHING. The land is hard and unsuitable for agricultural use.

law of the minimum   See LIEBIG'S LAW OF THE MINIMUM.

Le Châtelier's principle   A principle of dynamic equilibrium stating that a change in one or more factors that maintain equilibrium conditions in a system will cause the system to shift in a direction that will work against or adjust to the change(s), with a resulting reestablishment of equilibrium conditions. For example, assume the concentrations of gaseous oxygen in the atmosphere and DISSOLVED OXYGEN in a stream are in equilibrium at a certain temperature. As oxygen dissolves in water, heat is released. If an outside influence (e.g., sunlight) raises the water temperature in the stream, this shifts the equilibrium back in the direction of lower dissolved oxygen and greater atmospheric oxygen, and oxygen escapes from the water. As a result, at higher water temperatures, equilibrium concentrations of dissolved oxygen are lower. Named for French chemist Henri Louis Le Châtelier (1850–1936).

leachate   Water that has migrated through and escaped from a waste disposal site. The fluid contains dissolved and suspended material extracted from the waste and soil.

leachate collection system   An arrangement of reservoirs and pipes underlying a waste disposal site. The system is designed to accumulate and remove leachate, water that migrates through the waste.

leaching   The process by which soluble materials are washed out and removed from soil, ore, or buried waste.

leaching field   The area of land into which a septic tank drains. The wastewater exiting the tank is dispersed over and percolates through a defined area of land.

lead (Pb)   A toxic metal present in air, food, water, soil, and old paint. Overexposure to this metal can cause damage to circulatory, digestive, and central nervous systems. Children less than six years old are considered the most susceptible. Atmosphere levels have dropped sharply with the phaseout of leaded gasoline. Lead in air, water, and food is regulated by the CLEAN AIR ACT; CLEAN WATER ACT; SAFE DRINKING WATER ACT; FOOD, DRUG, AND COSMETIC ACT, and other environmental statutes. See PICA and TETRAETHYL LEAD.

lead agency   1. The federal administrative agency responsible for supervising the preparation of an ENVIRONMENTAL IMPACT STATEMENT when more than one agency is involved in the process. 2. The federal or state agency represented by the ON-SCENE COORDINATOR in a response action to a spill or leak of a hazardous substance.

leaking underground storage tank trust fund (LUST trust fund)   A $500 million fund established by the SUPERFUND AMENDMENTS AND REAUTHORIZATION ACT of 1986 to pay for any required cleanup of leaks from petroleum underground storage tanks. The fund was created because the COMPREHENSIVE ENVIRONMENTAL RESPONSE, COMPENSATION, AND LIABILITY ACT excludes petroleum. The money was collected from a 0.1 cent tax on each gallon of petroleum motor fuel sold until mid-1990, when the $500 million limit was reached.

legislative history   The committee reports, congressional debates, committee prints, and any other supplemental documents pertaining to an enacted federal statute. The material is used by administrative agencies formulating regulations based on the law and by the courts to review the intent of the law.

legume   A plant that produces a seed in a pod, such as bean plants. This type of plant is important in the symbiotic fixation of atmospheric nitrogen during which nitrogen gas from the air is converted to a form of nitrogen found in protein. The conversion requires the combined efforts of the plant and bacteria belonging to the genus *Rhizobium*. Legumes are in turn important sources of nitrogen for humans.

Lemna gibba   The genus and species name of a small, stemless, free-floating plant used in experiments to determine the toxicity of pollutants to aquatic plant life. Commonly called duckweed.

lentic water   The standing water of ponds, lakes, swamps, or marshes. Compare to LOTIC WATER.

Leq   A calculated value of a continuous sound level that equals, in total energy, a combination of different sound levels during a time period. It is expressed as

$$L_{eq} = 10 \log \sum_{i=1}^{N} 10^{L_i/10} t_i$$

where N is the total number of sound measurements, L_i is the sound level in DECIBELS of the ith sample, and t_i is the fraction of the total time for the ith sample. For example, if three 10-minute sound measurements are 50, 40, and 75 decibels, then the equivalent 30-minute continuous sound is 70.2 decibels. Note that the value of the Leq is influenced greatly by high sound levels.

lethal concentration–50 percent (LC$_{50}$)   The concentration of a chemical substance that will cause death in 50 percent of test organisms within a certain time period following exposure. The expression usually refers to air concentrations, but it is also used to define water concentrations causing a 50 percent death rate in aquatic organism toxicity tests. Usually expressed in units such as parts per million or micrograms per cubic meter, if air concentrations. Commonly used water concentration units are milligrams per liter or micrograms per liter. It is also possible to define the lethal concentration to 25 percent of the test organisms (LC$_{25}$), and so on. Also called the mean or median lethal concentration.

lethal concentration, low (LC$_{LO}$)   The lowest concentration of a chemical substance that has caused death in humans or animals within a certain period of time following exposure. The term usually refers to air concentrations, but it is also used to define the lowest water concentrations causing death of aquatic organisms during toxicity tests. See LETHAL CONCENTRATION–50 PERCENT.

lethal dose (LD)   The absorbed amount of a chemical agent sufficient to cause death. The inclusion of a subscript number with the LD designation indicates the percentage of the exposed population that dies within a certain time

period as a result of the exposure. For example, LD$_{50}$ means 50 percent fatality. See LETHAL DOSE–50 PERCENT.

lethal dose–50 percent (LD$_{50}$)   The amount of chemical, usually expressed as milligrams of the chemical per kilogram body weight, that causes death in one-half of the exposed organisms. Also called the median lethal dose.

lethal dose, low (LD$_{LO}$)   The lowest dose of a chemical introduced by any exposure route other than inhalation, that has caused death in humans or animals within a certain period following exposure. A commonly used unit is milligrams of the chemical per kilogram body weight of test subjects. See LETHAL CONCENTRATION, LOW.

leukemia   A term for cancers of the blood-forming tissues, characterized by the overproduction of white blood cells and their precursors. Excessive exposure to IONIZING RADIATION or benzene is associated with an increased risk of leukemia.

licensed material   Any material that contains 0.05 percent or more of uranium or thorium, enriched uranium, or by-products from nuclear reactors and that must be licensed by the federal government for use, transport, or processing.

lichen   Plants that result from a symbiotic relationship between algae and fungi. These composite growths are found on the surfaces of trees, stones, or other structures exposed to the air, and represent the most common example of mutualistic relationships between microorganisms. The uptake of air pollutants by lichens has been employed as an indicator of the level of air pollution.

Liebig's law of the minimum   The ecological principle stating that the existence, abundance, or distribution of a population is limited by the essential physical or chemical factor that is in shortest supply relative to the level required by the organisms. The one physical or chemical requirement restricting the growth of a population is called the LIMITING FACTOR. First proposed by Justus von Liebig in 1840. Liebig was a German chemist (1803–1873).

life-span study (LSS) The tracking of individuals and monitoring their health status throughout their lifetimes in an attempt to study long-term effects of exposure to risk factors such as low doses of ionizing radiation.

life table A tabulation of vital statistics on a population in which members are grouped by age. Information gathered for the various age groups includes such things as survivorship, mortality rates, and life expectancy.

lifetime risk The probability of an individual dying from a specific cause or contracting a certain disease (like cancer) as a result of exposure to a dangerous substance or radiation during 70 years of life.

lifetime risk ratio The probability of a person dying from a specific cause or contracting a certain disease as a result of exposure to a dangerous substance or radiation divided by the probability of an unexposed person dying from the same cause or contracting the disease.

lift A tier of landfill CELLS.

lift station A pumping facility that raises municipal sewage to a higher elevation to allow for further gravity flow. Such facilities are required in areas with a flat topography.

light detection and ranging (LIDAR) A technique employing high-intensity laser light to detect and track air pollutants released from industrial facilities. The pulses of laser light are reflected by the aerosols in the atmosphere.

light-and-dark bottle technique A method used to determine the extent of photosynthesis in an aquatic ecosystem. Duplicate portions of a water sample are collected. One portion is incubated in a clear bottle, and the other is incubated in a dark, light-impermeable bottle. Following incubation for a prescribed time period, the net uptake of carbon dioxide in each is measured and compared.

light-water reactor (LWR) A nuclear reactor that uses ordinary water as the coolant and moderator. The core of the reactor, which contains the uranium fuel and control rods, is totally immersed in the water. Heat generated by the fission of the uranium fuel raises the temperature of the water, which is then pumped to heat exchange units for the production of steam and subsequent generation of electricity. The process results in a continuous transfer of heat from the reactor to the outside. The water also functions as a moderator to reduce the energy level of neutrons released by the fission process in order to allow the neutrons to promote additional fission events. The light-water reactor is the most common type of nuclear reactor operated in the United States. Compare HEAVY-WATER REACTOR.

lignin A major macromolecular polymer contained in the woody structure of plants. The polymer is a random arrangement of phenylpropane subunits. Biodegradation of lignin is slow and takes place only in aerobic environments. Lignins are important in the formation of humic materials and the building of top soil.

lignite A type of coal, with the lowest energy content of coals that are widely used. Also called brown coal.

limestone A sedimentary geological deposit formed by the consolidation of carbonate minerals that were formerly the shells of living organisms. Heating of this mineral results in the formation of lime. The mineral is used in the control of sulfur dioxide release in industrial exhausts. See FLUE GAS DESULFURIZATION.

limestone scrubbing See FLUE GAS
DESULFURIZATION.

limited water-soluble substances
Water pollution chemicals that are soluble in water at less than one milligram of
substance per liter of water.

limiting factor The dominant factor
that restricts the continued reproduction
or spread of a particular species. The
factor may be physical constraint, such
as light, or a chemical resource, such as
an essential nutrient.

limits of tolerance Minimum and
maximum amounts of a required physical
or chemical environment factor within
which a particular species can exist; an
optimum level, supporting the maximum
number of organisms, lies within the tolerance limits.

limnetic zone The open water of a
pond or lake supporting PLANKTON growth.
Compare to PROFUNDAL ZONE.

limnology The study of inland bodies
of fresh water and their flora and fauna.

lindane A commercial CHLORINATED
HYDROCARBON insecticide that consists of
several isomers of hexachlorocyclohexane, with the gamma isomer being the
primary active agent. The chemical is a
6-carbon, cyclic compound; however, it
is not AROMATIC. Lindane is persistent in
the environment.

line source A roadway source of ve-
hicular air pollutants. It is assumed that
pollutants are released at a constant rate
along each segment of the roadway.

linear accelerator A device for ac-
celerating charged subatomic particles in
a straight line. The device is used to study
the structure of atoms.

linear alkyl sulfonate (LAS) A
common surfactive agent used in detergents.

linear dose model A projection that
postulates that the increase in risk associated with exposure to a dangerous substance or radiation is in direct proportion
to the dose of that agent.

linear energy transfer (LET) A
measure of the loss of energy as IONIZING
RADIATION passes through tissue, expressed as the frequency of ionizing
events, or energy locally imported to the
tissue, per unit length of biological material. X-rays and gamma rays are characterized by a low LET, since ionizing
events are widely spaced as the radiation
passes through cells, whereas alpha particles and fast neutrons are described as
having a high LET, since ionizing events
are spaced close together as they pass
through cells. See also QUALITY FACTOR,
RELATIVE BIOLOGICAL EFFECTIVENESS.

linear growth See ARITHMETIC
GROWTH.

linearized multistage model A
mathematical method of extrapolating the
observed probability of tumor formation
in animal tests using high doses to the
expected incidence of cancer in humans
exposed to much lower doses. The model
incorporates the assumption that a clinical cancer results after progressing through
several stages and that cancer risk is proportional to dose (a linear relation) at low
exposures. One form of the model, for a
fixed time of exposure is

$$P(d) = 1 - \exp\left\{ -\sum_{i=0}^{n} q_i d^i \right\}$$

where $P(d)$ is the cumulative probability
of cancer at dose d, n is the number of
stages, and q is an empirically derived
nonnegative coefficient.

liner A low-permeability material, such
as clay or high-density polyethylene, used
for the bottom and sides of a landfill.
The liner retards the escape of LEACHATE
from the landfill to underlying groundwater.

lipid   A group of organic compounds that are insoluble in water, but that dissolve in organic solvents; includes fats, oils, and fatlike compounds.

liquefied natural gas (LNG)   Natural gas, mainly methane and ethane, condensed to a liquid state to greatly reduce its volume for international transport in specially designed refrigerated tanker ships. At its destination the LNG is revaporized and introduced into a pipeline network. The facilities for liquefication, transport, and revaporization are industrial operations, and the public does not have direct access to LNG. Compare LIQUEFIED PETROLEUM GAS.

liquefied petroleum gas (LPG)   A liquid hydrocarbon fuel composed of condensed propane and butane gases: the fuel is stored and transported in pressurized tanks for distribution to public consumers that are not serviced by natural gas facilities. Compare LIQUEFIED NATURAL GAS.

liquid chromatography   A technique for the separation and analysis of higher-molecular-weight organic compounds. See CHROMATOGRAPHY.

liquid scintillation counter   See SCINTILLATION COUNTER.

liquid-metal fast breeder reactor (LMFBR)   See BREEDER REACTOR.

liquor   A liquid solution containing dissolved substances. A concentrated solution of process chemicals or raw materials added to an industrial process. Compare SLURRY.

listed hazardous waste   Chemical substances or processes that produce chemical substances defined as hazardous waste by the United States Environmental Protection Agency and published in the CODE OF FEDERAL REGULATIONS, Title 40, Parts 261.31–261.33. A facility producing hazardous waste is then regulated as a "generator"; these rules appear in the Code of Federal Regulations, Title 40, Part 262. Special provisions apply for SMALL QUANTITY GENERATORS.

listed waste   See LISTED HAZARDOUS WASTE.

listing site inspection (LSI)   A more extensive investigation at a potential SUPERFUND SITE following the PRELIMINARY ASSESSMENT AND SITE INSPECTION (PA/SI). The results of the PA/SI and the LSI are used to calculate a site score using the HAZARD RANKING SYSTEM. The score determines whether the site is included on the NATIONAL PRIORITIES LIST.

liter (l)   A metric volume unit equivalent to 1000 cubic centimeters. One liter equals 1.057 quarts or 0.0353 cubic feet.

lithosphere   The solid portion of the earth's crust and mantle. Constitutes, together with the ATMOSPHERE, hydrosphere and BIOSPHERE, the earth's ecosystem.

lithotroph   A type of bacteria capable of obtaining metabolically useful energy from the oxidation of inorganic chemicals, chiefly ammonium, nitrite, iron, and various forms of sulfur. These bacteria obtain their carbon from carbon dioxide as do the green plants. Compare AUTOTROPH, HETEROTROPH.

litter fence   A movable fence used at sanitary landfills to catch blowing debris.

littoral   An interface region between the land and a lake or the sea.

littoral zone   The area of a lake or pond close to the shore; includes rooted plants. Compare PROFUNDAL ZONE.

ln   See NATURAL LOGARITHM.

L_n   An expression indicating how often a certain sound level is equaled or exceeded, when *n* equals percent of the measurement time; a summary statistic to describe noise levels. For example, $L_{70} = 55$ dBA means that 70 percent of the noise measurements in an area were at or above 55 dBA (DECIBELS, A-WEIGHTING NETWORK).

load The amount of chemical material or thermal effluent released into a receiving stream, either by human or natural sources.

load allocation (LA) The portion of the pollution LOAD of a stream attributable to human NONPOINT SOURCES. The amount of pollution from each point source is the WASTELOAD ALLOCATION.

loading 1. The air concentration of a pollutant in a gas duct before entry into an air pollution control device. 2. Synonym for the pollution LOAD of a stream.

loading capacity The greatest amount of chemical materials or thermal energy that can be added to a stream without exceeding water quality standards established for that stream.

loading, acute toxicity test In the laboratory testing of the toxicity of pollutants to aquatic biota, the ratio of the biomass of the test species, in grams of wet weight, to the volume (liters) of solution used.

loam A type of rich soil consisting of a mixture of clay, silt, sand, and organic material.

local emergency planning committee (LEPC) A committee formed under the STATE EMERGENCY RESPONSE COMMISSION as required by TITLE III of the SUPERFUND AMENDMENTS AND REAUTHORIZATION ACT. The duties of the committee include the preparation of an emergency response plan for the local emergency planning district it represents, typically a county, which includes emergency re-

sponse to accidental releases of toxic chemicals.

lodging The bending over of a cereal crop that has grown too tall, damaging the grain, reducing yields, and/or making the crop difficult to harvest. This problem is exhibited by native plant varieties given large fertilizer applications. The new HIGH-YIELDING VARIETIES possess dwarf genes that prevent lodging.

log-normal distribution A distribution of airborne particles that is a normal, bell-shaped curve if the logarithm of the midpoint of various particle size ranges is plotted against the percent of the total mass in that size range (mass function). A log-normal distribution also exists if graph of the particle sizes is a straight line when plotted against the mass percent of the total particulate that is less than (or greater than) that size, on log-probability paper.

A LOG-NORMAL PARTICLE DISTRIBUTION

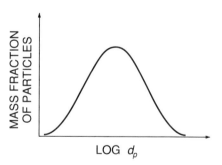

log phase That time period during which growth (cell division) of bacteria is occurring at a maximum rate. When bacteria are placed in an environment suitable for their growth, a pattern of increase in cell numbers that is typical of bacteria is observed. The various phases of growth are lag phase (a pause is cell division), log phase, stationary phase (cells stop dividing), and death phase (cells are dying). Same as exponential phase.

BACTERIAL PHASES
(log scale)

a = lag phase
b = log phase
c = stationary phase
d = death phase

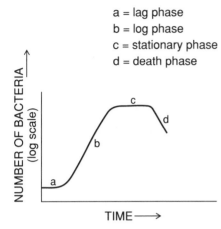

logarithm (log)   The value of the exponent that a fixed number (the base) must have to equal a given number. It is calculated as $b^x = y$, where b is the base and x is the logarithm. The base for the common logarithm is 10. For example, the logarithm of 100 is 2 since $10^2 = 100$. Also written as $\log_{10} 100 = 2$. The NATURAL LOGARITHM base is approximately equal to 2.718282.

logistic curve   A graph resembling a flattened S (sigmoid curve), obtained when certain phenomena (e.g., population size for a species) are plotted against time. The growth rate is high at the beginning, reaches a maximum during the early stages, starts to slow near the center, and declines to zero at the end.

long-range transport   The movement of air contaminants as far as hundreds of miles downwind before their removal by physical or chemical processes, especially precipitation; such movement is considered to be a significant contributor to ACID DEPOSITION.

loss-of-coolant accident (LOCA)   Potentially, one of the more serious accidents involving a nuclear reactor. Water used to transfer heat from, and therefore cool, the reactor core is lost from the vessel housing the reactor, resulting in melting of the core contents.

lotic water   The flowing water of rivers and streams. Compare LENTIC WATER.

loudness   The perceived intensity of a sound. It depends on the SOUND PRESSURE LEVEL (DECIBEL level) and the FREQUENCY of the sound; since the human ear is more sensitive to higher frequencies.

low fence   A hearing damage criterion that equals a 25-decibel threshold for the sound frequencies 500, 1000, and 2000 hertz.

lower detectable limit   See DETECTION LIMIT.

lower explosive limit (LEL)   The air concentration of a gas or vapor below which the chemical will not burn or explode in the presence of a flame or other ignition source. Compare UPPER EXPLOSIVE LIMIT.

lower flammable limit (LFL)   Same as LOWER EXPLOSIVE LIMIT.

lowest achievable emission rate (LAER)   The emission rate required for an air pollutant emitted by a new or modified source in an area that does not meet the national ambient standard for that pollutant (NONATTAINMENT AREA). The CLEAN AIR ACT requires that it be set as the most stringent emission limitation in use for that pollutant and type of facility.

lowest-observed-adverse-effect level (LOAEL)   The lowest dose of a chemical or physical agent that causes a significant increase in the frequency or severity of an adverse effect in an exposed population compared with an unexposed (control) population.

lowest-observed-effect level (LOEL)   The lowest dose of a chemical or physical agent that causes a significant biological change in an exposed population compared with an unexposed (control) population. The observed change is not necessarily deleterious. Compare LOWEST-OBSERVED-ADVERSE-EFFECT LEVEL.

low-flow augmentation The release of reservoir water into a stream during periods of low flow to maintain acceptable DISSOLVED OXYGEN levels. This practice is not permitted as a water quality maintenance technique in the United States.

low-level waste (LLW) Radioactive waste material with a radiation intensity of less than 10 nanocuries per gram. Such materials are commonly produced by hospitals and research laboratories and are not as dangerous as waste generated in nuclear reactors. See HIGH-LEVEL WASTE.

Lurgi process A method for COAL GASIFICATION developed in the 1930s that mixes pulverized coal, oxygen, and steam under high pressure, producing methane gas.

lymph nodes Small glands, located throughout the body (as in the neck, armpits, and groin), containing large numbers of lymphocytes, which are white blood cells used to defend the body against bacteria and viruses. Tumor cells that have migrated from an original site of growth are frequently observed in these structures.

lymphoma Cancers of the lymphatic tissue; such tumors are characterized by the overproduction of cells that compose LYMPH NODES.

lysimeter A device for measuring EVAPOTRANSPIRATION from soil covered with vegetation.

M

macroconsumer A class of heterotrophic organisms that ingests other organisms or PARTICULATE ORGANIC MATTER; includes HERBIVORES, CARNIVORES, and OMNIVORES. Compare to MICROCONSUMER.

macronutrients Chemical nutrients required in relatively large quantities by an organism. Plant macronutrients include nitrogen, potassium, phosphorus, calcium, sulfur, and magnesium. Compare MICRONUTRIENTS.

macrophage See ALVEOLAR MACROPHAGE.

macroscopic Capable of being seen with the naked eye. Items that can only be seen with the aid of a microscope are termed MICROSCOPIC.

Mad Hatter's disease A neurologic disorder resulting from chronic exposure to inorganic mercury compounds. Affected individuals may experience exaggerated emotional responses, tingling sensations of the skin, mental disturbances, or hallucinations. The term derives from the occupational exposure syndrome caused by overexposure to the mercuric nitrate used to soften fur in the making of felt hats.

magma Molten rock found in the mantle, beneath the crust of the earth. When forced toward the surface, magma cools and solidifies to become IGNEOUS ROCK.

magnet A naturally occurring or treated object almost always containing iron that exhibits the elemental natural force magnetism. Magnets exert a MAGNETIC FIELD around them and can affect objects within the field. The ends of magnets are called the north and south poles, because if a bar magnet is allowed to turn freely it will align itself with the earth's magnetic field (from its iron core). The like poles of magnets repulse each other; a north pole and south pole attract each other.

The passage of an electric current through a conducting coil of wire surrounding an iron or steel core will cause the core to become a magnet while the current is flowing; this is the phenomenon used to convert electrical energy to mechanical energy in electric motors. If a magnet is put in motion near a conducting coil of wire, then a current (ELECTRICITY) is induced in the coil, which is the operating principle of an electric generator.

magnetic field The area around a magnetic pole or moving charge under the influence of the forces exerted by the magnet.

magnetohydrodynamic generation A theoretical concept for the recovery of a greater amount of electricity from power plants that burn coal to generate electric energy. Potassium would be added to the hot gases produced by the burning of coal. The extremely high heat (about 2000° C) would result in the ionization of the potassium, and the ions would then be forced to move through a magnetic field, resulting in the production of an electric current. Such a design, which is theoretical, could double the efficiency of currently used coal-burning power plants.

main A relatively large pipe in a distribution system for drinking water or in a collection system for municipal wastewater.

makeup water Water added to the flow of water used to cool condensers in electric power plants. This new water replaces condenser water lost during passage of the cooling water through cooling towers or discharged in BLOWDOWN.

malathion An insecticide belonging to the organophosphate class. The agent is known as a soft pesticide because it is degraded relatively quickly when placed in the environment. Likewise, malathion has been shown to be less toxic to mammals than some of the more dangerous pesticides such as DDT and dieldrin.

malignant Describing a tumor that produces cells that can migrate to new sites in the body where additional tumors can subsequently develop.

malignant neoplasm A fast-growing tumor that invades other tissues and undergoes METASTASIS.

Malthusian Of or related to an analysis developed by English economist and clergyman Thomas Malthus in the late eighteenth century, in which he proposed that the human population would soon exceed available agricultural resources, with a resulting catastrophic famine and loss of human life as population levels adjusted to the available food supply. This is also referred to as "boom-to-bust" swings in population levels. (Malthus, in his later work, adopted a more optimistic view.)

mammalian selectivity ratio (MSR) A parameter employed to judge the relative toxicity of insecticides by comparing the amount needed to kill rats with the amount needed to kill houseflies. It is calculated as the ratio of the oral dose (milligrams per kilogram body weight) required to kill 50 percent of experimental rats to the topical dose (in the same units) needed to kill 50 percent of exposed houseflies. A ratio less than 1 indicates that the agent kills rats more effectively than it kills houseflies, and vice versa.

mangrove swamp A tidal swamp forest populated by plant species capable of growth and reproduction in areas that experience periodic tidal submergence in seawater with a resulting increase in saline conditions. These forests develop along coastal regions in tropical climates. More than 50 species of plants may be present; however, mangrove swamps are dominated by trees referred to as red mangrove, *Rhizophora mangle;* black mangrove, *Avicennia germinans;* and white mangrove, *Laguncularia racemosa.* Typically, these trees have large, exposed root systems.

manifest system   The regulations applicable to transporters of HAZARDOUS WASTE under the RESOURCE CONSERVATION AND RECOVERY ACT; found in Title 40, Part 263 of the CODE OF FEDERAL REGULATIONS. Transporters are required to use a set of shipment forms to track wastes in transit from their origin to their final site of disposal.

man-made mineral fibers (MM-MFs)   Fibrous materials prepared from substances such as glass wool, rock wool, slag wool, and ceramic fibers. These materials have been used as replacements for asbestos in insulation and building materials. Although questions remain concerning the relationship of these fibers to human health, they have coarser diameters and do not become airborne as easily as asbestos fibers which have been implicated in lung cancer and are being phased out in the United States.

Manning formula   An equation relating the volumetric flow rate *(Q)* of water in an open channel as follows:

$$Q = \frac{1.0 \text{ or } 1.49}{N} AR^{2/3} S^{1/2}$$

where *A* is the cross-sectional area of the water, *S* is slope of the channel sides, *N* is a channel roughness coefficient, and *R* is the HYDRAULIC RADIUS. Use 1.0/*N* for SI units, 1.49/*N* for American units. Also known as the Chézy-Manning formula.

manometer   An instrument for measuring pressure, consisting of a transparent tube filled with a liquid such as water or mercury. The level of the liquid in the tube indicates the pressure.

mantle   The division of the earth's interior between the core and the crust. It is composed mainly of silicate rock and is around 2900 kilometers (1800 miles) thick.

marasmus   The deficiency condition caused by inadequate intake of calories and protein. In the absence of famine,

the condition is usually seen in poor areas in infants less than one year old, following weaning and/or gastrointestinal infection that is accompanied by severe or prolonged diarrhea. Compare KWASHIORKOR.

margin of exposure (MOE)   The ratio of the NO-OBSERVED-EFFECT LEVEL to the estimated dose received. For example, if the no-observed-effect level is 20 milligrams per kilogram body weight per day and the dose received is 5 milligrams per kilogram body weight per day, then the MOE is 4. The higher the ratio, the safer the exposure.

margin of safety (MOS)   In setting an environmental standard, a factor applied to the maximum dose of a chemical substance showing no adverse effect in toxicity tests; the factor accounts for uncertainty in the toxicology data. The factor lowers the allowable amount of exposure or dose, which lowers the standard in terms of the acceptable concentration in an environmental medium (air, water, food). Also used to describe the difference between the NO-OBSERVED-EFFECT LEVEL and the estimated dose received by a population.

mariculture   The cultivation of marine organisms for use as a food resource.

Marine Protection, Research, and Sanctuaries Act (MPRSA)   A 1972 law that includes provisions requiring citizens of the United States to obtain a permit from the Environmental Protection Agency before disposing of materials in the oceans. Subsequent amendments to the Act have limited the types of waste that may be permitted for ocean disposal. For example, the phaseout of sewage sludge and industrial waste dumping is to be completed by 1992. Commonly called the OCEAN DUMPING ACT. See OCEAN DUMPING PERMIT.

marsh   A coastal region where the soil has a high moisture content because of periodic flooding caused by the tides.

The vegetation is normally dominated by grasses.

marsh gas   Gas produced during the decomposition of organic material buried in wetland soils. The primary gas is methane.

mass   The quantity of matter in a substance. The SI unit of mass is the kilogram, but mass is also expressed in milligrams, grams, and similar units.

mass balance   An approach used to estimate pollutant releases to the environment, based on the conservation of matter—that matter cannot be created nor destroyed. If the input mass to a process and the output, or product, mass are known and if the chemical reactions within the process are taken into account, the difference between the known mass in and known mass out is considered to be the amount lost to the environment.

mass burn   The incineration of municipal solid waste without prior materials recovery or processing but often accompanied by an energy recovery system, such as steam and electricity production.

mass burn waterwall incinerator   See MASS BURN and WATERWALL INCINERATOR.

mass density   Mass per unit volume; commonly expressed as kilograms per cubic meter or pounds per cubic foot.

mass flow rate   For a liquid or gas, $\dot{m} = Q\rho$, where $\dot{m}$ is the mass flow rate, Q is the VOLUMETRIC FLOW RATE, and ρ is the fluid DENSITY. The unit is mass per unit time.

mass flux   See MASS VELOCITY.

mass number   The number of protons plus the number of neutrons in the nucleus of an atom. This number is also the approximate atomic weight of an atom. See ATOMIC WEIGHT.

mass spectrometer (MS)   A sensitive device used for the analysis of organic materials in environmental samples. Extracts prepared from water, sediment, biota, or other materials are injected into the instrument, and organic compounds extracted from the sampled material are fragmented by a stream of electrons. The positive ions that are produced as a result are separated on the basis of their mass. A detector within the system responds to each ion and produces a corresponding peak on chart paper. The height of each peak indicates the relative abundance of each element in the original organic compound of interest. The identity of organic materials recovered from environmental samples can be determined by a comparison of the pattern of peaks (mass spectrum) produced by known compounds with the spectrum obtained from the unknown chemicals extracted from samples collected from within the environment. The comparison is done by computer.

mass spectrum   See MASS SPECTROM-ETER.

mass velocity   The MASS FLOW RATE in a duct or other enclosure per unit area. Expressed as

$$M_v = \frac{\dot{m}}{A}$$

where M_v is the mass velocity, $\dot{m}$ is the mass flow rate, and A is the cross-sectional area of the enclosure; used in gas absorber design. Typical units are kilograms per second per square meter or pounds per hour per square foot. Also called mass flux.

matching   The selection of persons for an epidemiological study so that the study groups are similar with respect to characteristics that are not being tested but that are related to the disease in question. The selection process is an attempt to allow the study results to be influenced only by the variable being studied. All such potential CONFOUNDING VARIABLES cannot be eliminated, however. For example, in a study of the effects of exercise on the risk of heart disease, with a frequently exercising group compared to a nonexercising group, one important variable related to heart disease that must be equally distributed in the study groups is age. When this is accomplished, the groups are said to be matched on age and any difference in heart disease noted in the groups cannot be attributed to age. Although the exercising group may show a lower risk of heart disease, frequency of exercise may not be the CONTROLLING VARIABLE if, for example, the groups are not matched on smoking histories.

material balance   See MASS BALANCE.

material safety data sheet (MSDS) A written summary of information about chemical substances required by the United States Occupational Safety and Health Administration to be distributed to all employees who may be exposed to the chemicals and to purchasers of the chemical materials. The required information for each chemical substance includes the address and telephone number for the manufacturer; common and scientific names for the chemical; important physical and chemical characteristics; fire, explosion, and reactivity data; adverse health effects that may be expected following overexposure; methods for safe handling, use, and disposal; and recommended methods for workplace control of exposure.

matrix   The material in which an environmental sample is embedded or contained, whether it is soil, water, dried biomass, or other substance.

matrix interference   The adverse influence of the environmental sample MATRIX on the ability to detect the presence or amount of a chemical substance in the sample.

maximum   acceptable   toxicant concentration (MATC)   The highest

concentration at which a pollutant can be present and not exert an adverse effect on the biota, used to experimentally determine the toxicity of the chemical.

maximum allowable concentration (MAC)   The predecessor workplace air standards to the THRESHOLD LIMIT VALUES.

maximum achievable control technology (MACT)   The level of air pollution control technology required by the 1990 amendments to the CLEAN AIR ACT for sources of certain AIR TOXICS. MACT is set by the United States Environmental Protection Agency based on the best demonstrated control technology or practices of the industry.

maximum contaminant level (MCL) The maximum permissible concentration of a drinking water contaminant set by the United States Environmental Protection Agency under the authority of the SAFE DRINKING WATER ACT. MCLs are set for certain inorganic and organic chemicals, turbidity, coliform bacteria, and certain radioactive materials.

maximum   contaminant   level   goal (MCLG)   A nonenforceable drinking water concentration defined by the United States Environmental Protection Agency for several listed contaminants that will cause no known or anticipated adverse human health effects, including a margin of safety. MCLGs are set for certain organic contaminants and for fluoride.

maximum holding time   The longest time period that water samples can be retained between the taking of the sample and the laboratory analysis for a specific material before the results are considered invalid. The times vary from none in the case of the test for residual chlorine levels to six months for the testing of radioactivity. Some types of analyses require that preservatives be added to the sample, and some require storage of samples at refrigerated temperatures.

maximum individual risk (MIR) The excess lifetime health risk (the risk above that of a nonexposed individual) for a certain disease (e.g., cancer) in an individual estimated to receive the highest exposure to a particular environmental pollutant. See MOST-EXPOSED INDIVIDUAL.

maximum permissible concentration (MPC) The amount of toxic chemical or radioactive material in air, water, or food that would be expected to result in a MAXIMUM PERMISSIBLE DOSE at the normal rate of consumption by an individual.

maximum permissible dose (MPD) That amount of exposure to toxic chemicals or radioactivity below which no adverse health effects are anticipated.

maximum sustainable yield The greatest amount of a renewable natural resource (e.g., forests or wildlife) that can be removed without diminishing the continuing production and supply of the resource.

maximum tolerated dose (MTD) Used in tests for carcinogenicity, the highest lifetime dose a test animal can receive without causing significant adverse effects other than cancer. Such large doses can overwhelm the normal metabolic detoxification and excretion responses of animals and not be representative of the risk posed by lower doses. The MTD, however, is used to overcome the problem of finding positive results in studies of cancers with perhaps a 1 in 100,000 risk in humans, in which only around 100 test animals are used for each exposure (dose) level.

maximum total trihalomethane potential (MTTP) The highest combined concentration of chloroform, bromodichloromethane, dibromochloromethane, and bromoform produced in water containing chlorine, chlorine dioxide, or chloroamines when held for seven days at 25° C or above. The measure is employed to assess the impact of chlorine addition on raw water supplies. See TRIHALOMETHANE.

MCF One thousand cubic feet.

mean free path The average distance that a molecule in a fluid (air or water) moves before colliding with another molecule.

mean relative growth rate (RGR) A measure used under test conditions to document the effects of pollutants on the growth rate of aquatic algae. Control and treatment conditions are compared during the active growth of the algae by the formula $RGR = (\log n_2 - \log n_1)/(t_2 - t_1)$, where n_1 is the number of algal cells at time 1 (t_1) and n_2 is the number of algal cells at time 2 (t_2).

mechanical integrity The demonstrated absence of significant leaks in the outer wall of the borehole (well casing), tubing, or packer of an underground injection well. Also, the absence of the downward vertical migration of the injected fluids through underground pathways adjacent to the bore hole of the well.

mechanical separation The use of air classification, magnetic separation, or other mechanized methods to separate solid waste into recyclable categories. See AIR CLASSIFIER.

mechanical turbulence Randomly fluctuating air motion caused by air moving over a rough surface or past objects. The eddies produced by this erratic motion help dilute polluted air with unpolluted air by a process called EDDY DIFFUSION. Usually less significant than the effects of THERMAL TURBULENCE.

median lethal concentration (LC$_{50}$) See LETHAL CONCENTRATION—50 PERCENT.

median lethal dose (MLD) See LETHAL DOSE—50 PERCENT.

Medical Literature Analysis and Retrieval System (MEDLINE) A computer database containing toxicological and medical information from over 3000 research journals. The system is operated by the National Library of Medicine, Bethesda, Maryland. The database is on-line, meaning that it can be searched from a terminal at a local library, with results available in about one-half hour.

medical waste Discarded paraphernalia associated with the practice of medicine, such as plastic tubing, needles, syringes, test tubes, and specimen containers.

Medical Waste Tracking Act (MWTA) A 1988 law attached to the RESOURCE CONSERVATION AND RECOVERY ACT that started a two-year medical waste management demonstration program; the act requires sources generating certain wastes at a rate greater than 50 pounds per month, excluding incinerator ash from on-site burning, to keep detailed records of the movement of those wastes from their facilities to the final disposal site. The regulated wastes must be labeled and placed in approved containers.

medium 1. A material or substance that acts as a carrier for some compound or chemical, such as air, water, soil, food. 2. In microbiology, a sterile mixture, either solid or liquid, of nutrients used for the cultivation of bacteria or fungi. Solid media are prepared using agar as the agent to solidify the liquid solution containing required nutrients.

mega- A prefix meaning one million. For example, one megawatt equals one million watts.

megahertz (MHz) One million HERTZ; a unit of FREQUENCY.

megaton An expression of explosive energy, applied to nuclear weapons, equal to one million tons of TNT.

meltdown A theoretical, worst-case accident at a nuclear reactor. A meltdown is caused by a loss of coolant, with a resulting melting of the reactor core and fuel elements. If the molten core of a reactor escaped from the containment structure housing the reactor, the potential for catastrophic consequences would be great. Large amounts of highly radioactive materials would be released into the air, water, and soil. Although accidents have badly damaged reactor cores, a meltdown has yet to occur.

melting point The temperature at which a solid changes to a liquid. The temperature will vary, and is consistent at equal temperatures and pressures, for each element or solid.

membrane filter A flat paperlike disk typically made of modified cellulose. The filters have a small pore size that allows for the retention of bacteria. These units are employed to remove bacteria from liquid samples for examination for public health purposes as well as for the sterilization of liquids. These membranes are available in a variety of sizes, with a diameter of 47–50 millimeters being the most common.

membrane filter method A procedure used to recover and count bacteria in samples of liquid substances, such as water. The liquid is drawn through a MEMBRANE FILTER using a slight vacuum, with the bacteria in the liquid being retained on the filter. The filter disk is then transfered to a MEDIUM suitable for the growth and incubation of the bacteria.

mercaptans A group of compounds that are sulfur analogs of alcohols and phenols. For example, the chemical formula of methyl alcohol is CH_3OH, and that of methyl mercaptan is CH_3SH. The compounds are noted for their odor and are used to impart odor to natural gas in public gas supplies. Air emissions from chemical manufacturers and pulp and paper mills may contain mercaptans.

mercury   One of a group of elements classified as heavy metals. Mercury has an atomic number of 80 and an atomic weight of approximately 200, and the element is the only metal that is a liquid at room temperature. Unlike most other metals, mercury tends to form COVALENT bonds with other inorganic materials as well as organic compounds. Elemental (metallic), inorganic (ionic), and organic forms of the mercury are released into the environment through natural and anthropogenic sources. The latter include mining, smelting, and industrial discharges. The chloralkali and paper industries were major sources at one time. The burning of coal and the refining of petroleum continue to be important sources. Once released into the environment, the element can undergo many transformations and conversions among the three major forms given above. The most common human exposures to elemental mercury are by the inhalation of vapor. The element is rapidly absorbed and distributed by the circulatory system. Absorption of mercury from the gastrointestinal tract is very limited for elemental mercury but very high for organic forms of the element. Mercury is excreted slowly from the body, with about one-half of an absorbed dose being lost in 30 to 90 days. That part of the human body most sensitive to mercury poisoning is the central nervous system. Symptoms can range from a mild tingling of the skin to death. See METHYL MERCURY and MINAMATA SYNDROME.

meromictic lake   A lake in which the waters are only partly mixed during the course of a year. The lake has a permanent stratification of the water. Compare DIMICTIC LAKE.

mesh size   A measure of particle size determined by the size of a sieve or screen (mesh) that particles can pass through.

mesophyll cells   Chlorophyll-containing leaf tissue.

mesophyte   A plant that requires only moderate amounts of water for growth. Compare HYDROPHYTE, XEROPHYTE.

mesoscale   The scale of air motion smaller than the SYNOPTIC (the scale of a weather map) and larger than the microscale (smaller, turbulent airflow). Appropriate horizontal distances are several tens of kilometers; the time scale is about one day. Air STAGNATIONS are mesoscale phenomena.

mesosphere   The division of the atmosphere above the STRATOSPHERE; it begins about 50 kilometers in altitude and extends to about 80 kilometers.

mesothelioma   Cancer of the membranes lining the abdominal and chest cavities.

mesotrophic   Describing a body of water with a moderate nutrient content. Compare EUTROPHIC and OLIGOTROPHIC.

metabolite   A chemical substance produced by the metabolic reactions of an organism. Because the body will metabolize most absorbed materials, tests for exposure to many materials are analysed for their metabolites rather than the original chemical substances absorbed. For example, the test for exposure to benzene is the presence of phenol, a benzene metabolite, in the urine.

metal fume fever   An acute condition caused by short-term, high exposure to the fumes of zinc, magnesium, or their oxides. The condition is characterized by fever, muscle pain,and chills. Symptoms begin several hours after the high dose is received.

metamorphic rock   Rock formed by the exposure of SEDIMENTARY or IGNEOUS material to high temperatures, high pressures, and chemical processes deep beneath the surface of the earth. For example, limestone, a sedimentary rock, is converted to marble, a metamorphic rock.

metastasis The migration of tumor cells to new sites in the body where additional cancerous growth can arise. See CARCINOGENESIS.

meteorology The study of the atmosphere and weather conditions in the atmosphere. Knowledge of this science is required for an understanding of the movement and activities of pollutants released into the atmosphere.

methane A gaseous hydrocarbon that is the main component of natural gas; the molecule contains a single carbon atom bonded to four hydrogen atoms. It is a primary component of firedamp and is explosive.

methemoglobin A hemoglobin molecule with the central iron atom in an oxidized (ferric) state. Methemoglobin cannot combine with and carry oxygen to the tissues, and thus, a high level of red blood cells in the methemoglobin state can lead to HYPOXIA, a deficiency of tissue oxygen. Environmental, dietary, or workplace exposure to NITRITE, NITRATE, aniline, and nitrobenzene, among other materials, produces methemoglobin in the blood.

methemoglobinemia The disease state characterized by inadequate tissue oxygenation caused by excessive levels of blood METHEMOGLOBIN.

method detection limit (MDL) The minimum concentration of a chemical substance present in a sample that can be measured and reported with a 99 percent probability (confidence level) that the measured concentration is above zero.

methyl mercury One of the common ALKYL MERCURY compounds in which an atom of mercury is bonded to a methyl (CH_4) group. Methyl mercury is produced by a variety of organisms (chiefly bacteria) in the natural environment following contamination by ionic forms of inorganic mercury. Methyl mercury is absorbed by a variety of animals and undergoes BIOACCUMULATION within the food chain. Humans exposed to significant levels of this compound can experience symptoms ranging from mild neurological disorders to paralysis or death. The compound has also been linked to birth defects.

metric ton (t) A unit of mass equal to 1000 kilograms or 2204.62 pounds.

microbar One millionth of a BAR. A unit used to express sound pressures. Standard atmospheric pressure equals about 1,000,000 microbars, or one bar.

microbe Short for microorganism. Small organisms that can be seen only with the aid of a microscope. The term encompasses viruses, bacteria, yeast, molds, protozoa, and small algae; however, microbe is used most frequently to refer to bacteria. Microbes are important in the degradation and decomposition of organic materials added to the environment by natural and artificial mechanisms. Also called germs.

microbial load The total number of bacteria and fungi in a given quantity of water or soil or on the surface of food. The presence of the bacteria and fungi may not be related to the presence of disease-causing organisms.

microbiology The study of organisms that can be seen only with the aid of a microscope. The science deals with the structure and chemical composition of various microbes, the biochemical changes within the environment that are caused by members of this group, the diseases caused by microbes, and the reaction of animals, including humans, to their presence.

microbiota The plants, animals, and microorganisms that can only be seen with the aid of a microscope.

microconsumer A class of heterotrophic organisms that utilizes waste material from other organisms or the tissues of dead animals or plants. Mainly composed of BACTERIA and fungi. Compare to MACROCONSUMER.

microcosm　A laboratory model of a natural ecosystem in which certain environmental variables can be manipulated to observe the response. The model test results are not always applicable to an actual ecosystem because the microcosm is, of necessity, a simplified collection of selected physical, chemical, and biological ecosystem components.

microfauna　Animals invisible to the naked eye, such as copepods and mites.

microflora　Plants invisible to the naked eye, such as diatoms and algae.

microgram (μg)　A mass unit equal to one-millionth of a GRAM.

micrograms per cubic meter (μg/m³)　An expression of the air concentration of a solid, liquid, or gaseous substance. The concentration of airborne solids or liquids must be expressed as a mass per unit volume, such as micrograms per cubic meter, but gas concentrations can also be expressed as volume/volume ratios, such as PARTS PER MILLION (PPM). At 25° C and one atmosphere of pressure, a gaseous concentration in ppm can be converted to micrograms per cubic meter by multiplying the number of ppm of the substance by the MOLECULAR WEIGHT of the substance and then multiplying by 40.9. For example, a carbon monoxide concentration (by volume) of 9 ppm is equal to 9 times the molecular weight of carbon monoxide (28) times 40.9, or about 10,300 micrograms per cubic meter.

microliter (μl)　A volume unit equal to one-millionth of a LITER.

micrometer (μm)　Synonym for micron.

micron (μ)　A unit of length equaling one-millionth of one meter. Airborne

particle diameters are commonly expressed in microns.

micronutrients　Chemical nutrients required in very small amounts by an organism. Plant micronutrients include copper, manganese, iron, zinc, vanadium, molybdenum, cobalt, boron, chlorine, and silicon. Compare MACRONUTRIENTS.

microscopic　Describing an object or organism visible only with the aid of a microscope. Objects that can be seen with the naked eye are called macroscopic.

microwaves　The area of the ELECTROMAGNETIC SPECTRUM between the infrared region and radio waves; energy with wavelengths between about 0.003 and 0.3 meter or with frequencies between 100,000 and 1000 megahertz. The most well known impact of microwaves is their interference with cardiac pacemakers. Lower-frequency microwaves (<3000 megahertz) are absorbed by internal body tissues. Overexposure can raise the body temperature, which could lead to burns or adverse effects on reproduction.

midnight dumping　The deliberate disposal of hazardous waste at a site other than a permitted disposal facility, often taking place at night.

migration velocity　See DRIFT VELOCITY.

millfeed　Mineral ores, such as uranium ore, that enter the refining process.

milliequivalents per liter (meq/l)　An expression of the concentration of a material dissolved in water; the expression is calculated by dividing the concentration, in milligrams per liter, by the EQUIVALENT WEIGHT of the dissolved material. For example, the equivalent weight of aluminum is 9.0. A water concentration of aluminum of 1.8 milligrams per liter equals an aluminum concentration of 0.2 milliequivalent per liter.

milligram (mg)   A unit of mass equal to one-thousandth of a GRAM.

milligrams per liter (mg/l)   An expression of water concentration of a dissolved material; one milligram per liter is equal to one PART PER MILLION.

milliliter (ml)   A unit of volume, equal to one cubic centimeter (cc or cm^3). One thousand milliliters equal one liter.

Millipore filter   A thin membrane of modified cellulose that is used as a filter in the bacteriological examination of water or wastewater. The filter is typically used to filter a given quantity of aqueous sample followed by transfer of the filter to the surface of a special medium to allow for the growth of the bacteria that have been retained by the filter. The filters are also used to filter-sterilize aqueous solutions. The only significant commercial source of the filters was the Millipore Corporation for many years. Although the filters are currently available from a variety of sources, workers in the microbiology area refer to all such filters as Millipore filters.

millirem (mrem)   A unit of IONIZING RADIATION dose equal to one-thousandth of a REM.

Minamata syndrome   Insidious neurological disorders resulting from the consumption of fish and shellfish contaminated by ORGANOMERCURIALS. The most famous case involved a substantial number of people near Minamata Bay, Japan. Fish and shellfish in the bay accumulated ALKYL MERCURY discharged into the bay by an industrial facility. (Although inorganic mercury can be converted to organic mercury in sediments, the Minamata Bay contamination resulted from the direct discharge of alkyl mercury compounds.) Consumption of the contaminated seafood resulted in several hundred poisonings during the 1950s

through the 1970s, ranging from mild neurological disorders to paralysis and deaths. See METHYL MERCURY.

Mineral Lands Leasing Act   The 1920 federal statute authorizing the Secretary of the Interior to issue leases for the extraction of coal, oil, natural gas, phosphate, sulfur, and other minerals from public lands. Later statutes required that environmental impact of mining on public lands be considered before a lease is issued or that specific controls be applied. These laws include the Mining and Mineral Policy Act of 1970 and the SURFACE MINING CONTROL AND RECLAMATION ACT of 1977.

mineralization   The conversion of an organic material to an inorganic form by microbial decomposition.

minimum moisture content   The amount of water in soil during the driest time of the year.

minimum tillage farming   A farming technique that reduces the degree of soil disruption. Crop residues are not plowed under after harvest, and special planters dig narrow furrows in the crop residue when new seeds are sown. Advantages of the technique include reductions in energy consumption by farm equipment, less soil erosion, and lower soil moisture losses during the fallow season. Disadvantages include the possibility of encouraging insect pests by leaving the crop residue in the field and the use of herbicides to control weeds in the place of mechanical cultivation.

minute volume   The amount of air moving through the lungs each minute as determined by the product of the breathing rate and the TIDAL VOLUME.

mist   Liquid AEROSOL, small droplets suspended in air.

mist eliminator   A device placed downstream from a SCRUBBER to remove particles that were introduced to the air stream by the turbulent gas-liquid contact in the scrubber. Also called an entrainment separator.

mitigation   Actions taken to lessen the actual or foreseen adverse environmental impact of a project or activity.

mixed funding agreement   Under the COMPREHENSIVE ENVIRONMENTAL RESPONSE, COMPENSATION, AND LIABILITY ACT, an agreement to clean up a waste site that involves payments from both the HAZARDOUS SUBSTANCES SUPERFUND and from companies that are held responsible for the hazardous waste at the site.

mixed liquor suspended solids (MLSS)   A measurement of solid material, mainly organic compounds and ACTIVATED SLUDGE, in the aeration tanks of a wastewater treatment plant.

mixed liquor volatile suspended solids (MLVSS)   That portion of MIXED LIQUOR SUSPENDED SOLIDS that will vaporize when heated to 600° C; this volatile fraction is mainly organic material and thus indicates the biomass present in the aeration tank. The material that does not vaporize in this test, mostly inorganic substances, is said to be fixed.

mixing height   The altitude, at a particular time, below which atmospheric dilution of pollutants can occur; the value is determined by the ENVIRONMENTAL LAPSE RATE present that day (or time of day). A low mixing height for an extended period will allow air contaminant concentrations to increase, possibly to unhealthy levels. See also TEMPERATURE INVERSION.

mixing ratio   The concentration of water vapor in the atmosphere, commonly expressed as grams of water vapor per kilogram of dry air. See also SATURATION MIXING RATIO and RELATIVE HUMIDITY.

mixture   Two or more elements or compounds present in various proportions. A mixture can be separated into its constituent parts by physical or mechanical means, and the materials composing the mixture retain their individual physical and chemical properties. Compare COMPOUND.

mixture rule   The United States Environmental Protection Agency regulatory provision stating that (1) with certain exceptions, any mixture of LISTED HAZARDOUS WASTE with nonhazardous solid waste is a HAZARDOUS WASTE, and (2) any mixture of CHARACTERISTIC HAZARDOUS WASTE with nonhazardous solid waste must be tested to determine if it is a hazardous waste.

MMCF   Million cubic feet.

MMCFD   Million cubic feet per day; used to express natural gas production, transport, or combustion rates.

mobile sources   Moving emitters of air pollutants such as automobiles, trucks, boats, and airplanes. See STATIONARY SOURCE, AREA SOURCE.

mobilization   The introduction to the air or water of a chemical material that formerly was not circulating in the environment. For example, certain metals can be chemically bound to soil or clay until they are dissolved into and then move with acidic groundwater. The carbon in fossil fuel resources can be said to have been mobilized dramatically in the last century, with most of it emitted to the air as carbon dioxide, following combustion of the hydrocarbon fuel.

model   1. A simplified representation of an object or natural phenomenon. The model can be in many possible forms: a set of equations or a physical, miniature version of an object or system constructed to allow estimates of the behavior of the actual object or phenomenon when the values of certain variables are changed. Important environmental models include those estimating the transport, dispersion, and fate of chemicals in the environment. 2. The action of using a model, for example, to model the atmospheric dispersion of the emissions from a smokestack.

moderator   A material employed in a nuclear reactor to decrease the energy level of fast, high-energy neutrons released in fission reactions and to increase thereby the possibility of a sustained fission reaction. When slow, low-energy neutrons collide with nuclei of the fissionable fuels used in nuclear reactors (e.g., uranium-235), the neutrons are captured by those nuclei, causing the production of an inherently unstable atom that undergoes fission. When fission takes place, fragments of the uranium are produced (fission products) along with a large amount of heat. The fission releases or ejects fast, high-energy neutrons. These neutrons must be slowed in order to increase the chance that they will collide with another atom of fissionable material to create a sustained nuclear chain reaction. The slowing of the neutrons is the function of a moderator. Regular (or light) water, heavy water, and graphite are the most common moderators in nuclear reactors.

modular incinerator   A prefabricated MASS BURN incinerator unit that is transported to a municipal waste incineration site. Each modular unit will typically have a capacity of 5 to 120 tons per day.

molar absorptivity   For a given chemical substance, the proportionality constant used in the calculation of absorbance for a given wavelength of light by a solution of the chemical. The BEER-LAMBERT LAW defines absorbance of a chemical solution as the product of the molar absorptivity of the absorbing chemical, the length of the light path, and the aqueous concentration of the chemical. The units for molar absorptivity are typically per molar concentration per centimeter, where molar concentration is the concentration of the chemical in water and centimeters indicates the length of the light path.

molar concentration   The number of MOLES of a chemical substance per unit volume of a medium, for example 0.2 mole (8 grams) of sodium hydroxide per liter of water.

molarity   The number of MOLES of a dissolved chemical substance per liter of solution.

mole   The SI unit for the amount of a substance that contains AVOGADRO'S NUMBER (6.02×10^{23}) of atoms or molecules. Frequently expressed as a grammole, equal to the molecular mass of a substance in grams. For example, the molecular mass of water (H_2O) is 18. Thus, one gram-mole of water equals 18 grams and contains 6.02×10^{23} molecules.

mole fraction   An expression of the concentration of a chemical substance in a solution (water) or mixture (air). Calculated by dividing the number of MOLES of the chemical substance by the total number of moles of the various substances making up the mixture or solution.

molecular sieve   A crystalline aluminosilicate material with uniform pore spaces that can be used to separate molecules by size. Technical applications include drying gases, ion exchange, catalysis, and gas chromatography.

molecular weight   The sum of the ATOMIC WEIGHTS of the ATOMS in a molecule. For example, the molecular weight of water (H_2O) is 18, the sum of the atomic weights of two hydrogen atoms ($1 + 1 = 2$) and oxygen (16).

molecule   A group of atoms held together by chemical bonds. They may be either atoms of a single element (O_2) or atoms of different elements that form a compound (H_2O). The smallest unit of a compound that retains the chemical properties of that compound.

monitoring   Sampling and analysis of air, water, soil, wildlife, and so on, to determine the concentration(s) of contaminant(s).

monitoring well   A well drilled in close proximity to a waste storage or disposal facility to check the integrity of the facility or to keep track of leakage of materials into the adjacent groundwater.

monoculture   The growing of a single plant species over a large area.

monofill   A solid waste disposal facility containing only one type or class of waste.

monolith   A solid mass of waste-containing material that has undergone solidification. See SOLIDIFICATION.

monomer   A compound that under certain conditions will join to other compounds of the same type to form a molecular chain called a polymer. For example, vinyl chloride monomers can polymerize to polyvinyl chloride (PVC).

montane   A forest ecosystem in mountainous areas of the tropics. The montane forest has far fewer plant species than does the TROPICAL RAIN FOREST, which is found at lower elevations below the mountains.

Monte Carlo method   A method that produces a statistical estimate of a quantity by taking many random samples from an assumed probability distribution, such as a normal distribution. The method is typically used when experimentation is infeasible or when the actual input values are difficult or impossible to obtain.

Montreal Protocol   A 1987 international agreement that establishes a schedule for reduced production of CHLOROFLUOROCARBONS and HALONS in participating countries. The full name is the Montreal Protocol on Substances that Deplete the Ozone Layer. See OZONE LAYER DEPLETION.

morbidity   Statistics related to illness and disease.

morbidity rate   See INCIDENCE.

mortality rate   The number of deaths in a given area during a specified time period, usually one year, divided by the number of persons in the area, multiplied by a constant, typically 1000. For example, if 200 deaths occurred in 1989 in Jonesville and if the population of Jonesville was 25,000 the 1989 mortality rate was 8 per 1000.

most exposed individual (MEI)   In a risk assessment of the off-site impact of a pollutant released by a facility, a hypothetical person receiving the highest dose of the pollutant. For example, in an analysis of air toxics risk, a person assumed to spend all of his or her time outside at the location predicted to have the maximum concentration of the modeled pollutant emissions. After the MEI is identified, the health risk attributable to the exposure is estimated. See MAXIMUM INDIVIDUAL RISK.

most probable number (MPN)   A statistical estimate of the levels of COLIFORM bacteria in a water sample, expressed as the number of coliforms per 100 milliliters. See MULTIPLE-TUBE FERMENTATION TEST.

mottling, of teeth   Discoloration of the teeth, which can be caused by extended intake of excessive fluoride. The condition is seen in certain populations using groundwater containing naturally occurring high fluoride levels.

mucociliary escalator   The mechanism that sweeps particles from the air-conducting tubes (bronchi and the smaller bronchioles) in the lungs. The bronchi and bronchioles are lined with hairlike projections called cilia; the cilia move in unison to force a constantly supplied sheet of mucus upward from just above the ALVEOLAR REGION toward the pharynx (throat). Trapped particles move upward with the mucus.

mucosa A mucous membrane lining those parts of the body communicating with the exterior. The mucosa in the gastrointestinal and respiratory tracts can therefore be irritated by certain ingested or inhaled environmental pollutants.

muffle furnace A device used to determine the organic content of a soil sample. A preweighed soil sample is heated in the furnace at temperatures sufficient to vaporize the HUMUS and the mass lost (the organic content) is calculated by subtracting the sample weight after the furnace treatment from the pretreatment weight. The device is also used to determine that fraction of PARTICULATE ORGANIC MATTER that can be volatilized by heating to high temperatures.

multiclone A set of individual CYCLONES connected in parallel to control air emissions of particulate matter.

multiple-tube fermentation test A method used to estimate the number of specific types of bacteria, such as COLIFORMS, in a water sample. Three different quantities of a sample, for example 10 ml, 1 ml, and 0.1 ml, are placed into three sets of five tubes each containing lactose broth. The 15 tubes are allowed to incubate at 37.5° C for 24 hours, and the number of tubes in each set of five tubes showing positive results (gas production) is determined. Most probable number tables are then consulted to provide a statistical estimate of the number of coliform bacteria in the water sample. The population levels of different types of bacteria can be determined by altering the media used and the incubation conditions. See STANDARD METHODS.

multiple use The policy of allowing public land to be used for varied purposes, such as timber production, camping and hiking, animal grazing, mineral extraction, and/or wildlife preservation. The national forests and the federally owned rangelands in the western United States and Alaska are managed under the multiple-use principle. The U.S. Forest Services manages the national forests (the 15% of these lands that are WILDERNESS AREAS do not have multiple-use management), and the BUREAU OF LAND MANAGEMENT administers the rangelands.

multistage cancer risk model The most frequently used model of carcinogenesis; the model assumes that a cell or cells pass(es) through two or more stages before becoming a detectable tumor. See CARCINOGENESIS.

municipal solid waste (MSW) Solid waste, including GARBAGE and TRASH, that originates in households, commercial establishments, or construction/demolition sites. Nonhazardous sludge from municipal sewage treatment plants or nonhazardous industrial waste can also be placed in this category of solid waste.

muon An elementary, subatomic particle with either a positive or negative charge and a mass 207 times the mass of an electron.

muskeg Large boggy areas found in Canada and Alaska, part of the North American boreal forest biome.

mustard gas A group of gases related to 2,2-dichlorodiethyl sulfide. The gas was used extensively during World War I as a vesicant (blistering gas).

mutagen Any agent that has the capability of causing a permanent change in the genes of a cell. See MUTATION.

mutagenicity The ability of an agent to cause permanent changes in the genetic material of a cell. See MUTATION.

mutation A significant change in the genetic material of a cell. These changes can be reflected in the physical or biochemical properties of the cell and can be transferred to offspring of that cell. Most mutations are deleterious but persist in a population because they are not expressed (are recessive) and thus carried in the GENOTYPE without affecting its appearance or viability. Excessive environmental exposures to a variety of agents, including x-rays, ultraviolet radiation, and an array of chemical compounds, can cause mutations.

mutualism   An interaction between two or more distinct biological species in which the members benefit from the association. Mutualism describes both symbiotic mutualism (a relationship requiring an intimate association between species in which none can carry out the same functions alone) and nonsymbiotic mutualism (a relationship between organisms that is of benefit but which is not obligatory, i.e., the organisms are capable of independent existence). Compare NEUTRAL-ISM.

mycorrhizal   An association between plant roots and fungi. There is an integration between the roots and the fungal mycelium (filaments) to produce a distinct morphological unit. The fungal mycelium can either form an external layer on the root or actually invade living cells of the root structure.

mycotoxins   Natural toxic materials produced by molds. Aflatoxin, a natural carcinogen, is a mycotoxin commonly found in rice, peanuts, wheat, and corn.

N

nanometer (nm)   An SI unit of length equal to 10^{-9} meter.

nappe   The stream of water flowing over a dam or weir; from the French for sheet.

National Air Monitoring System (NAMS)   A national network of air monitoring stations designed by the United States Environmental Protection Agency and individual state environmental agencies to assess the ambient air quality in major urbanized areas and used by the EPA to track long-term air quality trends. About 1500 stations monitor for PARTICULATE MATTER, about 300 for SULFUR DIOXIDE, about 200 for NITROGEN DIOXIDE, about 200 for CARBON MONOXIDE, about 200 for OZONE, and about 150 for LEAD. The NAMS stations are part of the STATE AND LOCAL AIR MONITORING SYSTEM.

National Air Toxics Information Clearinghouse (NATICH)   A computerized database containing air toxics information from federal, state, and local agencies; the database includes ambient monitoring data, source testing results, EMISSION INVENTORIES, acceptable ambient concentrations, agency contacts, permit limits, and selected United States Environmental Protection Agency (USEPA) air toxics risk assessments. Operated by the USEPA Office of Air Quality Planning and Standards.

national ambient air quality standards (NAAQS)   The ambient air concentration standards set for particulate matter, sulfur dioxide, nitrogen dioxide, ozone, carbon monoxide, and lead to protect human health (primary standards) or welfare (secondary standards).

National Audubon Society   A large American environmental interest group that emphasizes natural resource and wildlife conservation. Named in honor of John James Audubon (1785–1851), who was one of the first American conservationists and who gained recognition for his paintings of birds. Founded in 1905, by 1989 it had about 550,000 members.

National Cancer Institute (NCI)   A federal agency under the National Institutes of Health, United States Department of Health and Human Services that conducts and supports research on cancer and the identification of carcinogens. Based in Bethesda, Maryland, the NCI is a participant in the NATIONAL TOXICOLOGY PROGRAM.

National Contingency Plan (NCP)
The outline of procedures, organization, and responsibility for responding to spills and releases of hazardous substances and oil into the environment. Prepared by the United States Environmental Protection Agency as required by sections of the COMPREHENSIVE ENVIRONMENTAL RESPONSE, COMPENSATION, AND LIABILITY ACT and the CLEAN WATER ACT. The plan applies to sudden, accidental releases and to nonsudden, gradual leaks. SUPERFUND SITE cleanups are performed in accordance with the NCP. Officially called the National Oil and Hazardous Substances Pollution Contingency Plan.

National Council on Radiation Protection and Measurements (NCRPM)
A private group of scientists that recommends safe occupational and public exposure levels to IONIZING RADIATION. Based in Bethesda, Maryland. Formerly the National Committee on Radiation Protection.

National Emission Standards for Hazardous Air Pollutants (NESHAP)
National technology-based limits set by the United States Environmental Protection Agency for air emissions of pollutants determined by the agency to pose a significant risk of death or serious illness upon long-term exposure. NESHAP limits have been set for asbestos, benzene, beryllium, inorganic arsenic, mercury, radionuclides, and vinyl chloride. Coke oven emissions have been designated as hazardous, but an emission standard has not been set. Over 25 other pollutants are under consideration for a NESHAP.

National Environmental Policy Act (NEPA)
A 1969 statute that requires all federal agencies to incorporate environmental considerations into their decision-making processes. The act requires an ENVIRONMENTAL IMPACT STATEMENT for any "major Federal action significantly affecting the quality of the human environment." See COUNCIL ON ENVIRONMENTAL QUALITY.

National Fire Protection Association (NFPA)
An international organization with voluntary membership whose functions are to promote and improve fire prevention and to establish safeguards against the loss of life or property by fire. The organization has produced the National Fire Code, which lists standards for recommended practices and materials handling. Based in Quincy, Massachusetts.

National Institute for Environmental Health Sciences (NIEHS)
A government research organization attached to the National Institutes of Health, United States Department of Health and Human Services, located in Research Triangle Park, North Carolina. The purpose of the NIEHS is to conduct or sponsor research on the adverse effects of environmental agents on human health. The research results are used by federal environmental regulatory agencies in their prevention and control programs.

National Institute for Occupational Safety and Health (NIOSH)
An agency of the United States Public Health Service that recommends occupational exposure limits for chemical and physical agents and certifies respiratory and air-sampling devices. Based in Cincinnati, Ohio, and Washington, D.C.

National Oceanic and Atmospheric Administration (NOAA)
A federal agency within the Department of Commerce responsible for mapping and charting the oceans, environmental data collection, monitoring and prediction of conditions in the atmosphere and oceans, and management and conservation of marine resources and habitats. Based in Suitland, Maryland, and Rockville, Maryland.

National Oil and Hazardous Substances Pollution Contingency Plan

(NCP) The full name of the NATIONAL CONTINGENCY PLAN.

National Pollutant Discharge Elimination System (NPDES) The program established by the Clean Water Act that requires all POINT SOURCES discharging into any "waters of the United States" to obtain a permit issued by the United States Environmental Protection Agency or a state agency authorized by the federal agency. The NPDES permit lists permissible discharge(s) and/or the level of cleanup technology required for wastewater.

National Primary Drinking Water Regulations (NPDWR) Regulations for public drinking water supply systems that include health-based standards for various contaminants, and monitoring and analysis requirements. Issued by the United States Environmental Protection Agency under authority of the SAFE DRINKING WATER ACT. See MAXIMUM CONTAMINANT LEVEL and MAXIMUM CONTAMINANT LEVEL GOAL. Compare NATIONAL SECONDARY DRINKING WATER REGULATIONS.

National Priorities List (NPL) A list of the hazardous waste disposal sites most in need of cleanup; the list is updated annually by the United States Environmental Protection Agency, based primarily on how a site scores using the HAZARD RANKING SYSTEM. Also called the Superfund list.

National Research Council (NRC) A group of volunteer professionals supported by the National Academy of Sciences, National Academy of Engineering, and the Institute of Medicine that, working through study committees, conducts independent research for the United States government on public policy issues in the areas of science and technology. Based in Washington, D.C.

National Response Center (NRC) The United States Coast Guard unit that receives reports of hazardous chemical spills and is responsible for notifying other agencies which will help plan, coordinate, and respond to the release. Based in Washington, D.C. The 24-hour telephone number of the center is 800-424-8802. See also SPILL CLEANUP INVENTORY.

National Response Team (NRT) An organization of the federal government under the leadership of the Environmental Protection Agency that includes representatives of 10 other federal agencies. The team serves as an umbrella organization at the federal level, and its functions include, among others, evaluating methods to respond to discharges or releases; recommending needed changes in the response organization; making recommendations relative to the training, equipping, and protection of response teams; evaluating response capabilities, reviewing regional responses to discharges; and coordinating the activities of federal, state, and local governments as well as private organizations in response to discharges.

National Science Foundation (NSF) An independent United States government agency based in Washington, D.C., that supports basic and applied research in science and engineering.

National Secondary Drinking Water Regulations (NSDWR) Regulations governing the operation of public water supply systems under the Safe Drinking Water Act. The regulations define secondary maximum contaminant levels, the maximum concentrations of certain substances in drinking water that affect its aesthetic quality. The NATIONAL PRIMARY DRINKING WATER REGULATIONS set standards protective of the public health.

National Stream Quality Accounting Network (NASQAN) A data system operated by the United States Geological Survey that compiles measurements of water pollutant concentrations taken at the downstream ends of all major water basins in the United States.

National Strike Force (NSF) An organization under the leadership of the United States Coast Guard that responds to spills of oil or hazardous substances. The Force operates through various teams organized in different regions of the country. They provide, among other services, communication support, advice, and assistance in the event of discharges; shipboard damage control; containment and removal of discharges; and diving activities related to damage assessment and surveys.

National Technical Information Service (NTIS) An agency that sells reports from government-funded studies to the public. Located in Springfield, Virginia.

National Toxicology Program (NTP) An organization within the United States Department of Health and Human Services (DHHS) charged with coordinating toxicology research. The DHHS agencies involved are the Food and Drug Administration, the Occupational Safety and Health Administration, the National Cancer Institute, the National Institute for Occupational Safety and Health, and the National Institute for Environmental Health Sciences. The NTP executive committee also includes the heads of the United States Environmental Protection Agency, the Consumer Product Safety Commission, and two non-DHHS agencies that conduct toxicological research. The NTP is similar in purpose to the INTERAGENCY REGULATORY LIAISON GROUP.

natural draft A gas flow created by the difference in pressure between hot gases and the atmosphere, such as the draft operating in a fireplace chimney, incinerator stack, or in a natural-draft tower. See also FORCED DRAFT, INDUCED DRAFT.

natural-draft tower A cooling tower that is designed to remove waste heat from a heated effluent. The air that receives the heat from the water rises and exits the tower by convection currents without the aid of blowers or fans.

natural increase A positive change in the number of individuals in a community as a result of a larger number of births than deaths.

natural logarithm (ln) The value of the exponent that the base, e must have to equal a given number. It is calculated as $e^x = y$, where x is the logarithm. For example, the natural logarithm of 5 is the power *(x)* to which e (approximately 2.718282) must be raised to equal 5, or $e^x = 5$; therefore x is about 1.60944.

naturally occurring radioactive material (NORM) Radioactive material in fluids brought to the surface during the production of oil or gas, usually in the PRODUCED WATER. The material can contaminate drilling pipe by forming radioactive scale, or, if discharged into surface waters, can potentially accumulate in an aquatic ecosystem.

natural radioactivity Ionizing radiation from sources that are not related to human activities, for example cosmic rays and radiation emitted by RADIOISOTOPES found naturally in the earth's crust.

Natural Resources Defense Council (NRDC) A private American environmental organization emphasizing the proper management of natural resources. A participant in numerous precedent-setting lawsuits concerning national environmental policies. For example, NRDC suits resulted in the listing of lead as a CRITERIA POLLUTANT (*NRDC v. Train,* 1976) and the adoption of BEST AVAILABLE TECHNOLOGY effluent standards for toxic water pollutants (*NRDC v. Train,* 1976; see FLANNERY DECREE). In 1989, membership numbered 95,000.

natural selection A natural process by which certain members of a POPULATION that are well-adapted to prevailing environmental conditions survive and reproduce at greater rates than those organisms not suited to that particular environment. Expressed by Charles Darwin, the process is often referred to as the survival of the fittest (i.e., the genetic makeup that best fits the environment is most successfully passed to offspring) and is considered to be the selection pressure that drives EVOLUTION.

natural sink   1. A habitat that serves to trap or immobilize chemicals such as plant nutrients, organic pollutants, or metal ions through natural processes. For example, a river that enters a swamp may carry a substantial amount of dissolved plant nutrients. The natural biological activity of the swamp may remove these nutrients to such an extent that the water exiting the swamp is relatively low in nutrient concentrations. The swamp has then served as a sink to trap the nutrients that are no longer available for subsequent plant growth downstream from the swamp. 2. A natural process whereby pollutants are removed from the atmosphere. A sink process can be physical (particulates removed by rain), chemical (the reaction of ozone with nitric oxide to form nitrogen dioxide and oxygen), or biological (the uptake of airborne hydrocarbons by soil microorganisms). Also called a scavenging mechanism.

navigable waters   Water to which the CLEAN WATER ACT applies; such waters include "the waters of the United States," which is any body of water with any connection to interstate waters or commerce and this includes almost all surface water and wetlands; there is no requirement, despite the name, for vessels to be able to navigate these waters.

near field   The area very close to a noise source in which the sound pressure level does not drop with the inverse square of the distance from the source (INVERSE SQUARE LAW). Compare FAR FIELD.

necrosis   Localized death of body tissue that results in the development of a lesion characterized by inflammation and pus accumulation. A festering sore.

negligible residue   An amount of pesticide remaining in or on raw agricultural commodities that would result in a daily intake of the agent regarded as toxicologically insignificant.

nekton   Animals in aquatic systems that are free-swimming, independent of currents or waves. Compare BENTHOS, PLANKTON.

nematode   Roundworms One of the most common kinds of animals, with about 10,000 species known. Members of this group live in almost all known habitats, ranging from polar regions to the tropics and from soil to the deep ocean. Nematodes are parasitic to man, livestock and plants. Some common human pathogens are hookworm, pinworm, intestinal roundworm and whipworm.

neoplasm   The growth of new or abnormal tissue that has no prescribed physiological function. A tumor. See BENIGN NEOPLASM and MALIGNANT NEOPLASM.

nephelometer   A device that measures the scattering of light by particles (or bacteria) suspended in air or water, compared to a reference suspension. It consists of a light source aimed at a sample cell and a detector placed at right angles to the light path through the sample. The light scattered at right angles is measured by a nephelometer, in NEPHELOMETRIC TURBIDITY UNITS, while the decrease in light transmitted *directly* through the sample cell is a measure of TURBIDITY, in JACKSON TURBIDITY UNITS.

nephelometric turbidity unit (NTU) A unit used to express the cloudiness (turbidity) of water as measured by a NEPHELOMETER. Nephelometric turbidity units are approximately equal to JACKSON TURBIDITY UNITS.

neritic   Of the shallow regions of a lake or ocean that border the land. The term is also used to identify the biota that inhabit the water along the shore of a lake or ocean.

Nessler reagent   An aqueous solution of mercury and potassium salts of iodine and sodium hydroxide that is used to test for the presence of ammonia in water or reaction mixtures.

net community productivity (NCP) The gain of biomass within a defined region over time. The total amount of carbon dioxide fixed by the photosynthetic plants within the area (PRIMARY PRODUCTION) minus that amount of carbon dioxide lost through metabolism at all trophic levels within the same region.

net precipitation   The potential for leachate generation from a waste disposal site. It is computed for a specific location by subtracting the annual evaporation from lakes in the region from the normal annual rainfall.

net primary productivity (NPP) The number of grams of carbon dioxide fixed by photosynthesis of plants per unit area or volume of water minus the number of grams of carbon dioxide produced during the respiration of those plants. See also PRIMARY PRODUCTIVITY.

net reproductive rate (R_0)   The relative ability of a population of animals to increase in one generation. If R_0 is 3, the population has the capacity to triple in number during one generation.

netting out   The exemption of certain facility modifications from a more-detailed air pollution permit process (a NEW SOURCE REVIEW) if the emissions of a particular pollutant from the proposed modification and emission reductions of the same air pollutant within the same source result in no net increase in facility emissions for that pollutant.

neurotoxin   A substance that can damage or destroy nerve tissue.

neuston   Small particles or microorganisms found in the surface film that covers still bodies of water.

neutralism   The absence of interactions between two species. Compare MUTUALISM.

neutrino   A very small subatomic particle released from unstable atoms that emit beta particles as they undergo nuclear decay. The particle does not carry a charge, as do the protons and electrons of an atom.

neutron   One of the elementary particles in the nucleus of all atoms except hydrogen. A neutron does not have a charge, and the atomic mass is approximately 1.

new source performance standards (NSPS)   1. Pollutant-specific national uniform air emission standards for new or modified stationary sources, set by the United States Environmental Protection Agency by facility type, based on available emission control technology. 2. Effluent limitations set by the USEPA for new POINT SOURCES of water pollution. The standards are applied to INDUSTRY CATEGORIES, such as petroleum refineries and phosphate manufacturers.

new source review (NSR)   The procedural steps defined by an environmental regulatory agency for the issuance of a permit for a new facility (or major modification of an existing facility) that will emit significant quantities of air pollutants. The review includes specification or approval of air pollution control devices or methods and air quality dispersion modeling of the estimated emissions of a facility to assess their impact.

newton (N)   The SI unit of force equaling a mass of one kilogram accelerated at one meter per second per second. Expressed as $1\ N = 1\ kg\ m\ s^{-2}$.

niche   In ECOLOGY, a term that includes both the HABITAT and role (functional status) of an organism within an ecosystem.

night soil   Solid human excrement.

nitrate   A chemical compound having the formula NO_3^-. Nitrate salts are used as fertilizers to supply a nitrogen source for plant growth. Nitrate addition to surface waters can lead to excessive growth of aquatic plants. High groundwater nitrate levels can cause METHEMOGLOBINEMIA in infants. See NITRITE.

nitric acid   A strong mineral acid having the formula HNO_3. This acid is one of the constituents of ACID RAIN.

nitric oxide (NO)   The gas formed by heating air to high temperatures (THERMAL NO_x) or by the oxidation of organic nitrogen contaminants in a fuel during combustion (FUEL NO_x).

nitrification   The oxidation of ammonia to nitrate by bacteria in soil or water.

nitrite   An oxidized nitrogen molecule with the chemical formula NO_2^-. Nitrite can be formed from NITRATE (NO_3^-) by microbial action in soil, water, or the human digestive tract. Excessive nitrate levels in rural well water, caused by fertilizer application, has led to cases of METHEMOGLOBINEMIA, typically in infants. Sodium nitrite preservative added to bacon, lunch meats, hot dogs, ham, and other foods can react with dietary amines (compounds found in cereals, fish, cheese, beer, and others) to form carcinogenic NITROSAMINES. However, relative to nitrites formed by normal body metabolism and from dietary intake of natural nitrates (in many vegetables), the health risk of nitrite preservatives is small.

nitrogen cycle   A model illustrating the conversion of nitrogen from one form to another through a combination of biological, geological, and chemical processes. The process is continuous, with N_2 in the atmosphere being converted to forms usable by biota and then ultimately returning to the atmosphere as N_2.

nitrogen dioxide (NO_2)   A brownish colored gas that is a major ingredient in PHOTOCHEMICAL SMOG. It is readily produced in the atmosphere from nitric oxide by the addition of an oxygen atom ($NO + O \rightarrow NO_2$). Nitrogen dioxide can be converted by atmospheric reactions to peroxyacetyl nitrate or to nitric acid, an ingredient in ACID DEPOSITION. See NITROGEN OXIDES.

nitrogen fixation   The conversion of nitrogen in the atmosphere (N_2) to a reduced form (e.g., ammonia and amino groups of amino acids) that can be used as a nitrogen source by organisms. The process is important since all organisms require a source of nitrogen for nutrition, and N_2 cannot be used by the great majority of the biota to satisfy that need. Biological nitrogen fixation is carried out by a variety of organisms; however, those responsible for most of the fixation are certain species of bluegreen algae, the soil bacterium *Azotobacter,* and the symbiotic association of plants of the legume variety and the bacterium *Rhizobium.* In industry, the HABER PROCESS is used to fix atmospheric nitrogen for use as fertilizer. See also SYMBIOSIS.

nitrogen oxides (NO_x)   Gases containing nitrogen and oxygen that include NO, NO_2, NO_3, N_2O, N_2O_3, N_2O_4, and N_2O_5. The first two, nitric oxide (NO), and nitrogen dioxide (NO_2), are the primary NO_x air pollutants. About 60 percent of United States NO and NO_2 emissions are from stationary sources (smokestacks) and most of the remaining 40 percent is from transportation exhaust, mainly automobiles.

nitrogenous BOD   The amount of molecular oxygen required for the microbial oxidation of ammonia and nitrite contaminants in a specified volume of wastewater. This type of oxygen demand can complicate the interpretation of data obtained from the determination of the

BIOCHEMICAL OXYGEN DEMAND (BOD) of treated sewage, although a chemical can be added to the BOD test to prevent ammonia oxidation. Ammonia and nitrite are oxidized by chemoautotrophic bacteria. See CHEMOAUTOTROPHS.

nitrogenous waste   Wastewater that contains inorganic forms of nitrogen, including ammonia and nitrite.

nitrosamines   A large and diverse family of synthetic and naturally occurring compounds having the general formula $(R)(R')N—N{=}O$. Almost all are CARCINOGENIC, with biochemical activation to a cancer-causing intermediate taking place in tissue fluids. Different members of the family of compounds and different concentrations of the same nitrosamine result in cancer development in different tissues: liver, lungs, esophagus, kidney, pancreas, among others. Several nitrosamines produced from nicotine by bacterial activity during the tobacco curing process are responsible, in part, for the cancer-causing properties of smoking and smokeless tobacco products.

noise   Unwanted sound. Noise differs from most environmental insults in that, if the source is stopped, noise disappears immediately and has no residual accumulation. It also has important subjective aspects. No direct measurements can quantify the loudness of a noise or its disturbance quality. The Occupational Safety and Health Administration workplace standard for an 8-hour exposure is a limit of 90 DECIBELS, A-WEIGHTING NETWORK (dBA). There are no enforceable federal community noise standards, but the Federal Aviation Administration and the Department of Urban Development have established a guideline, which is a DAY-NIGHT SOUND LEVEL of 65 dBA. See NOISE-INDUCED HEARING loss, entries that start with the word SOUND.

noise-induced hearing loss (NIHL)   A PERMANENT THRESHOLD SHIFT attribut-

able to excessive noise exposure; hearing loss beyond the normal decline with age.

noise reduction coefficient   The arithmetic average of the SABIN ABSORPTION COEFFICIENTS for a certain material at the frequencies of 250, 500, 1000, and 2000 hertz, rounded to the nearest 0.05. The coefficient is used for comparing the sound absorption characteristics of materials.

nominal variable   A sample characteristic expressed as a class or descriptive category, such as marital status, ethnic group, socioeconomic class, and religious preference. These are qualitative, not quantitative, data, and only NONPARAMETRIC TESTS are appropriate for data collected as nominal variables. See also ORDINAL VARIABLE, PARAMETRIC TESTS.

nomograph   A graphical solution to a multivariable equation. Parallel vertical scales, one for each variable, are arranged such that a straight line across the scales produces values for each variable that together solve the equation.

nonattainment area (NAA)   A geographical area that does not meet a NATIONAL AMBIENT AIR QUALITY STANDARD for a particular pollutant. The extent of the area is defined by air quality monitoring data, air dispersion modeling, and/or the judgment of the state environmental agency and the United States Environmental Protection Agency. The CLEAN AIR ACT requires STATE IMPLEMENTATION PLANS to contain the steps necessary to become an ATTAINMENT AREA.

nonbinding preliminary allocation of responsibility (NBAR)   An allocation by the United States Environmental Protection Agency of percentages of total cleanup costs for a hazardous waste site to POTENTIALLY RESPONSIBLE PARTIES, under the authority of the COMPREHENSIVE ENVIRONMENTAL RESPONSE, COMPENSATION, AND LIABILITY ACT. The system is used to encourage settlements.

nonbiodegradable Describing organic compounds, usually synthetic, that are not decomposed or mineralized by microorganisms.

nonconventional pollutants Under the CLEAN WATER ACT, water pollutants not listed as CONVENTIONAL POLLUTANTS, toxic pollutants, or thermal discharges. These include chloride, iron, ammonia, color, and total phenols.

nondestructive testing (NDT) In geophysical surveying, methods used to detect subsurface water, subsurface containers, or the areal extent of groundwater contamination without soil borings. The testing involves the use of acoustic sounding, infrared radiation, x-rays, magnetic field perturbation, and electrical resistivity, among other methods.

nondiscretionary (duty) See CITIZEN SUIT PROVISION.

nondispersive infrared analysis (NDIR) An analytical method that uses the molecular absorption of infrared radiation to measure the concentration of certain chemical compounds in the ambient air. A broad (nondispersive) band of infrared radiation is used, in contrast to the particular wavelength tuning applied in methods like ULTRAVIOLET PHOTOMETRY. NDIR is the United States Environmental Protection Agency reference method for ambient measurements of carbon monoxide.

nonfissionable An atom that is not capable of undergoing nuclear fission, or breaking into two or more pieces, when bombarded by neutrons. Commonly used to refer to a specific isotope of some element, other isotopes of which will normally undergo nuclear fission. For example, uranium-235 represents an isotope of uranium that is capable of undergoing fission, whereas uranium-238 is an isotope of uranium that is nonfissionable.

nonhazardous oil field waste (NOW) Wastes generated by drilling of and production from oil and gas wells that are not classified by United States Environmental Protection Agency regulations as hazardous wastes. Typical NOW wastes include drilling muds, cuttings, drilling fluids, and PRODUCED WATER.

nonionizing radiation Electromagnetic radiation that is not sufficiently energetic to produce charged ions when it strikes an object. Examples are ultraviolet light, visible light, and radio waves. See IONIZING RADIATION.

nonmethane hydrocarbons (NMHC) Hydrocarbon compounds present in ambient air that participate in photochemical reactions leading to an accumulation of PHOTOCHEMICAL OXIDANTS, especially OZONE. Methane is excluded because it is relatively nonreactive and does not contribute to the atmospheric reactions forming ozone. Compare NONMETHANE ORGANIC COMPOUNDS.

nonmethane organic compounds (NMOC) Airborne NONMETHANE HYDROCARBONS, plus any oxygenated hydrocarbons, such as aldehydes and ketones, that participate in photochemical reactions producing OZONE or other PHOTOCHEMICAL OXIDANTS.

nonparametric statistics See NONPARAMETRIC TESTS.

nonparametric tests Statistical procedures that yield information about populations but not about population PARAMETERS. This type of test uses data from NOMINAL or ORDINAL VARIABLES and does not require the assumptions about sample population distributions that must be made or acknowledged using PARAMETRIC TESTS.

nonpersistent pollutant A substance that can cause damage to organisms when added in excessive amounts to the environment but is decomposed or degraded by natural biological communities and removed from the environment relatively quickly. Compare PERSISTENT POLLUTANT.

nonpoint source A diffuse, unconfined discharge of water from the land to a receiving body of water. When this water contains materials that potentially can damage the receiving stream, the runoff is considered to be a source of pollutants. Runoff from city streets, parking lots, home lawns, agricultural land, individual septic systems and construction sites that finds its way into lakes and streams constitutes an important source of water pollutants. Estimates of the fraction of water pollution attributable to nonpoint sources range up to 90 percent of the total.

nonpolar solvent A solvent with no positive POLARITY or negative polarity. This type of solvent is a good dissolver of other nonpolar materials. For example, oils dissolve in benzene, both nonpolar materials. Compare POLAR SOLVENT.

nonreactive See INERT.

nonrenewable energy Those sources of energy such as oil or natural gas that are not replaceable after they have been used. In contrast, sources of energy like firewood or hydroelectricity production are renewable.

nonrenewable resources Those natural resources such as coal or mineral ores that are not replaceable following their removal. In contrast, resources like food crops and timber forests can be replenished.

nonthreshold pollutant Any chemical or physical agent for which any human exposure is assumed to produce an increased risk of an adverse effect. Ionizing radiation and chemical compounds that are suspect human carcinogens are considered by United States regulatory policy to be nonthreshold pollutants. See

THRESHOLD DOSE, THRESHOLD EFFECT, THRESHOLD HYPOTHESIS.

no-observed-adverse-effect level (NOAEL) See NO-OBSERVED-EFFECT LEVEL.

no-observed-effect level (NOEL) In a toxicology study, the highest dose at which no adverse effect is observed. Also called the no-observed-adverse-effect level.

normal solution A solution containing one gram EQUIVALENT WEIGHT per liter.

nosocomial Describing a disease or infection contracted in a hospital because of either excessive exposure to infectious organisms or the general compromised condition of the patient.

notice letter 1. A formal notice from the United States Environmental Protection Agency to POTENTIALLY RESPONSIBLE PARTIES that a REMEDIAL INVESTIGATION/FEASIBILITY STUDY or a cleanup action is to be undertaken at a site at which hazardous substances have been released or pose a substantial threat of release, in accordance with the COMPREHENSIVE ENVIRONMENTAL RESPONSE, COMPENSATION, AND LIABILITY ACT. 2. A letter written to an environmental administrator under the CITIZEN SUIT PROVISIONS of various environmental laws, informing the administrator that the plaintiff alleges a failure to perform a nondiscretionary act. The citizen suit cannot begin until 60 days after the notice letter.

notice-and-comment rule making The process used by the United States Environmental Protection Agency to write rules (regulations or standards) that implement federal environmental laws. The rules are proposed in the FEDERAL REGISTER, followed by a public comment period and, often, public hearings at locations across the country. The agency responses to public comments appear in another *Federal Register* and the rules may be revised and reproposed for further comments, or they are issued (promulgated) as final with only one round of comments.

not-in-my-backyard syndrome (NIMBY) An expression of public opinion that, while waste materials must be treated and disposed of, this should be done somewhere *else*.

nuclear fission See FISSION.

nuclear fusion See FUSION.

nuclear reactor A device to promote and control nuclear FISSION for producing heat and then steam to generate electricity or for producing certain RADIOISOTOPES. All reactors have a core containing nuclear fuel, which serves as the energy source, and control rods, which regulate the rate of fission. The fuel is usually a mixture of URANIUM-238 (about97%) and URANIUM-235 and/or PLUTONIUM-239 (about 3%). The fuel is formed into pellets and packed into metal tubes called fuel rods. The core is immersed in coolant (commonly water). The system is energized by removal of the control rods. Heat generated by fission is used to make steam, which is used to turn generators that produce electricity. Some reactors, operated for the United States government, produce plutonium, tritium and other RADIONUCLIDES used in nuclear weapons. Other than their safe operation, one of the primary environmental concerns related to nuclear reactors is the accumulation of waste materials that are very radioactive and remain dangerous for hundreds of years. These radioactive wastes accumulate in the fuel rods as the reactor operates and within the reactor structure itself. Consequently, used fuel rods and reactors that are decommissioned are sources of large amounts of dangerous radioactive wastes.

Nuclear Regulatory Commission (NRC) A five-member U.S. government commission with supporting staff responsible for issuing licenses for the construction and operation of nuclear power plants; the Commission succeeded the Atomic Energy Commission in January 1975.

Nuclear Waste Policy Act (NWPA) A 1982 federal statute that established a schedule to identify a site for and construct an underground repository for civilian high-level radioactive waste. Initially, three sites were recommended: Deaf Smith County, Texas; Hanford Reservation, Washington State; and Yucca Mountain, Nevada. A 1987 amendment to the NWPA provided for further studies at the Yucca Mountain site only.

nuclear winter Predicted consequences of a major war involving the use of a large number of nuclear weapons. The detonation of the weapons and the resulting fires have been predicted to cause a significant accumulation of dust and smoke in the atmosphere, of sufficient concentration and duration to result in a dramatic reduction in the amount of solar radiation reaching the surface of the earth and a resultant catastrophic lowering of surface temperature. These theoretical predictions have been questioned strongly and remain a subject of much debate.

nucleons Common name applied to the particles in the nucleus of an atom. The most common are protons and neutrons.

nucleus, atomic The central, positively charged region of all atoms, made up of one or more nucleons. All nuclei contain protons and neutrons, with the exception of hydrogen, which has only a single proton, and are surrounded by a cloud of electrons. The nucleus contains almost all of the mass of an atom.

nuclide A general term for the various configurations of protons and neutrons in atomic nuclei. Based on the number of protons in the nucleus, over 100 different ELEMENTS have been identified. An element often has several ISOTOPES, with different numbers of neutrons. The elements and their various isotopes number about 1000 nuclides.

null hypothesis   The hypothesis to be tested; often the statement that there is no difference between two populations based on a measured or observed variable. Statistical tests are used to determine if a significant difference exists, that is, whether the null hypothesis should be rejected.

nutrient   Any substance that an organism obtains from its environment for use as an energy source, growth factor (such as a vitamin), or basic material for the synthesis of biomass. The term is often used to identify substances used for growth by plants.

nutrient cycle   The cyclic conversions of nutrients from one form to another within the biological communities. A simple example of such a cycle would be the production and release of molecular oxygen (O_2) from water (H_2O) during photosynthesis by plants and the subsequent reduction of atmospheric oxygen to water by the respiratory metabolism of other biota. The cycle of nitrogen is much more complex, with the nitrogen atom undergoing several changes in oxidation state (N_2, NO_3^-, R—NH_2, and NH_4^+, among others) during the cycling of this element through the biological community, and into the air, water, or soil, and back.

nutrient sink   See NATURAL SINK.

O

Occupational Safety and Health Act (OSHAct)   The 1970 federal statute that established the OCCUPATIONAL SAFETY AND HEALTH ADMINISTRATION, for the purpose of ensuring, to the extent feasible, a safe and healthy workplace. The act provided for a detailed record-keeping system of employee exposure to potentially harmful agents and of workplace injuries and illnesses, periodic workplace inspections, and the creation of the National Institute of Occupational Safety and Health to conduct studies of occupational hazards and recommend standards.

Occupational Safety and Health Administration (OSHA)   An agency of the United States Department of Labor responsible for issuing and enforcing regulations to protect the safety and health of workers. Authorized by the OCCUPATIONAL SAFETY AND HEALTH ACT.

Ocean Dumping Act   See MARINE PROTECTION, RESEARCH, AND SANCTUARIES ACT.

ocean dumping permit   Under the Ocean Dumping Ban Act of 1988, which amended the 1972 MARINE PROTECTION, RESEARCH, AND SANCTUARIES ACT, no new permits for the dumping of sewage sludge or industrial waste are allowed, and all holders of existing permits must cease ocean dumping of these materials by December 31, 1991.

ocean floor sediment   Unconsolidated materials that settle and accumulate on the floor of the deep ocean. These materials can be fine muds and clays, quartz grains, dust, glacial debris derived from the land masses, oozes comprised of microscopic shells of plants or animals, and substances precipitated directly from seawater.

ocean thermal energy conversion (OTEC)   The use of the solar energy absorbed by the ocean to produce electricity. The temperature difference between the warm surface water and the cooler, deeper water is as much as 20°C in the tropics, and the movement caused by this thermal gradient can drive an evaporation-condensation cycle of a fluid to turn a turbine generator. Low efficiency and saltwater corrosion are two current technical problems with OTEC.

octane number   A measure of the tendency of gasoline to knock, or ignite prematurely, in an INTERNAL COMBUSTION ENGINE. A fuel with a low knocking potential, isooctane, is arbitrarily given an octane number of 100, and a higher-knocking fuel, n-heptane, is given a value of zero. The octane number of a fuel is determined by equating its knocking tendency to the knocking of a mixture of isooctane and n-heptane; its octane number is the volume percentage of the isooctane in the equated isooctane/n-heptane mixture. The higher the octane number, the greater the antiknock property and the smoother an internal combustion engine operates.

octanol-water partition coefficient (K_{ow})   A ratio derived from the laboratory measurement of the solubility of a chemical compound in water relative to its solubility in n-octanol. The ratio is expressed as micrograms of substance per milliliter of n-octanol divided by the micrograms of the same substance per milliliter of water. The coefficient can be related to the water solubility, the SORPTION to soil organic matter, and the BIOCONCENTRATION factors of certain organic compounds.

octave bands   A series of frequency intervals used to analyze the makeup of a sound. Each interval is the range represented from a given frequency to twice the given frequency. Often, 11 octave bands are used, starting at 22 hertz (Hz) and ending at 44,800 Hz. Therefore, the first octave band is 22–44 Hz, the next, 44–88 Hz, and so on, up to the eleventh band, represented by 22,400–44,800 Hz. The center band frequency for each octave band is the geometric mean of frequencies at each end of the range, for example, the center band frequency of the 22–44 Hz octave band is the square root of the product of 22 and 44, or approximately 31.5 Hz.

odds ratio   An expression of relative risk used in epidemiological CASE-CONTROL STUDIES. The ratio is calculated as the disease rate in the exposed group divided by the disease rate in the nonexposed group.

odor fatigue   The loss of odor sensitivity that occurs after a period of continuous exposure to an odor.

odor threshold   The lowest concentration of a vapor or gas that can be detected as an odor by a stated percentage of a panel of test individuals.

Office of Management and Budget (OMB)   An agency in the executive branch with responsibility for drafting the annual budget of the federal government. The OMB also reviews major regulatory proposals, in accordance with EXECUTIVE ORDER 12291.

Office of Solid Waste and Emergency Response (OSWER)   The administrative unit of the United States Environmental Protection Agency responsible for solid and hazardous waste management, including responses to chemical releases.

offset   An air quality management rule that requires a facility locating or expanding in an area that does not meet a NATIONAL AMBIENT AIR QUALITY STANDARD for an air pollutant to reduce its own emissions or obtain emission reductions from other facilities in an amount greater than the additional emissions of the nonattainment pollutant associated with the new or expanded facility. The emission reductions must "offset" the new emissions for the pollutant that violates an ambient standard.

off-gas   The normal gas emissions from any process vessel or equipment.

Oil and Hazardous Materials Technical Assistance Data System (OHMTADS)   A system of the United States Environmental Protection Agency that provides technical information to units responding to spills or releases of hazardous substances.

oil shale   See KEROGEN.

oil skimmer   A device that collects and removes oil from a water surface. Ropes, belts, rotating drums, and similar devices are used as adhering surfaces for the oil, and the oil is pressed out or scraped off into a holding tank.

old field   Cropland that is no longer used to produce an agricultural crop and that has been allowed to revert to natural plant cover.

old growth   Forests that either have never been cut or have not been cut for many decades. Forests characterized by a large percentage of mature trees.

olfactometer   A device that allows a selected group of evaluators (odor panel) to compare the odor from a sample of ambient air to a series of different concentrations of a reference odorant. Each panel member inhales air delivered through a mask or sample port of the olfactometer and then selects the reference dilution that best matches the strength of the sample odor. The olfactometer-odor panel approach attempts to introduce a degree of objectivity to the identification and control of odors in community air.

oligotrophic   Describing a body of water with low content of plant nutrients. Compare EUTROPHIC and MESOTROPHIC.

omnivore   An animal that eats both plants and animals, such as humans. Compare CARNIVORE, HERBIVORE.

once-through cooling   See OPEN-CYCLE COOLING.

oncogene   A gene that if altered or damaged is associated with the development of certain cancers.

oncogenic   Describing any virus, chemical, or radiation that has the capacity to give rise to tumors. The term is most frequently used to describe viruses. See CARCINOGENIC.

one-hit theory   The assumption that a single molecular interaction between a chemical or IONIZING RADIATION and a cell, or "hit," at the cellular level can cause irreversible changes in the cell, leading to a tumor.

on-scene coordinator (OSC)   A federal official designated by the United States Environmental Protection Agency or the United States Coast Guard to direct and coordinate federal responses to oil spills and the releases of hazardous substances. The OSC is responsible for removal actions following spills. The prevention or minimization of the release of oil or hazardous substances is the responsibility of the REMEDIAL PROJECT MANAGER.

opacity   For smokestacks, the degree of light transmission through the plume indicating the concentration of PARTICULATE MATTER in the exhausting air. See RINGELMANN CHART, SMOKE READER.

open-cycle cooling   The practice of withdrawing surface or well water to cool the condensers of an electric power plant or other industrial equipment, followed by release of the heated water to the ocean, a river, or a lake.

open dump   A landfill operated without the environmental safeguards required by current law. All open dumps were forced by the RESOURCE CONSERVATION AND RECOVERY ACT of 1976 to upgrade to SANITARY LANDFILL status or close.

open-pit mining   See STRIP MINING.

open system   An environment or defined area that is characterized by a significant input of chemical elements used to support the growth of plant and animal communities and a corresponding loss of biomass or chemical elements from the area to the outside. Compare CLOSED SYSTEM.

open water loop Any process in which water is routed through a facility and then not reused, but discharged into a surface water body after any appropriate treatment. Compare CLOSED WATER LOOP.

optical coefficient An expression of the fraction of light energy striking a surface that is absorbed, reflected, or passes through the material.

oralloy Uranium that has been enriched with the fissionable isotope, uranium-235. The mixture is capable of supporting a sustained chain reaction.

order of magnitude A factor of 10 times.

ordinal variable A characteristic expressed by rank or order. For example, a sample group might be classified by shades of the color red, with the number 1 being the lightest red and 10 the darkest red. Although the ranking is often in numerical order, the data do not represent actual quantities; there is no indication of the measured difference between the ranked samples. Used only in NONPARAMETRIC TESTS. See also NOMINAL VARIABLE.

organ of Corti The cells in the ear that translate vibrations into nerve impulses to be sent to the brain. Different cells react to different sound frequencies. The aging process and excessive noise can damage these cells and cause permanent hearing impairment. The cells responding to higher-frequency sounds are almost always harmed first.

organic Of or related to a substance that contains CARBON atoms linked together by carbon-carbon bonds. All living matter is organic. The original definition of the term organic related to the source of chemical compounds, with organic compounds being those carbon-containing compounds obtained from plant or animal sources, whereas INORGANIC compounds were obtained from mineral sources. We now know that compounds containing bonds between carbon atoms can be made in the laboratory and industrially by man.

organic carbon 1. Carbon atoms in an organic compound that are linked to other carbon atoms by a COVALENT BOND. Used to distinguish such compounds from inorganic forms of carbon, such as those found in the carbonates and cyanides. 2. The amount of organic material in soil or water.

organic carbon partition coefficient (K_{oc}) A measure of the extent to which an organic chemical is ADSORBED to soil particles or sediment. The measure is expressed as the ratio of the amount of adsorbed carbon per unit mass of total organic carbon (milligrams of carbon adsorbed per kilogram organic carbon) to the equilibrium concentration of the chemical in solution (milligrams dissolved carbon per liter, or kilogram, of solution). Higher values of K_{oc} indicate greater organic carbon adsorption and greater retention of the organic chemical in the soil. Lower values indicate less retention and greater transport of the chemical in the ground or surface water.

organic load The amount of organic material added to a body of water. The amount of material, usually added by human activities, that must be mineralized or degraded within a particular environment.

organic nitrogen Nitrogen that is bound to carbon-containing compounds. This form of nitrogen must be subjected to mineralization or decomposition before it can be used by the plant communities in aquatic and terrestrial environments. This is in contrast with inorganic nitrogen, which is in the mineral state and more readily utilized by plant communities.

organic phosphorus Phosphorus bound to carbon-containing compounds. See ORGANIC NITROGEN.

organic solvent A liquid, usually a hydrocarbon like benzene or toluene, used to dissolve materials such as paints, varnishes, grease, oil, or other hydrocarbons.

organic waste Carbon-containing materials that are discarded into the environment. The term is often used as a euphemism for domestic sewage.

organochlorine See   CHLORINATED HYDROCARBONS.

organomercurials Common name applied to organic molecules complexed with MERCURY. In general, these compounds are toxic to humans with the nervous system being the most sensitive. The most well-known member of the class of compounds is METHYL MERCURY.

organophosphate A family of organic insecticides that includes tetraethyl pyrophosphate, diazinon, malathion, and parathion. The compounds, which are esters of phosphoric acid, contain phosphorous as an integral component of the molecules. Members of this family of insecticides interfere with nerve transmission through the inhibition of cholinesterase. In humans, they cause headache, weakness, and dizziness in lower doses and paralysis, convulsions, and coma in higher doses. Members of the family react readily with clay particles, ions, and other components of soil. They are relatively nonpersistent in the environment.

orifice meter A device for measuring gas flow rates in which the gas flows through a hole (orifice) in a plate positioned in a pipe. The difference in static pressure measured upstream and downstream of the plate is used to compute the gas flow rate.

orographic lifting The upward movement of air when currents in the atmosphere encounter mountains. The rising air expands and cools, condensing moisture in the air as clouds and resulting in precipitation. The air descending on the other side of the mountains has lost its moisture and, as it descends, is compressed and warms.

orphan site See ABANDONED SITE.

osmosis The diffusion of a solvent (typically water) across a membrane (either natural or artificial) separating two solutions of different concentrations. The semipermeable membrane allows the passage of water but prevents the passage of substances dissolved in the water. The water movement is from the more dilute solution toward the more concentrated solution, and will continue until the two solutions are equal in concentration. If pressure is applied to the more concentrated side, the flow of water will reverse, from the concentrated side to the more dilute side. See REVERSE OSMOSIS.

osmotic lysis The rupture of a cell placed in a dilute solution. For example, when a red blood cell is placed in distilled water, water tends to move into the cell because of the osmotic pressure generated as a result of the concentration of the materials inside the cell. As the amount of water increases within the cell, the cell membrane can no longer withstand the pressure, and ruptures. The process is not unlike putting so much air in a balloon that it pops.

osmotroph An organism that obtains nutrients through the active uptake of soluble materials across the cell membrane. This class of organism, which includes the bacteria and fungi, cannot directly utilize particulate material as nutrients. Compare to PHAGOTROPH.

outage 1. The difference between the volume capacity of a container and its actual content. 2. The time period when industrial equipment is shut down for routine maintenance. 3. An interruption of electric power delivery, resulting from a power plant failure or a break in the distribution system.

outfall   The location where wastewater is released from a POINT SOURCE into a receiving body of water.

outgassing   The loss of vapors or gases by a material, usually as a result of raising the temperature of and/or reducing the pressure on the material.

outwash   Sand and gravel deposited by meltwater from glaciers. Individual deposits tend to be of uniform particle size.

overburden   The rock and dirt that overlies a mineral deposit and that must be removed before the mineral deposit can be removed by surface mining.

overdraft   The sustained extraction of groundwater from an aquifer at a rate greater than the aquifer is recharged, resulting in a drop in the water table.

overfishing   The removal of a sufficiently large number of certain fish from a body of water such that breeding stocks are reduced to levels that will not support the continued presence of the fish in desirable quantities for sport or commercial harvest.

overland flow   The discharge of wastewater in such a way that it flows over a defined land area prior to entering a receiving stream. The movement over vegetated land fosters the removal of plant nutrients from the wastewater. See TERTIARY TREATMENT.

overpack   An external, secondary container used to enclose a packaged hazardous material, hazardous waste, or radioactive waste.

overturn   See FALL TURNOVER, SPRING TURNOVER.

oxidant   See PHOTOCHEMICAL OXIDANT.

oxidation   1. In a strict chemical sense, the loss of electrons by atoms or compounds. The term is often applied to a variety of other analogous chemical reactions, such as the addition of oxygen to a compound during chemical reactions and the removal of hydrogens during biochemical reactions. Oxidations are always accompanied by simultaneous reduction reactions in which some atom or molecule gains electrons. 2. The mineralization or decomposition of organic compounds by microorganisms. See also REDUCTION.

oxidation pond   A pond into which an ORGANIC waste (sewage) is placed to allow decomposition or mineralization by aerobic microorganisms.

oxidation-reduction reaction   A coupled reaction in which one atom or molecule loses electrons (through OXIDATION) while another atom or compound gains electrons (through REDUCTION).

oxides of nitrogen (NO$_x$)   See NITROGEN OXIDES.

oxides of sulfur (SO$_x$)   See SULFUR OXIDES.

oxidizing agent   Any material that attracts electrons, thereby oxidizing another atom or molecule. The oxidizing agent is itself reduced; that is, it gains electrons. The material donating the electrons is the REDUCING AGENT. Chlorine and oxygen are good oxidizing agents. See also OXIDATION.

oxygen demand   The need for molecular oxygen (O$_2$) to meet the needs of biological and chemical processes in water. The amount of molecular oxygen that will dissolve in water is extremely limited; however, the involvement of oxygen in biological and chemical processes is extensive. Consequently, the amount of oxygen dissolved in water becomes a critical environmental restraint on the biota living in water. The metabolism of large organisms like submerged plants and fish, the microorganisms engaged in decomposition, and spontaneous chemical reactions all require (demand) a portion of a limited resource, molecular oxygen. See BIOCHEMICAL OXYGEN DEMAND, CHEMICAL OXYGEN DEMAND.

oxygen-demanding waste   Any organic material that will stimulate the metabolism of bacteria with a corresponding use of DISSOLVED OXYGEN when discharged into a natural waterway. Often used as a euphemism for domestic sewage.

oxygen depletion   The removal of DISSOLVED OXYGEN from a body of water as a result of bacterial metabolism of degradable organic compounds added to the water, typically by human activities.

oxygen sag curve   A graph of the measured concentrations of DISSOLVED OXYGEN in water samples collected (1) upstream from a significant POINT-SOURCE of readily degradable organic material, (2) from the area of the discharge, and (3) from some distance downstream from the discharge, plotted by sample location. The amount of dissolved oxygen is typically high upstream, diminishes at and just downstream from the discharge location (causing a sag in the line graph) and returns to the upstream levels at some distance downstream from the source of pollution.

oxyhemoglobin   The hemoglobin of red blood cells that is bound to molecular oxygen. Hemoglobin of red blood cells that is functioning normally.

ozonation   The use of ozone gas (O_3) as a disinfectant to reduce the microbial load and to kill dangerous pathogenic bacteria. The treatment can be applied to a public drinking water supply before it enters the distribution system or to wastewater prior to its discharge into a receiving stream.

ozone   Triatomic oxygen (O_3). A very reactive gas produced by photochemical reactions or lightning in the troposphere and by the absorption of ultraviolet radiation in the lower stratosphere. In sufficiently high concentrations at ground level, the gas acts as an irritant to the eyes and respiratory tract. See OZONE LAYER, PHOTOCHEMICAL AIR POLLUTION, and PHOTOCHEMICAL OXIDANTS.

ozone hole   A large area over Anarctica recently discovered to have a seasonal drop in stratospheric ozone concentration of as much as 50 percent. It is linked to the formation of stratospheric ice clouds that release chlorine atoms from CHLOROFLUOROCARBONS during the Antarctic winter. The chlorine is present in quantities that cause the extensive ozone

OXYGEN SAG CURVE

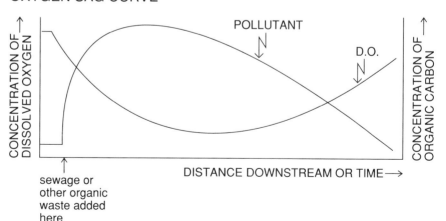

depletion when the spring sunshine returns. See OZONE LAYER DEPLETION.

ozone layer An area of the stratosphere, about 12 to 30 miles in altitude, where the intensity of short-wavelength ultraviolet light from the sun (between 130 and 200 nanometers) is sufficiently high to convert normal diatomic oxygen (O_2) to ozone (O_3). The ozone thus formed provides a measure of protection to plant and animal life on the surface of the earth because it absorbs ultraviolet radiation at wavelengths that are mutagenic (about 250 to 270 nanometers). See OZONE LAYER DEPLETION.

ozone layer depletion The destruction of ozone molecules in the OZONE LAYER of the stratosphere by chemical reactions with materials released by human activities. The main ozone-consuming chemicals are the CHLOROFLUORO-CARBONS (CFCS) and the HALONS, both of which are groups of chemicals that are extremely stable in the troposphere, with typical atmospheric lifetimes of 60 to 100 years. If the CFCs or halons migrate to the stratospheric ozone layer, the ultraviolet radiation there is strong enough to break the molecules apart, releasing chlorine atoms (CFCs) or bromine atoms (halons) which react with and destroy ozone. See also MONTREAL PROTOCOL.

P

packed bed absorber See PACKED TOWER.

packed tower An air pollution control device in which contaminated air is moved through a tower containing materials (packing) that have a large surface area; the air contaminants are then absorbed into a liquid flowing over the packing materials. The flow is usually countercurrent; that is, the liquid falls downward

and the air is forced upward through the tower. See also SPRAY TOWER, TRAY TOWER.

packed tower aeration A process for the removal of organic contaminants from groundwater. The groundwater flows downward inside a tower filled with materials (packing) over a large surface area. Air is introduced at the bottom of the tower and is forced upward past the falling water. Individual organic contaminants are transferred from the water to the air, according to the gas and water equilibrium concentration values of each contaminant. See also AIR STRIPPING and HENRY'S LAW CONSTANT.

packer A device lowered into a well to produce a fluid-tight seal.

packing Corrosion-resistant materials, such as porcelain, stainless steel, or polypropylene, used in PACKED TOWERS to maximize gas-liquid contact for air pollutant removal efficiency. Packing types vary by shape and include Berl saddles, Lessing rings, Pall rings, Raschig rings, and Intalox saddles.

paint filter liquids test (PFLT) A test used to determine the presence of FREE LIQUIDS in an untreated waste sample or in a sample of a waste that has undergone STABILIZATION/SOLIDIFICATION. The test is performed by placing the sample in a supported funnel lined by a filter; if no liquids pass through the filter and flow from the funnel within 5 minutes, no free liquids are present.

palisade parenchyma Along with the spongy parenchyma, chlorophyll-containing cells between the upper and lower epidermis, or outer layers, of a leaf.

palustrine Describing marsh or wetlands.

paper tape sampler An air sampling device for suspended particulate matter. Sampled air is directed toward a filter tape. The tape is moved periodically, and the series of small circular dust spots produced is evaluated by measuring the transmittance of white light through the spots compared to the transmittance through an unsoiled filter. See COEFFICIENT OF HAZE.

paper trail   See ADMINISTRATIVE RECORD.

paraffins   Straight- or branched-chain hydrocarbons with carbon-carbon single bonds; the members of this group are relatively nonreactive. A large component of crude oil. Also called aliphatic hydrocarbons. When the carbon atoms are arranged in a ring structure, they are called cycloparaffins.

paralytic shellfish poisoning (PSP)
A pathological condition in humans caused by the consumption of certain marine mussels or clams that have fed on planktonic dinoflagellates belonging to the genus *Gonyaulax*. The mussels or clams become contaminated with a neurotoxin produced by the dinoflagellates, and subsequent consumption by humans results in respiratory or gastrointestinal distress. The condition occurs in conjunction with the phenomenon known as the red tide, a bloom of dinoflagellate populations in marine waters.

parameter   The true value of a population characteristic. The estimate of a parameter, called a statistic, is a measurement of a sample of the population. For example, the average height of 12-year-old girls, based on measuring the height of 10 girls, may be stated as 54 inches, a statistic. The parameter corresponding to the true average height of all 12-year-old girls is unknown but is approximated by the sample results.

parametric tests   Statistical tests that attempt to draw inferences about population PARAMETERS, for example the mean or standard deviation. These tests use quantitative data (measurements) and require certain assumptions about the sam-

ple population data distributions. Compare NONPARAMETRIC TESTS.

paraquat   The best known example of the bipyridyl herbicides, 1,1'dimethyl-4,4'-bipyridylium. Large exposures to the agent result in damage to the lungs, liver, and kidneys, with the cause of death usually associated with respiratory distress even when the herbicide is taken by mouth. Evidence of lung damage can become apparent up to two weeks after the exposure, at which time the agent has been eliminated from the body. Paraquat adsorbs strongly to soil particles and resists microbial degradation.

parasite   An organism (the parasite) whose natural habitat is either on or inside another, larger, organism (the host). The presence of the parasite results in damage to the host.

parasitism   A biological interaction between species in which a parasite gains nourishment by living on or inside a host organism. The interaction is beneficial to the parasite and detrimental to the host.

parathion   One of the early organophosphate insecticides developed in the 1940s. The agent, *O,O*-diethyl-*O-p*-nitrophenylphosphorothioate, is a stable, AROMATIC compound that binds tightly to sediments. The relatively high toxicity to man is associated with the inhibition of nerve function.

parent material   The rock from which a soil is derived; bedrock.

Part A   Under the RESOURCE CONSERVATION AND RECOVERY ACT, the first part of the two-part permit application for a hazardous waste TREATMENT, STORAGE, OR DISPOSAL facility; the form asks for general information about the facility. INTERIM STATUS facilities are said to operate under a Part A permit. See PART B.

Part B   The detailed, technical, second part of the permit application for a hazardous waste TREATMENT, STORAGE, OR

DISPOSAL facility, under the RESOURCE CONSERVATION AND RECOVERY ACT. Approval of Part B constitutes a final permit to operate the facility. See Part A.

partial pressure   The pressure exerted by an individual gas in a gaseous mixture. If the mixture of gases is assumed to behave as an IDEAL GAS, the sum of the individual partial pressures in a mixture is equal to the total pressure of the mixture of gases. For example, the atmosphere contains roughly 78 percent nitrogen, 21 percent oxygen, and many other gases total to the remaining one percent. Of a total atmospheric pressure of 760 millimeters of mercury (mm Hg), the partial pressure of the nitrogen is about 593 mm Hg (78 percent of 760 mm Hg), the oxygen partial pressure is about 160 mm Hg (21 percent), and the other gases total about 7 mm Hg.

particle size distribution   A presentation of the relative masses of airborne particles in different size ranges present in an air sample. For example, the air might contain particles of 0–2 microns in diameter, 15 milligrams (8 percent of total); 2–4 microns, 20 milligrams (11 percent of total), and so on.

particulate control device   Any of several types of air pollution control equipment that remove particulate matter from exhaust gases. See ELECTROSTATIC PRECIPITATOR, BAGHOUSE, SCRUBBER, PACKED TOWER, TRAY TOWER, SPRAY TOWER.

particulate matter   In water pollution, particulate matter describes solid material in either the solid or dissolved states. Insoluble particulate matter includes particulate substances that either settle from water that is allowed to stand or are removed by passing the water through a filter. Sand, clay, and some organic matter constitute insoluble particulate matter. Dissolved substances that will neither settle if water is allowed to stand nor be removed by passage through a filter, but which will be recovered if the

water is allowed to evaporate, are called dissolved particulate matter. Salt is an example of this type. In air pollution, particulate matter is used to describe either solid particles or liquid droplets that are carried by a stream of air or other gases.

particulate matter, 10-micron diameter (PM₁₀)   Airborne particles with diameters equal to or smaller than 10 microns. The size range is defined in the ambient air quality standard for particulate matter established by the United States Environmental Protection Agency. Particles of this size can penetrate deep into the respiratory system and can potentially damage the gas-exchange surfaces of the lung or be absorbed into the blood, with possible systemic toxic effects.

particulate organic matter (POM)   Material of plant or animal origin that is suspended in water. The amount of this type of material suspended in water can be estimated by first removing the suspended material from water by filtration, followed by either a direct measurement of the amount of carbon retained on the filter or by estimating the amount of carbon present from the weight lost upon heating of the filter in excess of 500°C. Generally, the greater the amount of particulate matter present, the more severe the water pollution problem.

particulate phosphate   That portion of the total amount of phosphate (PO_4^{-3}) suspended in water that is attached to particles and will not pass through a filter. The aggregates can be either inorganic or organic. This form of phosphate must be solubilized before it can be used as a plant nutrient.

partition coefficient   A measure of the distribution of some chemical between two IMMISCIBLE solvents. See OCTANOL-WATER PARTITION COEFFICIENT.

parts per billion (ppb)   A unit of measure commonly employed to express the number of parts (e.g., grams) of a chemical contained within a billion parts of gas (air), liquid (water), or solid (soil). Conversions of various ppb expressions to metric units are given in the Concentration Appendix at the back of this dictionary. See CONCENTRATION.

parts per million (ppm)   A unit of measure commonly employed to express the number of parts (e.g., grams) of a chemical contained within a million parts of gas (air), liquid (water), or solid (soil). Conversions of various ppm expressions to metric units are given in the Concentration Appendix at the end of the dictionary. See CONCENTRATION.

pascal (Pa)   The SI unit of pressure equal to one NEWTON per square meter. Expressed as $1 \text{ Pa} = 1 \text{ N m}^{-2}$.

Pasquill-Gifford stability class   One of six atmospheric stability categories used to estimate DISPERSION COEFFICIENTS for GAUSSIAN PLUME MODELS. For a particular stability class at a given distance downwind from a source of air pollution, a dispersion coefficient is input, with other variables, to the model to estimate the air concentration of the released pollutant. The classes represent different combinations of solar insolation and wind speed present during a pollutant release.

passive smoking   The involuntary inhalation of tobacco smoke from another person's cigarette, cigar, or pipe. Studies have estimated that a nonsmoker with a smoking spouse has about twice the risk of lung cancer as a nonsmoker living with a nonsmoker.

passive solar system   A system that captures sunlight to heat a structure. Solar energy is captured directly through the use of large windows, special structural materials, or greenhouses, and that heat is distributed within the structure without the aid of mechanical devices such as fans. Compare ACTIVE SOLAR SYSTEM.

pathogen   A microorganism, such as a bacterium or fungus, that has the ca-

pacity to cause disease under normal conditions.

peak load   The maximum electricity demand in an area, expressed on a daily, seasonal, annual, or other time basis.

peat   The residue of partly decomposed plant material in which various plant parts such as stems can easily be discerned. The residue is mined for use as a soil builder in gardening or as a low-grade fossil fuel. Peat is considered to be an early stage in the conversion of plant residues to coal.

pelagic   Of marine biota that live suspended in the water. Compare to BENTHIC.

penetration   An expression of the fraction of particulate matter in a gas stream that is not removed by a collection device.

pentachlorophenol (PCP)   A synthetic compound very similar to hexachlorobenzene, with the only difference being the substitution of one of the chlorine atoms (—Cl) with a hydroxyl group (—OH). (See HEXACHLOROBENZENE for a diagram.) The substance has been used in the United States since 1936 as an insecticide, herbicide, and fungicide. The treatment of wood to prevent attack and decomposition by fungi and insects is one of the major uses for this chemical. Because of its widespread use and because the material is persistent in the environment, PCP can routinely be recovered from soil and water samples at concentrations less than one part per billion. Such concentrations do not represent a health hazard. Adverse reactions to the agent have been reported as a result of accidental exposures to relatively high concentrations. Commercial preparations of the chemical have been found to be contaminated with other POLYCHLORINATED compounds, such as DIBENZODIOXINS and DIBENZOFURANS, which increases the danger to individuals exposed to commercial sources of pentachlorophenol.

percent destruction A term commonly used to describe the efficiency of an incineration system used to burn dangerous or hazardous substances. Efficiencies such as 99.99 or 99.9999 percent are prevalent. A 99.99 percent destruction means that 0.01 percent (1 pound per 10,000 pounds) of the substance added to the furnace escapes into the environment.

percent error An expression of the error of a measurement as a fraction of the measurement itself. For example, a measurement could be expressed as 50 centimeters, plus or minus 2 percent, meaning that the true value is thought to be within 49 $(50-0.02\times50)$ and 51 $(50+0.02\times50)$.

perforated tray absorber A mechanical device designed to remove unwanted materials from an exhaust gas. A series of plates containing holes are arranged so that a liquid that reacts with or traps the pollutant flows over the trays as the exhaust gas is forced through the holes. This arrangement provides for maximum contact between the exhaust gas and the liquid.

performance curve A graph depicting the performance of a pollution control device for various conditions, for example the collection efficiency of a cleaning device for various sizes of airborne particulate matter.

performance standards Environmental or workplace health and safety standards written in terms of the result to be achieved but without detailed requirements for specific actions. Compare SPECIFICATION STANDARDS.

period For electromagnetic radiation, the inverse of FREQUENCY, or the time required for one wave cycle to pass a given point. Frequency is given in units of inverse time (1/second); a period in time units (seconds).

periodic table An arrangement of the chemical elements by ATOMIC NUMBER (increasing number of protons) to show the relationship between elements. Horizontal rows are called periods; vertical rows are called groups.

peripheral nervous system The part of the nervous system exclusive of the brain and spinal cord.

periphyton Microscopic plant species attached to submerged objects in aquatic ecosystems. Compare NEKTON, PLANKTON.

permafrost Soil layers that remain frozen throughout the year. The upper layers of soil in the tundra region of the arctic thaw during the summer months; however, the lower layers remain frozen throughout the year, forming permafrost.

permanent hardness Water hardness that cannot be reduced or removed by heating the water, a reflection of the presence of dissolved calcium, magnesium, iron and other divalent metal ions. These ions will react to form insoluble precipitates. See HARDNESS. Compare TEMPORARY HARDNESS.

permanent threshold shift (PTS) An irreversible reduction in hearing ability for a certain sound frequency. Exhibited by an enduring increase in the HEARING THRESHOLD level. Normal hearing loss with age (PRESBYCUSIS) and NOISE-INDUCED HEARING LOSS begin with a permanent threshold shift in the higher sound frequencies. See TEMPORARY THRESHOLD SHIFT.

permeability The ease with which water and other fluids migrate through geological strata or landfill LINERS. Compare IMPERMEABILITY and POROSITY.

permeant An animal that can move freely among different types of environments, or an animal that can move freely from one group of animals to another group.

permissible exposure limit (PEL) An employee's allowable workplace exposure to a chemical or physical agent; the levels are enforced by the OCCUPATIONAL SAFETY AND HEALTH ADMINISTRATION.

permit-by-rule The shortened permit process under the RESOURCE CONSERVATION AND RECOVERY ACT for facilities holding and in compliance with a permit issued under one of the following laws: the SAFE DRINKING WATER ACT (Underground Injection Control permit), the MARINE PROTECTION, RESEARCH, AND SANCTUARIES ACT (OCEAN DUMPING PERMIT), or the CLEAN WATER ACT (National Pollutant Discharge Elimination System permit).

peroxyacetylnitrate (PAN) A component of PHOTOCHEMICAL AIR POLLUTION with the chemical formula $CH_3C(=O)NO_2$; excessive exposure to this compound can cause eye irritation and injuries to vegetation. See PEROXYACYL NITRATES.

peroxyacyl nitrates (PAN) Air pollutants created by photochemical reactions involving sunlight, reactive hydrocarbons, and oxides of nitrogen. Their general chemical formula is $RC(=O)NO_2$, the R standing for a hydrocarbon group. The most common species is PEROXYACETYLNITRATE.

persistence The relative ability of a chemical to remain chemically stable following release into the environment. Persistent chemicals resist biodegradation and thus are of greater concern in the treatment of wastes.

persistent See PERSISTENCE.

persistent pesticide A chemical used to control unwanted plants or insect pests and that is only slowly decomposed, often by physical rather than biological processes, after its application.

persistent pollutant Some substance that does not undergo ready decomposition or degradation when added to the environment. The term is usually applied to materials such as dieldrin, DDT, and PCBs (polychlorinated biphenyls), which remain in the environment for years following their release.

person-gray A unit that reflects the total radiation exposure of a defined group (population) of people. A population dose, in person-grays, is found by multiplying the number of persons within each subgroup of the population receiving a certain radiation dose by that dose (in GRAYS) and summing the products obtained from each of the other subgroups. For example, a population of 8 individuals, 2 of which receive a dose of 4 grays (8 person-grays) and 6 of which receive a dose of 5 grays (30 person-grays), is exposed to a total of 38 person-grays.

person-rem Same as PERSON-GRAY except that the amount of radiation is given in REMS.

persons at risk The number of individuals that can possibly experience a health-related event during a time period.

person-years-at-risk (PYAR) A unit that reflects both the number of individuals who have been exposed to some dangerous substance or radiation and the length of time during which adverse side effects have been expected from that exposure. The unit is computed by multiplying the total number of individuals exposed times the number of years since the exposure. If there is a latent period between the exposure and the appearance of the specific effect, that time must be subtracted from the number of years since exposure.

perturbation Some action or event that disrupts the normal functioning of the environment.

pesticide A chemical agent used to kill an unwanted organism. Most of the agents are not highly selective in their action; however, they can be placed into categories depending on their target species. The most common pesticides are the insecticides (chemical agents applied to the environment to kill insects). Another class of materials used in great quantities is the herbicides (chemical agents applies to the environment to kill unwanted plants). Fungicides, which are frequently used as preservatives to prevent rotting, and rodenticides, which are used to control mice, rats, and similar animals, represent two lesser used classes of chemicals. The pesticides were introduced to control vector-transmitted diseases, to increase agricultural productivity, to exercise urban pest control, and to preserve materials such as wood.

pesticide residues   Small amounts of insecticides, fungicides, or herbicides remaining in or on food when it is consumed.

Peterson dredge   A device used to collect sediment samples for the identification of bottom-dwelling animals in lakes and streams. The device has the appearance of a closed metal cylinder sectioned in half through the long axis. The two sections are hinged together in such a way that they resemble large jaws when locked open for deployment by wire. When the device is on the bottom and the line is pulled to retrieve the apparatus, the apparatus is levered in such a way that the two halves close to trap a sample of sediment. Weights can be attached to the outside of the cylinder sections to provide a deeper bite into the sediment. The Peterson dredge is useful in the sampling of sediments that have a high content of sand and gravel. Also called the Peterson grab.

Petri dish   The customary device for the isolation and/or cultivation of bacteria in the laboratory. The standard dish is 100×15 millimeters and consists of two overlapping halves. A solid media (agar) is placed in the bottom portion of the dish for INOCULATION of bacterial cultures.

petroleum   A liquid fossil fuel consisting of various HYDROCARBONS. Also called

oil. The mixture is found in various geological deposits and can be refined to produce such products as gasoline, fuel oil, kerosene and asphalt.

petroleum exclusion   The exclusion of oil and natural gas from the definition of a hazardous substance under the COMPREHENSIVE ENVIRONMENTAL RESPONSE, COMPENSATION, AND LIABILITY ACT. More specifically, the exclusion applies to crude oil and crude oil fractions (except certain fractions listed as hazardous substances), natural gas, natural gas liquids, liquefied natural gas, and synthetic fuel gas.

pH   A unit used to express the strength of an acidic or basic solution; calculated as the negative logarithm of the hydrogen ion concentration. Values commonly range from 0 to 14, with less than 7.0 being acidic and greater than 7.0 being basic. A pH of 7.0 is considered neutral. Because the units are derived from common logarithms, a difference of one pH unit indicates a tenfold (10^1) difference in acidity; a difference of two units indicates a hundredfold (10^2) difference in acidity.

phagocytic   Describing a cell capable of engulfing particles. For example, white blood cells and cells that reside in lymph nodes can engulf and destroy bacteria and other microorganisms that enter the body.

phagotroph   An organism that obtains nutrients through the ingestion of solid organic matter. This class of organism includes all animals from the simplest, single-celled animal (for example, protozoa) to the higher forms. Organisms have some type of device to ingest particles, a digestive system, and a system to discard waste products.

Phase I   Under the RESOURCE CONSERVATION AND RECOVERY ACT (RCRA), the first phase of the assumption of responsibility by a state for administering the RCRA hazardous waste (SUBTITLE C) program. Phase I state authorization refers to regulations identifying hazardous wastes and the standards for generators, transporters, and TREATMENT, STORAGE, OR DISPOSAL facilities with INTERIM STATUS. Compare PHASE II.

Phase II   Under the RESOURCE CONSERVATION AND RECOVERY ACT, the second phase of the assumption of responsibility by a state for the hazardous waste (SUBTITLE C) program, which covers the detailed, technical requirements for issuing final permits to TREATMENT, STORAGE, OR DISPOSAL facilities for hazardous wastes.

phenol   An ingredient or chemical intermediate for many plastics, drugs, and explosives; the chemical formula is C_6H_5OH. Commonly known as carbolic acid. Environmental exposures primarily are through industrial employment. The compound has an adverse effect when discharged into sewage treatment systems.

phenology   The scientific study of the changes in biological organisms that coincide with the yearly seasons, such as the timing of plant emergence and the progression of the fall tree colors from north to south.

phenotype   The genetically based traits of an organism that are actually observable as some morphological, behavioral, or biochemical characteristic of the organism. Because some traits are not expressed (are recessive), they may not be reflected in the plant's or animal's appearance. The total genetic makeup of the organism is its GENOTYPE.

phenyl group   The chemical group C_6H_5 that is contained in many organic compounds, both natural and synthetic. For example, polychlorinated biphenyls (PCBs) contain two phenyl groups.

pheromone   A substance produced by an animal that serves as a chemical signal to other members of the same species. For example, the sex attractant produced by many animals or a chemical substance used to mark territory are pheromones.

phon   A unit used to express the perceived loudness of a sound. For a particular sound, the unit is equal to the DECIBEL level of a 1000-hertz sound evaluated by a group with healthy hearing to be of the same loudness as that particular sound. FLETCHER-MUNSON CONTOURS are given in phons. Compare SONE.

phosphates   General term used to describe phosphorus-containing derivatives of phosphoric acid (H_3PO_4). The chemical containing the phosphate group (PO_4^{-3}) can be either organic or inorganic and either particulate or dissolved. An important plant nutrient.

photic   Related to either the presence of or the effects of light.

photic zone   That area of a body of water into which light penetrates. The upper portion of a lake or sea within which light is sufficiently abundant to support the growth of phytoplankton. Compare APHOTIC.

photochemical   Describing a chemical reaction that is driven by sunlight.

photochemical air pollution   A type of air pollution resulting from the production of chemicals in the atmosphere by reactions between sunlight and airborne substances released by automobiles and industrial facilities. The most important pollutants that undergo these reactions are volatile organic compounds and oxides of nitrogen. These are converted by sunlight to nitrogen dioxide, ozone, and peroxyacyl nitrates (PAN). These gases can mix with particulate matter and produce concentrations of nitrogen dioxide high enough to impart a tint to the air, popularly called smog.

photochemical cycle   A complex series of chemical reactions leading to an accumulation of ozone or other PHOTOCHEMICAL OXIDANTS in the troposphere. Generally, the cycle involves the absorption of sunlight by nitrogen dioxide (NO_2), causing it to split into NO + [O] (atomic oxygen). The atomic oxygen is very reactive, and it can combine with a diatomic oxygen molecule (O_2, the usual form of oxygen in the atmosphere) to form ozone (O_3). The O_3 can react with the NO to form NO_2 and O_2 (the photolytic cycle). If reactive hydrocarbons are present, atmospheric reactions can convert them to free radicals, which can oxidize the NO formed by the initial splitting of NO_2. If the NO is oxidized to NO_2, it cannot react with and destroy the O_3, thus allowing ozone concentrations to rise.

photochemical oxidants Components of PHOTOCHEMICAL AIR POLLUTION that include ozone, peroxacyl nitrates (especially peroxyacetylnitrate), and oxygenated hydrocarbons, such as the aldehydes. The photochemical oxidant present in the highest concentrations in the ambient air is ozone. The United States Environmental Protection Agency initially set a NATIONAL AMBIENT AIR QUALITY STANDARD for all photochemical oxidants but has redesignated it as an ozone standard.

photochemical smog See PHOTOCHEMICAL AIR POLLUTION.

photoelectric effect An electric current induced when light strikes certain metals, the principle behind the operation of automatic exposure meters in cameras and photovoltaic devices.

photoionization detector (PID) An analytical instrument that determines the amount of a specific organic material present in a gas stream by exposing the gas to ultraviolet energy that will be absorbed by that material. If present, the organic compound will be ionized, producing a current proportional to the amount of the material. See GAS CHROMATOGRAPH.

photolysis The breakdown of a material by sunlight. For example, nitrogen dioxide (NO_2) is split into nitric oxide (NO) and atomic oxygen (O) by the ultraviolet energy in sunlight. Photolysis is

also an important degradation mechanism for contaminants in surface water and in the terrestrial environment.

photolytic cycle An atmospheric reaction that forms OZONE in the troposphere. The cycle involves the splitting, by the absorption of sunlight, of a nitrogen dioxide molecule (NO_2) into NO + [O] (atomic oxygen). The atomic oxygen atom [O] can then combine with a diatomic oxygen molecule (O_2) to form ozone (O_3). In the absence of reactive hydrocarbons, the ozone that is formed is destroyed by reacting with the NO to form NO_2 and O_2. See also PHOTOCHEMICAL CYCLE.

photomultiplier tube An electronic device in some analytical instruments that enhances the photon signal emitted by a reaction or interaction used to detect the presence of a particular chemical substance. Photons are emitted by certain chemical reactions or by an interaction between a specific wavelength of electromagnetic energy and the substance being measured.

photon A quantum of electromagnetic radiation. A unit of intensity of electromagnetic radiation, including light. A photon has properties that relate to both particles and waves. It has no charge or mass; however, it does have momentum.

photosensitization The increased susceptibility of the skin to ultraviolet light caused by exposure to photosensitizing agents such as coal tar and its derivatives, hexachlorophene, and certain plants.

photosynthate Carbohydrates and other organic molecules produced, and often released by algae, during the process of photosynthesis.

photosynthesis A process in green plants and some bacteria during which light energy is absorbed by chlorophyll-containing molecules and converted to chemical energy (the light reaction). During the process, carbon dioxide is reduced and combined with other chemical elements to provide the organic intermediates that form plant biomass (the dark reaction). Green plants release molecular oxygen (O_2), which they derive from water during the light reaction.

photovoltaic Producing an electric current as the result of light striking a metal; the direct conversion of radiant energy into electrical energy.

photovoltaic cell See SOLAR CELL.

phreatophyte A plant with roots extending to the water table and with, therefore, a relatively high transpiration rate.

phthalates Agents added to plastics to improve their flexibility. Over 25 different compounds are produced for commercial use, with di(2-ethylhexyl) phthalate (DEHP) and di-n-butyl phthalate (DBP) being the most common. Because they are used in every major category of consumer products and because they are only slowly degradable, phthalates are distributed throughout the environment. The level of acute toxicity of the compounds is very low; however, some evidence indicates a possible capability to induce tumors.

phylogeny The evolutionary development of a group or species of organisms.

phytoplankton Microscopic plants, such as algae, suspended in aquatic environments.

phytotoxicant A chemical that can damage or kill plants in aquatic environments.

phytotoxicity The ability of chemicals to damage or kill plants in aquatic environments.

pica The tendency of children to eat nonfood items, such as chips of old paint.

pico- (p) An SI unit prefix meaning 10^{-12}.

picocurie A unit of radioactivity equal to 1×10^{-12} curie.

piezometer A well used for the determination of water table elevation or the water level in a tightly cased well, the latter called the POTENTIOMETRIC SURFACE.

piezometric height See HYDRAULIC HEAD.

piezometric surface See POTENTIOMETRIC SURFACE.

pig 1. A container used to store or ship radioactive materials. The container is constructed of materials that act as shielding to prevent the transmission of high-energy radiation. 2. A device forced through unlined cast-iron water pipes by hydraulic pressure to scrape iron oxide scale from the inside surfaces.

pioneer community The first plant species to colonize a land area that is essentially free of plant life. The nonvegetated areas may be present because of previous farming, surface mining, or severe fires. These initial plant communities usually give way to different species that arrive later. See SUCCESSION, PRIMARY SUCCESSION.

pitchblende A mineral ore containing commercial quantities of the elements uranium and radium.

pitot tube A device used to measure the velocity of a gas or liquid. Two concentric tubes are oriented in the same axis as the material flow, and the tubes measure the total pressure and the static pressure of the flowing fluid. The difference between the total pressure and static pressure is the velocity pressure, which can be related to the velocity of the fluid.

pK_a The negative logarithm of the acid DISSOCIATION CONSTANT. Lower pK_a values indicate stronger acids.

pK_b The negative logarithm of the base DISSOCIATION CONSTANT. Lower pK_b values indicate stronger bases.

pK_s The negative logarithm of the SOLUBILITY PRODUCT CONSTANT for a chemical compound dissolving in water.

placard   An identifying sign required on all four sides of any highway vehicle or railcar carrying 1000 pounds or more of a hazardous material or any quantity of substances belonging to the United States Department of Transportation hazardous material classes: poison gas, explosives (A), radioactive (III), or flammable solids (W).

Planck's constant (h)   The constant relating the energy content of electromagnetic radiation to its frequency; the constant is equal to 6.626×10^{-34} joule-second.

Planck's law   The radiation energy emitted per unit area by a blackbody, by wavelength (λ), at absolute temperature T is expressed by

$$E_\lambda = \frac{2\pi hc^2 \lambda^{-5}}{\exp\,(hc/\lambda kT) - 1}$$

where h is PLANCK'S CONSTANT, c is the speed of light, and k is Boltzmann's constant. Planck's law also expresses the energy value (E) of a photon of electromagnetic radiation by $E = h\nu$, where ν is the radiation frequency.

plankton   Microscopic plants and animals that live suspended in water. See PHYTOPLANKTON and ZOOPLANKTON. Compare BENTHOS, NEKTON.

plant nutrients   The primary mineral ingredients of fertilizer: phosphate (PO_4^{-3}), nitrate (NO_3^-), and ammonium (NH_4^+), together with an extensive array of chemical elements used in lesser amounts to support the growth of plants. See TRACE ELEMENTS.

plasticizers   Chemical additives (often phthalates) used to increase the flexibility of plastics.

plate boundaries   According to the theory of plate tectonics, the locations where the rigid plates that comprise the crust of the earth meet. The plates move slowly on the molten material beneath in the process called continental drift. As the plates meet, the boundaries can be classified as divergent (places where the plates are moving apart, as at the mid-ocean ridges of the Atlantic Ocean), convergent (places where the plates are colliding, as at the Himalayas), and transform (places where the plates are sliding past each other, as in California).

plate count   A methodology used to determine the microbial population of a sample of soil, water, or biological material. The source material is suspended or mixed with a sterile solution, and a portion of the mixture is applied to the surface of a suitable agar medium in a Petri dish. Theoretically, each bacterium present in the sample applied to the dish will divide until a visible colony develops. By counting the colonies following incubation and applying the necessary mathematical manipulations, the number of bacteria in the original sample can be determined.

plate tectonics   A concept stating that the earth's crust is composed of crustal plates moving on the molten material below. Seven major and many minor plates are recognized. The seven major plates constitute the great majority of the earth's crust and are named for the continents or oceans (Pacific, Eurasian, African, Australian, North American, South American, and Antarctic). These and the minor plates are moving slowly but relentlessly. The boundaries of the plates are the foci for earthquakes and volcanic activity. Deep ocean ridges are formed where plates diverge or move apart;

transform faults like those in California are formed when plates slide past one another; and ocean trenches and volcanic activity are produced when one plate overrides another and pushes it into the mantle. The latter boundary is characteristic of areas like the region bordering the western Pacific Ocean.

plenum A chamber in a ventilation system used for air distribution via connecting ducts.

plug flow See CONTINUOUS-FLOW SYSTEM.

plugging Stopping the flow of water, gas, or oil into or out of a well. Also, cementing or otherwise blocking the casing of a dry oil, gas, or water well.

plume A relatively concentrated mass of emitted chemical contaminants spreading in the environment. In surface water, the effluent added to a receiving stream near a point source. For example, when a heated-water discharge is added to a stream, the heated water does not mix immediately with the stream water. The mass of hot water remains detectable for some distance downstream. In groundwater, the leachate leaking downgradient from a site of buried waste material. In air pollution, visible or invisible gases, vapors, or particulate matter emitted from a smokestack or moving downwind from an urban area. See URBAN PLUME.

plume reflection The assumption in the GAUSSIAN PLUME MODEL, used for air quality dispersion estimates, that an expanding plume downwind from an elevated source (stack) will, when it reaches the ground, reflect back upward instead of being absorbed.

plume rise The movement of an identifiable mass of gas (plume) upward for a period after its exit from a stack; the movement is caused by (1) the vertical momentum of the exiting gas which, when spent, allows the plume to bend over with the wind, and (2) the buoyancy effect of the lower-density hot gas mixture. Plume rise is estimated to calculate effective stack height (physical stack height plus plume rise), which is a required input to models that estimate air concentrations downwind from the stack.

plutonium-239 (^{239}Pu) A radioactive element, not found in natural uranium ores, produced in nuclear reactors by the bombardment of uranium-238 with FAST NEUTRONS. The atomic number of plutonium is 94. The 239 isotope is a fissionable material and is used as the primary nuclear fuel in nuclear weapons.

pneumoconiosis A general term for diseases of the lung resulting from chronic overexposure to dusts.

poikilotherm See ECTOTHERM.

point of compliance (POC) For a hazardous waste treatment, storage, or disposal facility, the location, specified by the operations permit of the facility, for downgradient wells. The wells are placed to detect the presence of any contaminants released from the facility into groundwater that will move into the uppermost aquifer in the area.

point source An identifiable and confined discharge point for one or more water pollutants, such as a pipe, channel, vessel, or ditch. Compare NONPOINT SOURCE.

point-of-use/point-of-entry (POU/POE) An approach to the management of the quality of drinking water that locates a water cleaning device at the faucet in an individual household; sometimes used in homes supplied by a private well that does not meet drinking water standards.

poise (P) A unit, equal to 1.0 gram per centimeter-second, of viscosity.

polar solvent A solvent, with a slight negative charge on one part of the molecule and a slight positive charge at another position, that dissolves other polar materials. The most common polar solvent is water. Compare NONPOLAR SOLVENT.

polarity   A property of a molecule that results in one side or end of the molecule having a slight negative charge while the other has a slight positive charge. When two atoms are bound together with a COVALENT BOND, they share a pair of electrons. For example, the water molecule is composed of one oxygen atom joined to two hydrogen atoms by covalent bonds. The oxygen atom exerts a slightly stronger attraction for the shared electrons than do the hydrogen atoms. Consequently, the oxygen atom tends to be slightly negative while the hydrogen atoms tend to be slightly positive in character. This creates a molecule with a positive pole and a negative pole. Solvents composed of molecules that exhibit polarity (water) are better able to dissolve ionic materials than are solvents consisting of molecules that do not exhibit polarity (carbon tetrachloride).

polarized light microscopy   A light microscope, equipped with polarizing filters (which screen out all light waves not vibrating or moving in the same plane), used to observe specific optical characteristics of geological materials. This type of microscopy is useful in the identification of asbestos.

polishing   The removal of low concentrations of dissolved, recalcitrant organic compounds from either water intended for human consumption or wastewater that has been subjected to PRIMARY and SECONDARY TREATMENT. The passage of water through a charcoal filtering device is a frequently employed polishing technique.

pollutant   A chemical or physical agent introduced to the environment that may lead to POLLUTION.

Pollutant Standards Index (PSI)   A single-digit representation of the air quality of an urban area. The number is obtained by combining data from air sampling stations for different air pollutants using a standardized method. PSI values falling within certain numerical ranges are then described by standard terminology: 0–50, good; 51–100, moderate; 101–200, unhealthy; and so on. The index was developed to communicate summary air quality information to the public.

pollution   The addition of one or more chemical or physical agents (heat, electromagnetic radiation, sound) to the air, water, or the land in an amount, at a rate, and/or in a location that threatens human health, wildlife, plants, or the orderly functioning or human enjoyment of an aspect of the environment.

pollution exclusion clause   A feature of COMPREHENSIVE GENERAL LIABILITY POLICIES since approximately 1970; the clause excludes bodily injury or property damage coverage for routine or gradual leaks of pollutants from a facility. See ENVIRONMENTAL IMPAIRMENT LIABILITY POLICY.

pollution indicator organism   A plant or animal species that is not normally present in an aquatic environment unless the body of water has been subjected to damage by pollution. For example, *Escherichia coli* is a bacterium that is not found in the aquatic environment unless the system has been contaminated by the addition of fecal material. The organism signals the presence of pollution.

polychlorinated biphenyls (PCBs)   A diverse mixture of AROMATIC compounds that in the past were used extensively in commercial and consumer products as insulating and cooling agents in electrical transformers, as plasticizers in waxes, and in the manufacture of paper and inks. The compounds are very stable, widely distributed in the environment, and bioaccumulated by mammals. Excessive exposures cause a severe acnelike eruption (chloracne) in humans, and the material has been shown to induce cancer development in mammals. As required by the TOXIC SUBSTANCES CONTROL ACT (1976), the U.S. Environmental Protection Agency has written extensive regulations for the manufacture, use, marking, storage, disposal, and cleanup of PCBs. They are found under Title 40, Part 761, of the CODE OF FEDERAL REGULATIONS.

polychlorinated dibenzofurans (PCDFs) A contaminant found in commercial preparations of POLYCHLORINATED BIPHENYLS and PENTACHLOROPHENOL that may be responsible for some of the physiological effects ascribed to those compounds. Exposure to the furans results in severe acnelike eruptions (chloracne) in humans; they are less toxic than the dioxins, which also contaminate the two classes of chemicals given above. See dibenzofurans.

polycyclic Describing chemical compounds composed of multiple units of the six-carbon AROMATIC nucleus, BENZENE, for example anthracene and related compounds.

polycyclic aromatic hydrocarbons (PAH) A group of AROMATIC ring compounds that are derivatives of anthracene, which consists of three benzene rings in a row. Other aromatic rings or organic groups are attached to the anthracene. They are found in coal, tar, and petroleum, and are emitted by combustion-related activities. Many different compounds can be formed through metabolic conversions involving the basic aromatic nucleus within biological systems, as well as during chemical syntheses, because of the reactive nature of anthracene and its derivatives. This class of compounds appears to be responsible in part for the cancer-causing properties of cigarette smoke.

polycyclic hydrocarbons See POLYCYCLIC AROMATIC HYDROCARBONS.

polymer Macromolecules composed of repeating units of some smaller molecule. Many natural materials (cellulose, proteins, starches, nucleic acids) and synthetic materials (nylon, plastic, rubber) are made from linking smaller units together. Cellulose is a natural polymer composed of glucose molecules; polyethylene plastic is an artificial polymer composed of ethylene molecules

polynuclear aromatic hydrocarbons (PAH) See POLYCYCLIC AROMATIC HYDROCARBONS.

polynuclear organic matter (POM) See POLYCYCLIC AROMATIC HYDROCARBONS.

polyvinyl chloride (PVC) A strong synthetic POLYMER, plastic, used in pipe, toys, electrical coverings, and many other products. Produced from vinyl chloride monomer.

population All of the members of a single SPECIES that inhabit a defined geographical area. The snow goose population of a marsh would consist of every member of the species *Chen caerulescens* found in that marsh.

population at risk (PAR) The number of persons that potentially can contract or develop a disease, adverse health effect, or condition; the proper denominator for use in calculating the rate of a disease or condition, which is the number of diseased persons divided by the population at risk.

population dose The sum of the radiation doses received by individuals in a given population, expressed as PERSON-GRAYS or PERSON-REMS. A measure of total population exposure to ionizing radiation.

population equivalent (PE) A way to express the strength of industrial waste in terms of the comparable amount of BIOCHEMICAL OXYGEN DEMAND (BOD) in the household wastewater produced by one person. An industrial waste that has a PE of 1000 is equivalent to the BOD of waste produced by 1000 people.

population risk A risk estimate equal to the product of INDIVIDUAL LIFETIME RISK and the size of the population exposed. Population risk is typically expressed as the number of excess cases of a disease, such as cancer, per year attributable to a given exposure. See UNIT RISK ESTIMATE.

porosity A description of the total volume of soil, rock, or other material that is occupied by pore spaces. A high porosity does not equate to a high permeability, in that the pore spaces may be poorly interconnected.

positive association The direct relationship between two variables, the values of which fluctuate together, in the same direction. For example, as the strength of incoming solar radiation increases seasonally, the atmospheric temperature increases.

positive crankcase ventilation (PCV) The routing of automobile crankcase emissions back into the cylinders for combustion. The process lowers the amount of hydrocarbons released into the atmosphere.

post-closure plan A document prepared by a hazardous waste TREATMENT, STORAGE, OR DISPOSAL facility outlining the groundwater monitoring and reporting, waste containment provisions, and security arrangements for the 30-year period following CLOSURE.

potable water Water suitable for human consumption.

potency See CANCER POTENCY FACTOR.

potency slope factor See CANCER POTENCY FACTOR.

potential energy The energy available in a substance because of position

(water held behind a dam) or chemical composition (hydrocarbons). This form of energy can be converted to other, more useful forms (for example, electrical energy).

potential temperature The temperature of a dry air parcel if it were compressed or expanded, without loss or gain of heat from its surroundings, to a pressure of 1000 millibars. The potential temperature is sometimes used to express ATMOSPHERIC STABILITY. If potential temperature increases with altitude, the atmosphere is stable; if it decreases with altitude, the atmosphere is unstable.

potentially responsible party (PRP) Any hazardous waste generator, owner or operator of a hazardous waste disposal facility, or transporter identified by the United States Environmental Protection Agency as potentially liable for the cost of clearing up a site under the COMPREHENSIVE ENVIRONMENTAL RESPONSE, COMPENSATION, AND LIABILITY ACT.

potentiometric surface The water level in a tightly cased well, or the level to which water would rise if a well were sunk at a particular point. For an artesian well, for example, the potentiometric surface is above ground level.

pounds per square inch (absolute) (psia) Common units for the expression of ABSOLUTE PRESSURE exerted by gases relative to zero pressure. Absolute pressure units are required in IDEAL GAS LAW calculations.

power (P) The rate at which work is done, expressed as work/time. Units of power are the WATT and HORSEPOWER. One watt of power is equal to one JOULE per second. One horsepower equals about 33,000 FOOT-POUNDS per minute or 746 watts.

pozzolanic Certain materials containing silicates or aluminosilicates that will solidify when combined with cement or lime. This type of material is used to trap (immobilize) certain hazardous waste contaminants before land disposal. Common sources of pozzolanic materials are fly ash, lime kiln dusts, and slag from blast furnaces.

precipitant An agent added to a liquid mixture to encourage the formation of solid materials that will settle from the mixture. For example, alum (aluminum sulfate) is added to sewage to promote the formation of FLOC, which facilitates the removal of organic materials from the wastewater.

precipitate The solid that settles from a liquid suspension. The solid produced by a chemical reaction involving chemicals that are in solution.

precipitation scavenging The removal of particles or certain gases from the atmosphere by rain or snow.

precision The repeatability of a series of test results; whether the testing method gives the same answer under the same set of circumstances.

precursor In air pollution, an air contaminant that reacts with sunlight and/or other compounds to produce new chemical materials, for example, the production of OZONE from the atmospheric reactions involving oxides of nitrogen and volatile organic compounds.

predation The process of one animal killing and eating another in order to obtain food.

predator An animal that kills and eats other animals in order to obtain food.

preheat The heating of a raw material or reactant before it is added to some process, or the heating of a gas exhaust before release into the atmosphere.

preliminary assessment and site inspection (PA/SI) The first data collection and evaluation at a site containing hazardous waste that may require re-mediation under the COMPREHENSIVE ENVIRONMENTAL RESPONSE, COMPENSATION, AND LIABILITY ACT. Based on the assessment results, no further action may be recommended or the site may be scheduled for a LISTING SITE INSPECTION, for possible inclusion on the NATIONAL PRIORITIES LIST.

preliminary assessment information rule (PAIR) A requirement of the regulations implementing the TOXIC SUBSTANCES CONTROL ACT under which the United States Environmental Protection Agency gathers certain information about a chemical substance, including the location and quantity manufactured, an estimate of worker exposure, and consumer uses. The agency uses this information for a preliminary assessment of the risk posed by the chemical to human health and the environment. This rule is being replaced by the COMPREHENSIVE ASSESSMENT INFORMATION RULE (CAIR).

premanufacturing notice (PMN) Under provisions of the TOXIC SUBSTANCES CONTROL ACT, toxicity test results and/or other appropriate data required to be submitted to the United States Environmental Protection Agency by a producer of any new chemical or existing chemical proposed for a significant new use. The Act requires the agency to determine whether the new chemical or new use will pose an "unreasonable risk" to human health or the environment.

presbycusis The normal loss of hearing ability in the higher frequencies due to age.

preservation The natural resources policy that stresses the aesthetic aspects of forests, rivers, wetlands, and other areas and tends to favor leaving such areas in an undisturbed state. Compare CONSERVATION.

pressure (p) Force per unit area. Enclosed fluids exert a force perpendicular to the surface of the containing vessel.

The shape of the container does not affect the fluid pressure. At equilibrium, the general expression for the relation between pressure at a point in a fluid and the elevation (depth) z is

$$\frac{dp}{dz} = -\rho g$$

where ρ is the fluid density and g is gravitational acceleration. This is called the hydrostatic equation. This expression indicates that an increase in elevation is accompanied by a drop in pressure.

For pressures p_1 and p_2 at the elevations z_1 and z_2, $p_2 - p_1 = -\rho g(z_2 - z_1)$. Elevation z_2 is higher than elevation z_1.

For stationary incompressible fluids (liquids), density is constant and the ABSOLUTE PRESSURE p_{abs} at a certain depth h below the point z_2 at the surface of the liquid is expressed as $p_{abs} = \rho g h + p_a$, where $h = z_2 - z_1$ and p_a is ATMOSPHERIC PRESSURE, equal to p_2 at point z_2.

For stationary compressible fluids (gases), because density varies with elevation z, the pressure change between two elevations is expressed as an integral:

$$p_1 - p_2 = \int_{z_1}^{z_2} \rho g \, dz$$

Pressure is typically expressed in pounds per square inch, newtons per square meter, or PASCALS.

pressure filter   A device used to remove fine particulate matter from water; the filter consists of a filter medium, such as sand or anthracite coal, packed in a watertight vessel.

pressure gradient   The change in pressure with distance, from lower to higher pressure, or vice versa. See ISOBARS, PRESSURE GRADIENT FORCE.

pressure gradient force   The force causing horizontal air flow from areas of relatively high pressure to areas of relatively low pressure. It is perpendicular to the lines of equal pressure (ISOBARS). Air moving under the pressure gradient force (wind) is deflected to the right by the

Coriolis force in the northern hemisphere. Centrifugal force and friction also affect wind direction and velocity.

pressure head   The hydraulic energy represented by the water pressure per unit weight. The numerical value is expressed in length units: water pressure in pounds per square foot divided by the unit weight of the water, in pounds per cubic foot, equates to pressure head, in feet. The sum of the pressure head and ELEVATION HEAD is the HYDRAULIC HEAD. See also HEAD, TOTAL.

pressurized-water reactor (PWR)   A type of nuclear reactor in which water is employed to transfer heat produced in the reactor CORE. The water that is in contact with the reactor core, called the primary loop, is maintained under high pressure to prevent boiling. Compare BOILING-WATER REACTOR.

presumptive test   The first of three steps in the analysis of water or wastewater for the presence of bacteria of fecal origin. Portions of a water sample are inoculated into lactose broth and incubated for 24 hours at 37° C. The presence of acid and gas after that time is a positive test, and the water is presumed to be contaminated. See CONFIRMED TEST, COMPLETED TEST.

pretreatment   Under the CLEAN WATER ACT, the required alteration and/or reduction of certain water pollutants in a waste stream before the wastewater is discharged into a PUBLICLY OWNED TREATMENT WORKS (POTW). The purpose of this requirement is to prevent discharges that will reduce the efficiency of the water treatment facility or to treat materials that are not treated or inadequately treated by the POTW.

pretreatment standards for existing sources (PSES)   See PRETREATMENT.

pretreatment standards for new sources (PSNS)   See PRETREATMENT.

prevalence The number of cases of a disease or pathological condition present in a given population at a certain time, expressed as a rate; for example, three cases of measles per 1000 persons during the month of April.

prevalence study See CROSS-SEC-TIONAL METHOD.

prevention of significant deterioration (PSD) A Clean Air Act regulatory program under which air quality in an area can only worsen by a fixed amount for particular pollutants above a defined baseline, even if the NATIONAL AMBIENT AIR QUALITY STANDARD for the pollutant is met. The United States is divided into Class I, II, and III areas, and the classification determines how much the air quality in the area can deteriorate, Class I allowing the smallest deterioration. See INCREMENTS, INCREMENT CONSUMPTION.

prey An animal that is potentially or actually killed and consumed by another animal in order to serve as the food source for the second animal.

primary air pollutant A pollutant in the ambient air that is hazardous in the chemical form in which it is released into the atmosphere, such as ambient carbon monoxide that was emitted by an automobile. Compare SECONDARY AIR POLLUTANT.

primary consumer In a food chain, a HETEROTROPHIC organism that feeds on plants. An animal that eats other animals is a secondary consumer.

primary industry categories The 34 types of facilities requiring the BEST AVAILABLE TECHNOLOGY ECONOMICALLY ACHIEVABLE (BAT) for toxic water pollutants under the Clean Water Act. They are listed as Appendix A to Title 40, Part 122, of the CODE OF FEDERAL REGULATIONS. See FLANNERY DECREE.

primary irritant A chemical that, upon overexposure, causes irritation to the skin, eyes, or respiratory tract at the site of contact but does not cause an adverse systemic effect. For example, ammonia inhalation causes irritation of the upper respiratory tract but no other systemic effect. Compare SECONDARY IRRITANT.

primary pollutant A general term for a pollutant in the environment that is in the same chemical form as it was when it was released. Compare SECONDARY POLLUTANT and PRIMARY AIR POLLUTANT.

primary producer See PRODUCER.

primary productivity The weight of plant biomass accumulated as a result of photosynthesis, per unit area or volume of water over a specific time interval. It can be expressed as GROSS PRIMARY PRODUCTIVITY or NET PRIMARY PRODUCTIVITY. See also SECONDARY PRODUCTIVITY.

primary recovery The oil and gas produced from a well that is forced to the surface by natural pressure. See SECONDARY RECOVERY, ENHANCED OIL RECOVERY.

primary settling tank A holding tank where raw sewage or other wastewater is retained to allow the settling and removal of particulate material. The material that separates from the suspension is often termed SLUDGE.

primary sludge The sludge produced by primary treatment in a wastewater treatment plant.

primary standards Standards set by the United States Environmental Protection Agency for the maximum amount of pollutants that can be present in air and water without adverse health effects on humans. The six air pollutants with primary standards are particulate matter, sulfur dioxides, carbon monoxide, nitrogen dioxide, ozone, and lead. The primary standards for drinking water are set for 20 materials, ranging from arsenic to fluoride and from pesticides to radionuclides. Compare SECONDARY STANDARDS.

primary succession The development of plant and animal communities in a land area that does not contain topsoil, for example in an area covered by lava that has solidified. This type of succession depends on the slow weathering of rock. Compare SECONDARY SUCCESSION.

primary treatment The removal of particulate materials from domestic wastewater, usually done by allowing the solid materials to settle as a result of gravity. Typically, the first major stage of treatment encountered by domestic wastewater as it enters a treatment facility. The wastewater is allowed to stand in large tanks, termed clarifiers or settling tanks. Also, any process used for the decomposition, stabilization, or disposal of sludges produced by settling. The water from which solids have been removed is then subjected to secondary treatment.

principal organic hazardous constituents (POHCs) Chlorinated organic compounds found in certain chemical waste mixtures that have the potential for forming PRODUCTS OF INCOMPLETE COMBUSTION when incinerated. Typical POHCs include 1,1,1-trichloroethane, 1,2-dichloroethane, trichloroethane, and trichlorofluoroethane (Freon-113).

priority pollutants A list of 129 chemicals in 65 classes of chemical materials defined as toxic pollutants by Section 307 of the 1977 CLEAN WATER ACT, which also requires technology-based effluent standards for the control of these chemicals. See also FLANNERY DECREE.

probabilistic risk assessment (PRA) The analytical estimation of the probability of an undesired consequence using the probabilities of events that will lead to the consequence, such as a study of the likelihood of a catastrophic release of radiation from a nuclear power plant.

procarcinogen A chemical that is converted to an active CARCINOGEN through a variety of biotransformations that follow its ingestion and metabolism by animals. Most of the chemicals known to induce cancer fall into this category. For example, aflatoxin produced by the mold *Aspergillus* in grain stored under moist conditions and subsequently ingested by an animal is converted to aflatoxin-2,3-epoxide within the tissues of that animal. The epoxide form of the aflatoxin then combines with DNA to cause the mutation needed to induce cancer development.

procaryotic Describing an organism composed of cells that do not contain membrane-bound organelles such as nuclei, mitochondria, or chloroplasts. The genetic material of this type of cell is not associated with large chromosomes within the nucleus, as is characteristic of higher plants or animals. The bacteria and blue-green algae comprise this group. Compare EUCARYOTIC.

process water Any water that comes in contact with a new material or product. The water is often released as wastewater following use.

produced water As crude oil is extracted from a well, the water that comes to the surface with the oil. The produced water can constitute a large fraction of the total fluids extracted, and it is either pumped back into an underground formation via an INJECTION WELL or discharged to surface water.

producer An organism that is involved in the fixation of carbon dioxide that results in the gain of organic biomass within an environment, primarily green plants. The green plants, together with a few types of bacteria, assimilate carbon dioxide and other inorganic nutrients into organic materials which, in turn, serve as food for the consumers. See CONSUMER.

productivity The rate of biomass accumulation within an environment as a result of PHOTOSYNTHESIS. The rate of plant growth. See PRIMARY PRODUCTIVITY, GROSS PRIMARY PRODUCTIVITY, and NET PRIMARY PRODUCTIVITY.

products of incomplete combustion (PICs) Potentially hazardous organic materials formed during the incineration of wastes containing chlorinated organic compounds, or PRINCIPAL ORGANIC HAZARDOUS CONSTITUENTS. PICs in the exhaust gases of an incinerator could include chloroform, vinyl chloride, benzene, tetrachloroethane, and 1,1-dichloroethylene, among others.

profundal zone In deep lakes, the water stratum found beneath the LIMNETIC ZONE. The water is often APHOTIC (dark) and part of the HYPOLIMNION (cool).

progeny The offspring of animals or plants. The term is also used to describe the products that result from the decay of radioactive elements.

program impact statement An ENVIRONMENTAL IMPACT STATEMENT prepared for a broad action or a series of cumulative actions to be taken by a federal agency, such as the development of a new technology or a series of mineral leases on public lands.

promoter In the two-stage model of carcinogenesis, a substance that converts an initiated cell to a tumor cell. See CARCINOGENESIS.

pronatalist Describing an individual or a policy that encourages an increase in the human birth rate in order to achieve a larger human population. Compare ANTINATALIST.

propagule 1. Seeds, cuttings, or other plants parts capable of separately and individually starting a new plant. 2. A piece of fungal biomass that is capable of starting the development of a fungal colony if the proper nutritional and environmental conditions are met. The piece can be a viable fragment of fungal hyphae or a spore.

proportional counter An instrument used to measure IONIZING RADIATION. The counter usually consists of a gas-filled cylinder with a wire running through its center. A voltage is applied to the wire. Any ionizing radiation will ionize the gas in the cylinder, allowing a current to flow. The output voltage is proportional to the number of ionizing events.

prospective study See COHORT STUDY.

protective factor (PF) An expression of the efficiency of a RESPIRATOR used to protect the wearer from harmful air concentrations. The factor is calculated as the ratio of the air concentration of a particular air contaminant outside the respirator to the concentration of the contaminant inside the respirator. A higher PF means greater protection from that material.

proton One of the elementary nuclear particles. It is located in the nucleus of atoms, has a mass number of one, and carries a positive charge. The number of protons in the nucleus of an atom is referred to as the atomic number.

protoplast An intact bacterial cell or plant cell from which the cell wall has been removed. Protoplasts are produced in the laboratory, and they must be protected from OSMOTIC LYSIS.

proximate analysis A chemical analysis that determines the fractions of volatile carbon, FIXED CARBON, water, and noncombustibles in a waste. This crude analysis is an attempt to classify the waste for possible incineration without undertaking a more extensive analysis of the precise chemical composition of the material. Compare ULTIMATE ANALYSIS.

psychrometer An instrument used to determine RELATIVE HUMIDITY that consists of two thermometers, one dry and the other with its bulb surrounded by a wetted wick. Relative humidity can be read from a psychrometric chart using the two thermometer readings.

publicly owned treatment works (POTW)   Facilities for the treatment of domestic sewage that are owned and operated by a public body, usually a municipal government.

pug mill   A mechanical device to blend dry solids with waste material in order to improve handling characteristics of the waste.

pulmonary edema   An abnormal accumulation of fluids within the lungs. The condition can result from a bacterial infection or from an irritation caused by exposure to certain chemical agents.

pumped storage   Water pumped upgrade into a reservoir or lake during periods of low electric power consumption. The water added to the reservoir is later released through the hydroelectric facility to generate electricity during times of high power demand.

purgeable organic   A volatile organic compound that has a boiling point less than or equal to 200°C and that is less than 2 percent water-soluble.

purgeable organic carbon (POC)   The amount of carbon composing a PURGEABLE ORGANIC that can be removed from soil or water.

purgeable organic halogens (POX)   Organic derivatives containing chlorine, bromine, or fluorine that are found within a PURGEABLE ORGANIC that can be removed from soil or water.

putrefaction   The partial degradation of organic materials in the presence of an insufficient oxygen supply. The result is the release of noxious oxidation products and gases.

putrescible   Describing any substance that is likely to result in the production of a rotten, foul-smelling product as it is decomposed by bacteria, such as the plant and animal waste material in municipal solid waste.

pyramid of biomass   A visual representation obtained by depicting the total weight of organisms residing in each trophic level in a given area as a horizontal bar. The mass of plants within the area is placed on the bottom, and the mass represented by the top carnivores is placed on the top, with other forms of biota placed at intervening levels. Once arranged in an order representative of the feeding structure in the community, the drawing resembles a pyramid.

pyramid of energy   A drawing that is similar to a PYRAMID OF BIOMASS except that the total caloric content of the organisms at each trophic level is used to construct the representation.

pyrethroids   A class of natural insecticides developed from pyrethrum extracted from plants belonging to the genus *Chrysanthemum*. Agents belonging to this group of materials are nonpersistent in the environment.

PYRAMID OF BIOMASS

GRAMS DRY MASS PER SQUARE METER

pyrite A mineral consisting of iron sulfide. Burning of the mineral (frequently done when high-sulfur coal is burned) results in the release of OXIDES OF SULFUR (SO_x) into the atmosphere which contributes to ACID RAIN.

pyrolysis The thermal destruction of some material (e.g., coal, oil, wood, or other organic substance) in the absence of molecular oxygen. Also termed destructive distillation.

pyrophoric Describing a material that can ignite spontaneously in air at less than 130°F.

p-value A value, calculated by various statistical tests, that is the probability assuming that there is no difference between two samples in terms of a particular characteristic or disease rate, that the observed difference between two sample estimates is due to sampling variability alone. For example, in a study of the association between cigarette smoking and bladder cancer, the NULL HYPOTHESIS is that cigarette smoking is not a significant risk factor for bladder cancer. If the study of two groups finds that the incidence of bladder cancer in smokers is 16 in 10,000 and the incidence in nonsmokers is 8 per 10,000, and the p-value is 0.01, then there is only a one in 100 chance of seeing such a difference between the two groups if there really is no link between smoking and bladder cancer risk.

Q

q* See CANCER POTENCY FACTOR.

quad (q) An energy unit representing one quadrillion (a million billion, or 1×10^{15}) BTUs that is sometimes used to discuss national or global energy production or consumption. Compare QUINT.

qualitative analysis The examination of a substance or sample to determine what chemical compounds or elements are present irrespective of the amounts of those compounds or elements. Compare QUANTITATIVE ANALYSIS.

qualitative variable See CATEGORICAL VARIABLE.

quality assurance/quality control (QA/QC) All methods and procedures used to obtain accurate and reliable results from environmental sampling and analysis. Includes rules for when, where, and how samples are taken; sample storage, preservation, and transport; and, during the analysis, the use of BLANKS, duplicates, and SPLIT SAMPLES.

quality factor (QF) A factor by which a dose of IONIZING RADIATION is multiplied to obtain a quantity that corresponds to the biological effect of the radiation. The factor adjustment reduces radiation exposures to a common scale, weighing them by their ionizing potential. See RELATIVE BIOLOGICAL EFFECTIVENESS.

quantitation limit The minimum amount of some chemical that can be detected and measured with a suitable degree of reliability using currently available instrumentation. Usually three to five times the INSTRUMENT DETECTION LIMIT.

quantitative analysis The examination of a substance or sample to determine the precise amounts of certain chemicals or elements that are present. Compare QUALITATIVE ANALYSIS.

quantitative variable A characteristic that can be measured and expressed in units that describe the precise quantity present; for example, length, mass, or time. Compare CATEGORICAL VARIABLE.

quantum A unit of energy conveyed by an electromagnetic wave. The magnitude of the unit is proportional to the frequency of the wave; for example, a unit of ultraviolet light (relatively high frequency) conveys more energy than a unit of infrared light (relatively low frequency). Also referred to as a photon.

quencher The inlet section of a vessel designed to remove particles from an exhaust gas. The hot exhaust gas is cooled before particles are removed.

quint (Q) An energy unit representing one quintillion (1 × 10^{18}) BTUs that is sometimes used to discuss national or global energy production and consumption. See QUAD.

R

rabbit A container used for substances to be subjected to exposure in a nuclear reactor.

rad See RADIATION ABSORBED DOSE.

radiant heat INFRARED ENERGY emitted by a surface; the amount of heat is proportional to the fourth power of the ABSOLUTE TEMPERATURE of the radiating body, according to the STEFAN-BOLTZMANN LAW.

radiation Emitted energy, as particles or as electromagnetic waves. Based on energy content, the two major divisions are IONIZING RADIATION and NONIONIZING RADIATION.

radiation absorbed dose (rad) The non-SI unit expressing the amount of radiation absorbed by any medium. One rad is equal to an energy absorption of 100 ergs per gram of medium (usually human tissue). The equivalent SI unit is the GRAY. One gray equals 100 rad.

radiation chemistry The study of the effects of alpha and beta particles and x-rays and gamma rays on the structure and properties of matter.

radiation inversion An atmospheric temperature INVERSION (cooler air beneath warmer air) caused by the loss of heat, or infrared energy, from the ground on a cold, clear night. This radiation heat loss causes the air closest to the ground to be cooler than the air above it. By mid-morning incoming solar energy warms the air near the ground sufficiently to reestablish the normal atmospheric temperature profile within the atmosphere. During inversion conditions, dispersive capability of the atmosphere is lowered, and air pollutants typically increase in concentration. See also SUBSIDENCE INVERSION.

radiation shielding Material, such as lead, placed between people and a source of ionizing radiation to reduce human radiation exposure.

radiation sickness Acute adverse health effects, such as gastrointestinal upset, decreased blood cell production, and loss of hair, caused by overexposure to IONIZING RADIATION.

radiation sterilization The use of ionizing radiation, such as gamma rays, to either render a plant or animal incapable of reproduction or to kill all microorganisms associated with some material or product.

radioactive Describing an unstable atom that undergoes spontaneous decay, releasing IONIZING RADIATION.

radioactive decay The spontaneous emission of mass or energy from the nucleus of an unstable element. The emission results in the production of a new element or a different ISOTOPE of the same element. These new products may in turn undergo further decay. Each step in the decay process takes place at a predictable rate normally expressed as a HALF-LIFE. See RADIOISOTOPE, IONIZING RADIATION, and RADIOACTIVE SERIES.

radioactive series A series of ELEMENTS produced by RADIOACTIVE DECAY of unstable atoms, with one decay product following another until a stable element is reached. For example, the uranium series begins with uranium-238 and ends with lead-206. The other two naturally occurring radioactive element series are the thorium and actnium series.

radioactive wastes Nonuseful residual materials and equipment produced by nuclear power generation, nuclear reactor decommissioning, nuclear weapons manufacture, medical applications, and research. Depending on the RADIOACTIVITY and the half-life of the materials, they are classified as LOW-LEVEL or HIGH-LEVEL WASTE.

radioactivity The property exhibited by certain unstable ELEMENTS of spontaneously emitting mass or energy from the nucleus. See RADIOACTIVE DECAY, and RADIOISOTOPE.

radiocarbon See CARBON-14.

radiochemical Describing the activities of RADIOCHEMISTRY.

radiochemistry The study of the properties and use of RADIOACTIVE materials.

radiography The production of images using radiation other than visible light. The term usually refers to the production of images of the human body using X-RAYS.

radioisotope An unstable form of an element that undergoes RADIOACTIVE DECAY, emitting energy in the form of GAMMA RAYS or mass in the form of ALPHA PARTICLES or BETA PARTICLES.

radionuclide See RADIOISOTOPE.

radiopoison A substance, such as boron, that stops a fission chain reaction by absorbing neutrons and thereby preventing neutrons from striking a nucleus that can undergo fission.

radiosensitivity The susceptibility of living organisms or certain body tissues to the adverse effects resulting from exposure to IONIZING RADIATION. Humans are quite sensitive, with sensitivity decreasing with increasing age. Sensitive human body tissues include the lining of the gastrointestinal tract and the bone marrow, among others.

radiosonde A device containing meterological instruments and a radio transmitter that is carried aloft by a balloon. Measurements of air temperature, pressure, and humidity are transmitted to the ground. If wind speed and direction as a function of altitude are determined by ground tracking of the position of the balloon at various times, called a RAWINSONDE.

radius of influence For a groundwater well, the horizontal distance from the well to the point at which the WATER TABLE is not influenced (lowered) by the withdrawal of water from the aquifer.

radon (Rn) A radioactive element with atomic number 86 and atomic weight 222. The element is a gas that is produced directly from the radioactive decay of radium, which is a member of the uranium decay series. It has been proposed that radon provides about one-half of the radiation to which the average American is exposed. The chemically inert gas enters homes through soil, water, and building materials. The threat is not uniformly distributed across the United States. For example, dwellers in Colorado and Illinois are exposed to about four times the amount of radon as are dwellers in Louisiana and South Carolina. Even in those states where the average exposure is high, specific localities differ markedly from one another. An important source of personal exposure to radon appears to be drinking water obtained from wells. The threat comes from the inhalation of the gas released from water during showering, bathing, cooking, and other water uses. Ingestion of water does not appear to present a threat.

radon daughters Elements produced by the radioactive decay of radon, including isotopes of polonium and bismuth. These compounds, which are also radioactive, contribute to the hazard associated with exposure to radon.

radwaste A term for RADIOACTIVE WASTES.

rain forest See TROPICAL RAIN FOREST.

ramp method, landfill A method for the placement of solid waste in a landfill in which the waste is handled in the same way as in the AREA METHOD, but part of the COVER MATERIAL is obtained from an excavation at the base of each layer of waste, which forms a ramp.

range 1. For a set of observations, the difference between the lowest and highest values. 2. The thickness of some material needed to absorb a specific type of radiation.

range of tolerance See LIMITS OF TOLERANCE.

range-finding test An analysis of the toxicity of a pollutant chemical to an aquatic organism during which the concentration range to be used in the more precise DEFINITIVE TEST is determined. The organisms are exposed to a wide range of concentrations of the specific chemical for a relatively short time so that a narrow range for use in future detailed analysis can be determined.

rapid sand filter A water treatment method that removes suspended or colloidal particles as drinking water passes through a sand-filtering medium. Generally used after a sedimentation treatment. The filters are cleaned periodically by backwashing.

rappers Devices used on an ELECTROSTATIC PRECIPITATOR to mechanically dislodge the dust on the collection plates. Magnetic impulse rappers are raised by an energized coil and drop by gravity,

striking rods attached to the collecting plates. Hammer/anvil rappers have hammers attached to a rotating shaft; as the hammers fall they strike anvils attached to the collecting plate.

Raschig rings A type of packing used in packing tower gas absorption devices.

Rasmussen report A study commissioned by the United States Atomic Energy Agency, named for study director Norman Rasmussen, that estimated the probability and consequences of an accident at a nuclear power plant, published in 1975. One conclusion was that the chance of injury to an individual from a serious accident leading to a meltdown was negligible. The report is also referred to as WASH-1400.

rate In public health statistics, the frequency of a health-related event, calculated as the number of persons experiencing the event during a time period divided by the number of PERSONS AT RISK, or the population at risk. The value is usually expressed in a way that allows the use of whole numbers; for example, if deaths in an area divided by the area population equals 0.008 per year, this is written as 8 deaths per 1000 persons per year.

rate constant (K) The proportionality constant appropriate to the rate of a chemical reaction. In the differential equation $dX/dt = -kX$, the concentration of X is decreasing at a rate proportional to the remaining concentration of X. Such a reaction is called a FIRST-ORDER REACTION, and the k in the equation is the rate constant.

rating curve A graph depicting the DISCHARGE of a river or stream as a function of its water elevation (stage) at a certain point along its bank.

rational method A simple procedure for calculating the direct precipitation peak runoff from a watershed, using the rainfall intensity, the area of the watershed, and a runoff coefficient appropriate for the type of watershed runoff surface.

raw sewage Wastewater that has not undergone any treatment for the removal of pollutants.

raw water Groundwater or surface water before it is treated for use as a public water supply.

rawinsonde A RADIOSONDE that is tracked from the ground to determine wind speed and direction at various altitudes. The term is derived from *ra*dar, *win*d, and radio*sonde*.

Rayleigh scattering The scattering of visible radiation by gaseous molecules in an unpolluted, particle-free atmosphere. The degree of scattering is greatest at the shorter wavelengths of visible light. See VISIBLE RANGE and EXTINCTION COEFFICIENT.

reactant An element or compound taking part in a chemical reaction.

reactive waste Solid waste exhibiting the HAZARDOUS WASTE characteristic of interacting chemically with other substances in ways defined by United States Environmental Protection Agency regulations in the CODE OF FEDERAL REGULATIONS, Title 40, Part 261.23.

reactivity A property of a chemical substance in relation to other substances; the likelihood of a substance interacting chemically with other substances in the environment or in the body. See REACTIVE WASTE.

reactor-year A unit equaling one nuclear reactor in operation for one year. Probabilities of serious nuclear accidents are often stated in reactor-years. For example, if a MELTDOWN occurred once in 20,000 reactor-years and if 5000 nuclear reactors were in operation, a meltdown would be expected every four years.

readily water-soluble substances In water pollution, chemicals that are soluble in water at a concentration equal to or greater than one milligram per liter.

ready biodegradability A property of those substances that produce positive, unequivocal results when used in DECOMPOSITION tests.

reaeration (of streams) The natural process by which flowing stream water is mixed with the atmosphere, resulting in the addition of DISSOLVED OXYGEN to the water.

reagent A laboratory chemical added to a test for the purpose of promoting a specific reaction (for example, sulfuric acid is added to promote the digestion of samples for the measurement of the amount of nitrogen present) or for the development of a colored compound in a specific procedure (for example, sulfanilic acid and dimethyl-α-naphthalamine are added to the laboratory test for the presence of nitrites in water, the intensity of the color produced being determined by the amount of nitrite present).

reasonable further progress (RFP) A necessary feature of a STATE IMPLEMENTATION PLAN (SIP). An SIP outlines the emission controls and other methods that a state will use to achieve the NATIONAL AMBIENT AIR QUALITY STANDARDS by a certain date, sometimes several years away. The schedule of controls is required to show incremental progress toward the clean air goal during the period before the deadline in such a way that avoids "bunching" controls just before the SIP compliance date.

reasonably available control measures (RACM) The provisions in a STATE IMPLEMENTATION PLAN (SIP) for reducing air emissions from existing sources and limiting new emission sources in an area that does not meet one or more of the NATIONAL AMBIENT AIR QUALITY STANDARDS.

reasonably available control technology (RACT) The minimum level of emission control technology required for existing air pollutant sources in NON-ATTAINMENT AREAS. The RACT controls apply to emissions of the pollutant(s) for which the area does not meet the NATIONAL AMBIENT AIR QUALITY STANDARD.

rebuttable presumption against registration (RPAR) See SPECIAL REVIEW.

recalcitrant Of a substance that is degraded at an extremely slow rate if at all when released into the environment. Consequently, this type of material tends to accumulate in water, soil, or biota.

receptor 1. A substance, compound, or location on the surface of a cell that binds a specific chemical. This binding process can stimulate the cell to respond in certain ways, or the binding can take place prior to the uptake of the substance into the cell. 2. A person, plant, animal, or geographical location that is exposed to a chemical or physical agent released to the environment by human activities.

recessive gene A gene that is not expressed unless paired with a similar recessive gene. When male and female gametes fuse in the reproductive process, individual gametes furnish one member of each pair of chromosomes that will determine the genetic makeup of the offspring. If the gametes are from plants and if plant height is governed by one gene, the possible genes may be *h* for short variety and *H* for tall variety. In the case where *H* is dominant and *h* is recessive, a plant with gene pairs *HH* or *Hh* would be tall, since *h* is recessive and is not expressed in the presence of *H* gene. The gene pair in a short plant would have to be composed of *hh*. Compare DOMINANT GENE.

recessive gene disorder A genetic disorder that is exhibited in an offspring only if that offspring is double recessive for the gene that controls the condition.

In that circumstance each parent would have contributed a RECESSIVE GENE to the child.

recharge basin A man-made lake or reservoir designed to allow infiltration of water into the ground to recharge an underlying aquifer.

recharge zone A land area into which water can infiltrate into an AQUIFER relatively easily. The infiltration replenishes the aquifer. The location is also called a recharge area.

recirculation cooling system A process design that reuses industrial cooling water after heat from the water is transferred to the atmosphere through evaporation of a small amount of the water.

reclamation Restoring land to the natural state following its destruction associated with some economic activity such as surface mining. The original contour of the land is restored as much as is feasible, top soil and fertilizer are applied, and vegetation native to the region is planted.

recommended daily allowance (RDA) A guideline set by the federal Food and Drug Administration for the amounts of nutrients (such as iron, calcium, protein, and vitamins B & C) needed daily for the maintenance of good health. Varying RDAs are set for adults and children over three years old, children under four years, infants to one year old, and for pregnant or lactating women.

record of decision (ROD) The document containing the choice of remedial action to be taken at a hazardous waste SUPERFUND SITE; the ROD is based on the REMEDIAL INVESTIGATION/FEASIBILITY STUDY.

recycling The practice of collecting and reprocessing waste materials for reuse.

red bag waste Infectious waste, the name of which is derived from the red plastic bags used for clear identification.

Red Book 1. The 1976 publication issued by the United States Environmental Protection Agency, *Quality Criteria for Water,* that is used as a basis for ambient water quality standards. 2. The Food and Drug Administration 1982 guidelines for animal toxicity testing, *Principles for the Safety Assessment of Direct Food Additives and Color Additives Used in Food.*

Red Data Book A collection of the available information relative to threatened and ENDANGERED SPECIES. Each volume contains colored loose-leaf information sheets arranged by species. The sheets are updated as the status of a species changes. Red sheets are used for those species that are endangered; amber for vulnerable; white for rare; green for out of danger; and gray for species that are indicated to be endangered, vulnerable, or rare, but with insufficient information to be properly classified. The book is maintained by the International Union for Conservation of Nature and Natural Resources.

red tide An unusual condition associated with a bloom or excessive growth of dinoflagellates in marine waters, resulting in a red, brown, or yellow tint in the water. The event causes the death of marine biota and the accumulation of toxins in mussels or clams. Consumption of the toxin-containing shellfish can cause paralytic shellfish poisoning or severe gastric distress in humans.

redox potential An expression of the oxidizing or reducing power of a solution.

reducing agent Any substance that loses electrons when involved in an oxidation/reduction reaction, or a substance that will combine with oxygen during a chemical reaction.

reduction A chemical reaction during which electrons are added to an atom or molecule. With organic compounds, the addition of electrons is frequently accompanied by the addition of hydrogen atoms. Also called hydrogenation of organic compounds.

reference daily intake (RDI) The international term equivalent to RECOMMENDED DAILY ALLOWANCE.

reference dose (RfD) The lifetime (chronic) daily exposure level to a non-carcinogen that will protect sensitive human populations from adverse effects; developed by the United States Environmental Protection Agency for exposure evaluations at SUPERFUND SITES. The agency is also developing subchronic reference doses, for shorter-term exposures, and developmental reference doses, for single or short-term exposures to fetuses, infants, or children.

reflux A method that uses solvents to extract certain classes of chemicals from solid materials. For example, the extraction of methanol-soluble chemicals from soil can be accomplished by placing the soil into a reflux apparatus which allows the methanol in a reservoir to be boiled, the vapor to be condensed, and the condensate to percolate through the soil sample and return to the boiling reservoir. As the methanol cycles through the system several times, the methanol in the reservoir accumulates soil materials that dissolve in the methanol.

reforestation The planting of trees on land from which the forest has been removed.

refraction The bending of light as it passes through transparent material.

refractory 1. A highly heat-resistant material that can be used as a lining for incinerators or furnaces. 2. Describing a material or substance that is difficult to metabolize within biological systems or react chemically during chemical pro-

cesses. See also WATERWALL INCINERA-
TOR.

refuging   In ECOLOGY, the aggregation
of large numbers of organisms in a small
area with nearby food supplies, bird colonies, for example.

refuse   All organic solid waste produced by a community. The combustible
fraction of solid waste, including garbage,
rubbish, and trash.

refuse-derived fuel (RDF)   An energy resource produced by the separation of combustible materials (refuse) from
municipal solid waste. This material is
then combined with coal and burnt as a
refuse-derived fuel, which is usually 5 to
25 percent refuse.

Regional Administrator (RA)   The
head of one of the 10 multistate regions
organized by the United States Environmental Protection Agency.

Regional Response Center (RRC)
Under the NATIONAL CONTINGENCY PLAN,
designated locations in each of 10 federal
regions which will provide communica-
tions and coordination for the response
to an oil or hazardous substance release.
See also REGIONAL RESPONSE TEAM.

Regional Response Team (RRT)
An organization under the joint leadership of the Environmental Protection
Agency and the United States Coast
Guard that serves as the organizational
unit to provide for planning and prepar-
edness activities related to spills or discharges of oil and hazardous substances
and for coordination and technical advice
during such spills or discharges in each
of the 10 federal regions outlined by the
EPA.

**Registry of Toxic Effects of Chem-
ical Substances (RTECS)**   A publication of the NATIONAL INSTITUTE OF OC-
CUPATIONAL SAFETY AND HEALTH that is
intended to identify all known toxic ma-

terials and to list toxicity and reference
information on each entry.

Regulatory Impact Analysis (RIA)
An estimate of the benefits and costs of
any federal regulation or standard that
may result in an annual impact of $100
million or more to the economy of the
United States; the analysis is used by the
Office of Management and Budget in its
review of the rule, as required by EXEC-
UTIVE ORDER 12291.

reheater   A device used to increase
the temperature of an exhaust gas after
particles have been removed. Scrubbing
devices that are used to remove particles
from an exhaust gas require that the
temperature of the gas-particle mixture
be lowered before cleaning can take place.
Following the scrubbing process, the gas
must be reheated before release in order
to prevent damage to downstream
equipment, to avoid the release of a visible plume through the smokestack, or
to enhance the rising and dispersion of
the released gas.

Reid vapor pressure (RVP)   The
American Society for Testing and Materials method for measuring the VAPOR
PRESSURE of VOLATILE ORGANIC COM-
POUNDS (VOCS). To control the evaporative losses of VOCs, the United States
Environmental Protection Agency requires refineries to produce gasoline with
a lower vapor pressure during the warmer
months.

**relative   biological   effectiveness
(RBE)**   A comparison of the effect of
some type of radiation to that caused by
200-keV x-rays. The relationship to
gamma rays is about 1, whereas the relationship to fast neutrons is approximately 10, meaning that the latter type
of radiation is about 10 times more effective than x-rays in producing a particular biological change. The RBEs are
used as the QUALITY FACTOR to adjust
absorbed doses of different types of ionizing radiation to a constant scale.

relative humidity   The ratio of the amount or weight of water vapor present in a specified volume of air to the maximum amount of water vapor that can be held by the same volume of air at a specified temperature and pressure. If the relative humidity is 50 percent in a room, the air contains one-half the amount of water vapor that it is capable of holding at the temperature and atmospheric pressure in the room. Also expressed as the MIXING RATIO divided by the SATURATION MIXING RATIO, times 100.

relative risk   The ratio of the risk of some adverse condition developing in individuals exposed to pollution to the risk of that same condition developing in an unexposed population.

release height   The height above the ground from which an air pollutant is emitted. A higher release height produces a lower ground level air concentration at points downwind of the source.

relict   A species or community that is unchanged from some earlier period of time.

rem   See ROENTGEN EQUIVALENT MAN.

remedial investigation/feasibility study (RI/FS)   A detailed, technical study of the type and extent of contamination at a SUPERFUND SITE, including the alternatives for its cleanup.

remedial project manager (RPM)   The federal official responsible for directing and coordinating federal remedial actions at sites on the NATIONAL PRIORITIES LIST. Remedial actions are taken to prevent or minimize the release of a hazardous substance into the environment. Removal actions are the responsibility of the ON-SCENE COORDINATOR.

remote sensing   1. The use of imagery generated by high-flying aircraft or earth-orbiting satellites to study the resources of the earth or aspects of environmental quality.   2. The use of ground devices to study atmospheric conditions aloft or to determine air pollutant concentrations in the upper urban atmosphere or a smokestack plume.

renewable energy   A source of energy that is replaced by natural phenomena, such as firewood or the water held behind by a dam used for hydroelectric purposes. Conversely, fossil fuels are a nonrenewable source of energy.

renewable resources   A supply of biological organisms that can be replaced after harvesting by regrowth or reproduction of the removed species, such as seafood or timber. Resources such as natural gas or mineral ores are nonrenewable.

replicate   Two or more duplicates of a test, analysis, experiment, exposure, and so on.

reportable quantity (RQ)   The size of a spill or release of a hazardous substance or material requiring a report to the United States Department of Transportation, the United States Environmental Protection Agency, or a state or local environmental agency.

repository   A site for the isolation of radioactive wastes from humans or the biological environment for an indefinitely long period of time. Such storage facilities would be located about 2000 feet below the surface in geological deposits that represent stable, dry environments with good heat-dissipation characteristics. The site would be designed to isolate high-level radioactive waste for about 10,000 years.

representativeness   How well a sample of air, water, soil, or food represents the whole from which the sample was taken.

reprocessing nuclear fuel   See FUEL REPROCESSING.

reproducibility   The ability of a test, analysis, or experiment to yield the same results when repeated under the same circumstances. The likelihood that duplicate tests will produce the same results.

Research Triangle Park (RTP)   An area in North Carolina containing the offices of many United States Environmental Protection Agency technical and scientific activities together with a variety of private laboratories and public institutions of higher education.

reservation   In public land management, a permanent dedication of land to, usually, a single purpose, such as a national forest, wildlife refuge, or wilderness area.

reserves   Natural resources that can be exploited in an economically feasible manner employing current technology. Compare RESOURCES.

reservoir   At the planetary scale, a storage depot for chemical substances, for example the atmosphere as a reservoir for nitrogen gas. See STOCKS.

reservoir rock   Porous rock containing oil or natural gas.

residence time   1. The time a chemical substance spends in a biological or environmental "container" such as a lake, the human body, or the atmosphere. Under equilibrium conditions, calculated by the mass of the material in the container divided by the flow rate in (or out). 2. The amount of time a specified volume of liquid or gas spends in a certain device. The residence time for sewage undergoing treatment is the amount of time a specified volume of sewage stays in the treatment facility.

resilience stability   The ability of a biological community or ecosystem to recover its original condition following a severe stress or perturbation. Compare RESISTANCE STABILITY.

resistance stability   The ability of a biological community or ecosystem to withstand a stress or perturbation without adverse change to its structure or function. Compare RESILIENCE STABILITY.

resistivity   For FLY ASH, the resistance of the ash to the conduction of electricity. Low resistivity reduces the collection efficiency of ELECTROSTATIC PRECIPITATORS by allowing the electrostatic charge attracting the particles of ash to the collection plate to drain away. High-resistivity ash holds a negative charge and can repel other particles (BACK CORONA); resistivity also causes the particles to adhere strongly to the plate, making them difficult to remove. See FLUE GAS CONDITIONING.

resistivity survey   A type of noninvasive survey of soil for the detection of buried metal objects and chemical contaminants capable of conducting an electric current. The method is especially useful for the evaluation of groundwater depth and site geology.

resolving time   The time that must elapse between two radiation impulses in order for them to be detected as two discrete events by an electronic radiation detection device.

Resource Conservation and Recovery Act (RCRA)   The 1976 federal statute, amended in 1980 and in 1984, providing for the comprehensive management of nonhazardous and hazardous solid wastes. The United States Environmental Protection Agency, in its implementing regulations, sets minimum standards for all waste disposal facilities, and, for HAZARDOUS WASTE, regulates treatment, storage, and transport. See MANIFEST SYSTEM, SANITARY LANDFILL, SECURE LANDFILL, SUBTITLE C, SUBTITLE D, and HAZARD AND SOLID WASTE AMENDMENTS.

resource recovery   The processing of solid waste to extract recyclable paper, glass, metals, or combustible material. Also, the recovery of energy (steam) by the direct incineration of waste.

resources The total amount of any rock, mineral, or fuel in the crust of the earth. See RESERVE.

respiration 1. For a person or animal, the act of breathing. 2. The metabolism of an individual cell, tissue, or organism resulting in the release of chemical energy derived from organic nutrients. 3. A specific series of reactions in a cell during which electrons removed during the OXIDATION of organic nutrients (substrates) are transferred to a terminal acceptor. When the terminal acceptor is molecular oxygen, the process is termed AEROBIC respiration. When the terminal acceptor is an inorganic substitute for molecular oxygen, such as sulfate or nitrate, the process proceeds without oxygen and is termed ANAEROBIC respiration. In either case, potential energy present in the organic nutrients is converted to a chemical form useful to the cell. Most organisms, including man, are capable of aerobic respiration, whereas only a few types of bacteria are capable of anaerobic respiration.

respiration/biomass ratio An expression of ecological TURNOVER, the ratio of total community RESPIRATION (often expressed in grams per square meter per day) to community BIOMASS (in grams per square meter). The ratio indicates the rate of energy flowing through the community compared to the mass of the community.

respirator Any device designed to deliver clean air to the wearer. The two general types are an AIR-PURIFYING RESPIRATOR and a SELF-CONTAINED BREATHING APPARATUS.

respirator fit test (RFT) A method used to determine if a working respirator provides the proper protection when worn by a particular worker. A quick, simple, and low-cost RFT is the qualitative test, which involves the wearer's response to a test atmosphere containing an easily detected material such as banana oil or irritant smoke. If the wearer can detect the material in the test atmosphere, the respirator does not fit properly. Quantitative fit testing is more accurate but more expensive and time-consuming, and is not widely used. The quantitative analysis involves highly trained technicians using instrumentation to determine the presence and extent of any respirator leaks.

restoration The establishment of natural land contours and vegetative cover following extensive degradation of the environment caused by activities such as surface mining. Same as RECLAMATION.

resuspended Describes particles that have been remixed with the air or water from which they have settled. For example, sediment particles will settle from river water if the water is allowed to stand. Those particles may be remixed with the water if turbulent conditions reoccur.

retention basin A permanent lake or pond used to slow stormwater runoff. See DETENTION BASIN.

retention time The interval of time that some waste, fluid or other material is in a treatment facility or process unit.

retort A bulb-shaped laboratory apparatus for heating some material in the absence of oxygen.

retrofit The addition of new parts or equipment that were not previously available or required, such as pollution control equipment, when the facility was built: a type of upgrading.

retrospective study See CASE-CONTROL STUDY.

reverse osmosis A process used in water purification in which pressure is applied to the more concentrated (contaminated) solution on one side of a semipermeable membrane. The result is the movement of water, but not contaminants, from the more concentrated side to the more dilute side, thus separating clean water from the contaminated water. See OSMOSIS.

Reynolds number (Re)   A dimensionless number that represents fluid flow in a pipe, duct, or around an obstacle, such as an airborne particle. The expression for fluid flow in a pipe or duct is

$$Re = \frac{\rho D v}{\mu}$$

where ρ is the fluid density, D is the pipe/duct diameter, v is the fluid velocity, and μ is the fluid viscosity. For fluid flow around a particle it takes the form

$$Re = \frac{d_p v_r \rho}{\mu}$$

where d_p is the particle diameter, v_r is the velocity of the particle relative to the fluid, ρ is the fluid density, and μ is the fluid viscosity. For fluid flow in a pipe or duct, a Reynolds number below about 2100 is considered to be streamline, smooth, or LAMINAR FLOW; above 4000 the flow is turbulent; 2100–4000 is a transition zone. For the flow of fluid around a particle, a Reynolds number less than 1.0 is laminar flow, or in the Stokes regime, and as the value increases above 1.0 turbulence increases. The difference between the conditions for laminar flow around particles and in pipes is explained by the impact of inertial forces as the fluid flows around a particle compared to the straight flow in a pipe or duct.

ribbon sprawl   The development of residential areas and businesses along major roadways, such as the Interstate Highway System, that carry traffic into a city.

Richardson number (Ri)   A dimensionless number for a fluid (usually air) at a certain place and time, expressed as follows:

$$R_i = \frac{(g/\Theta)(d\Theta/dz)}{(du/dz)^2}$$

where Θ is the POTENTIAL TEMPERATURE, z is the elevation, μ is the wind velocity,

g is the acceleration of gravity. The value of the number indicates the dominant type of turbulence present, either mechanical or convection (thermal). See MECHANICAL TURBULENCE and THERMAL TURBULENCE.

richness   1. The total number of species in an area, usually expressed as the number of species divided by the total number of individuals, or the number of species per unit area. See SPECIES RICHNESS INDEX. 2. The relative abundance of the fissionable form of uranium (^{235}U) in a mixture of uranium containing both of the naturally occurring ISOTOPES of the chemical (^{235}U and ^{238}U).

rickets   A medical condition, normally seen in children, characterized by poor bone structure resulting from a deficiency of vitamin D.

ring compound   Normally used to refer to an organic compound containing an AROMATIC structure typified by BENZENE, in which six carbon atoms are joined together in a closed circle or ring.

Ringelmann chart   An illustration containing shades of gray that can be used to estimate the density of smoke. The scale varies from 0, clear stack gases, to 5, 100 percent black smoke. See also SMOKE READER.

riparian   1. Describing the legal doctrine that says a property owner along the banks of a surface water body (lake, river) has the primary right to withdraw water for reasonable use. 2. Related to plant communities located on the banks of rivers.

risk assessment   An analytical study of the probabilities and magnitude of harm to human health or the environment associated with a physical or chemical agent, activity, or occurrence. The assessment involves estimates of the types, quantities, and locations of the release of harmful substances or energy; a dose-response evaluation linking exposure to possible harm; a characterization of the exposure of humans, wildlife, or ecosystem components to the release; and a summary using the foregoing analysis to produce the overall risk assessment. The final step is also known as RISK CHARACTERIZATION. See HAZARD IDENTIFICATION, SOURCE/RELEASE ASSESSMENT, DOSE-RESPONSE ASSESSMENT, and EXPOSURE ASSESSMENT.

risk characterization An estimation of the nature, magnitude, and likelihood of adverse effects on human health or the environment using the results of the previous analyses in a RISK ASSESSMENT.

risk estimate The probability of an adverse health effect associated with an exposure to a physical or chemical agent. See CANCER POTENCY FACTOR.

risk extrapolation model Any of a variety of methods used to apply the measured results of animal toxicology testing, especially those involving a potential CARCINOGEN, to the risk of adverse effects in humans. The EXTRAPOLATION involves the species differences, the estimation of the risk at much lower doses than those used in the testing, a longer human exposure time, and possibly a different route of exposure (human inhalation risk from animal ingestion data). See SCALING FACTOR, SURFACE AREA SCALING FACTOR, LINEARIZED MULTISTAGE MODEL.

risk factor Any exposure, habit, or personal characteristic linked to an increased risk of an adverse health effect or condition.

risk-specific dose (RSD) The daily dose or exposure level that implies a given risk level; for example, the dose of chemical A, in grams per kilogram body weight per day, corresponding to a 1 in 100,000 lifetime excess cancer risk.

R-meter A type of IONIZATION METER designed to measure radiation in ROENTGENS.

rod deck absorber A device used to remove pollutants from an exhaust gas. The exhaust gas is brought into contact with a liquid slurry used to remove pollutants from the gas. Mixing of the liquid and gas is enhanced by decks of cylindrical rods positioned perpendicular to the gas and liquid flows.

roentgen (R) The quantity of radiation (x-rays or gamma rays) that produces one electrostatic unit charge (or 2.08×10^9 ion pairs) in 1 cm^3 of dry air at 0°C and 1 atmosphere of pressure. Roentgens per unit time is the normal expression used when the unit is used to report radiation exposures. One roentgen equals 2.58×10^{-4} coulomb per kilogram of air.

roentgen equivalent man (rem) A non-SI unit of radiation dose that incorporates both the amount of ionizing radiation absorbed by tissue (RAD) and the relative ability of that radiation to produce a particular biological change, called the RELATIVE BIOLOGICAL EFFECTIVENESS (RBE). Expressed as rem = rad × RBE, where rad is an absorbed dose of 100 ergs per gram of tissue (0.01 GRAY). RBE is also called the QUALITY FACTOR. The unit is frequently applied to total body exposure for all types of ionizing radiation. Approximately 0.1 to 0.2 rem per year represents the level of exposure of individuals to natural background radiation. Five rem per year is the maximum occupation exposure allowed. An exposure of 100 rem in a single incident is thought to increase the risk of leukemia, and an exposure to 600 rem over 48 hours is a sufficient dose to kill 100 percent of the individuals exposed. The corresponding SI unit is the SIEVERT (Sv); 1 sievert equals 100 rem.

rollback model A control strategy by which excessive environmental concentrations or pollutants are to be reduced by lowering the emissions of all sources by an equal percentage. The percentage reduction is calculated using the amount by which the ambient concentration exceeds a standard. For example, if the ambient air standard for pollutant A is 8 and the measured ambient concentration is 10, then a 20 percent reduction by all sources will be required (10 minus 8, divided by 10, times 100).

room constant   The sum of the surface areas of the various materials in a room, weighted by their respective SABIN ABSORPTION COEFFICIENTS. The room constant includes walls, doors, floor and ceiling, together with all furniture, cabinets, draperies, and so on. Room constant units are area-sabins, for example 3 m^2-sabins.

room-and-pillar mining   A method employed in the underground mining of coal in which portions of the coal deposit are left undisturbed as columns that support the roof of the mined cave to prevent collapse.

root-mean-square sound pressure (rms sound pressure)   An expression of the average sound pressure (P_{rms}) from a series of sound wave AMPLITUDE measurements. The value is represented by

$$P_{rms} = \left[\frac{1}{T} \int_0^T P^2(t) \, dt \right]^{1/2}$$

where P is an instantaneous pressure measurement and T is the averaging time. This method is used because sound wave amplitude values fluctuate around a zero pressure, positive to negative, and squaring the pressure measurements avoids an average pressure of zero. Sound-level meters perform these calculation internally.

rotameter   A device used to measure gas flow rates; the meter consists of a tapered tube containing a float that rises as gas flows through the tube and that remains suspended as long as the gas flow is constant. The height of the float is read, using calibrated marks on the tube, and converted to a flow rate.

rotary kiln   A long inclined rotating drum, designed to manufacture cement or lime, used for thermal destruction of hazardous waste. Typical kiln operating temperatures and furnace residence time ensure excellent DESTRUCTION AND REMOVAL EFFICIENCY of the waste.

rotating biological contactor (RBC)   See BIODISC.

rotating biological filters   See BIODISC.

rotenone   A commercial preparation containing extracts of the tuber roots of plants belonging to the genus *Derris*. The most common use is to paralyze the fish in a pond or other body of water for their removal or collection in ecological studies. The extracts have also been used to control head lice.

route of exposure   The way in which an organism is exposed to a chemical agent, such as inhalation, ingestion, or skin contact.

r-selected   A type of reproductive strategy of a species that allocates a relatively high proportion of energy available in nutrients to reproduction and a relatively low proportion to the maintenance and survival of individual members of the species. See R-STRATEGIST. Compare K-SELECTED.

r-strategist   A type of animal whose reproductive behavior is characterized by "boom-to-bust" swings in population levels. Typically, animals with this type of reproductive strategy produce large numbers of offspring in a short period of time to take advantage of favorable environmental conditions. These explosions in numbers of animals are routinely followed by dramatic declines in numbers once the conditions change slightly. Likewise, the dramatic increases in population levels normally exceed the capacity of the habitat to support the species. The result is a precipitous decline in numbers of organisms present. See R-SELECTED.

rubbish   Nonkitchen solid waste, such as yard litter, junk, or other material. Also called trash.

rules   Standards and regulations issued by administrative agencies to implement laws.

run   The time period during which an emission sample is taken, as in a sampling run for a smokestack. The term is also used as a slang expression to describe the performance of a laboratory analysis: "The Ames test was run."

runoff   That portion of rain water or snow melt that enters surface streams rather than infiltrating into the ground.

rutherford (Rd)   A unit that describes the amount of some radioactive substance that decays at a rate of 10^6 disintegrations per second.

S

sabin   The SI unit used to compare the sound absorption ability of materials. A sabin is the unit area of a totally absorbent surface, that is, a surface that does not reflect any sound. One m^2-sabin is one square meter of perfect absorption. Sound-absorbing materials are given SABIN ABSORPTION COEFFICIENTS, which are expressed as a fraction of a sabin.

sabin absorption coefficients   Fractions of a sabin assigned to different materials indicating their sound absorbance relative to a totally absorbent surface. A sabin absorption coefficient of 0.15 indicates that 15 percent of the sound energy striking the surface will be absorbed and 85 percent will be reflected. Sabin absorption coefficients vary with sound frequency. The average absorption coefficient, over several frequencies, for a

material is the NOISE REDUCTION COEFFICIENT. Also see ROOM CONSTANT.

Safe Drinking Water Act (SDWA)
A federal statute enacted in 1974, and subsequently amended, that required the United States Environmental Protection Agency to set chemical and radioactivity standards for public drinking water supplies. The Act also began an UNDERGROUND INJECTION CONTROL program to protect underground drinking water sources and established a special SOLE SOURCE AQUIFER designation process. See MAXIMUM CONTAMINANT LEVEL, SECONDARY MAXIMUM CONTAMINANT LEVEL, and NATIONAL PRIMARY DRINKING WATER REGULATIONS.

Safety Control Rod Ax Man (SCRAM)
The individual responsible for the emergency lowering of control rods into the first nuclear reactor operated by the United States during the World War II Manhattan Project. The control rods were suspended by ropes; in the event of a malfunction, this individual was assigned the responsibility of cutting the ropes with an ax to allow the control rods to fall into the reactor and thereby stopping the fission reaction. The acronym SCRAM is used today to signify any sudden shutdown of a nuclear reactor by the emergency insertion of control rods to stop the fission chain reaction.

sag pipe   See INVERTED SIPHON.

salinity   The amount of salts dissolved in water. The term is usually reported in grams per liter (parts per thousand), and the unit symbol ‰ is normally used to signify the parameter. Although the measurement takes into account all of the dissolved salts, sodium chloride normally constitutes the primary salt being measured. As a reference, the salinity of seawater is approximately 35 ‰.

salinization   The accumulation of salts in soil to the extent that plant growth is inhibited. This is a common problem when crops are irrigated in arid regions; much of the water evaporates and salts accumulate in the soil.

Salmonella The genus name of bacteria that cause typhoid fever and that are associated with some types of food poisonings.

salmonellosis The bacterial disease caused by the presence of bacteria belonging to the genus *Salmonella*. The disease is a type of food poisoning characterized by a sudden onset of gastroenteritis involving abdominal pain, diarrhea, fever, nausea, and vomiting. A variety of foods, such as sweets, meats, sausages, and eggs, can be the mode of infection. Pet turtles and birds can also transmit the bacteria.

salt A chemical class of ionic compounds formed by the combination of an ACID and a BASE. Most salts are the result of a reaction between a metal and one or more nonmetals.

saltwater intrusion The movement of seawater into coastal areas normally flooded with fresh water. The term is applied to the flooding of freshwater marshes by seawater, the migration of seawater up rivers and navigation channels, and the movement of seawater into freshwater aquifers along coastal regions.

sample A representative portion withdrawn from a larger whole to determine some characteristic, such as the concentration of some constituent in a body of water, the atmosphere, or a waste stream. For example, if one wishes to investigate the level of lead contamination in a lake, a 500-milliliter portion of water can be withdrawn from the lake and the amount of lead in the 500 milliliters determined. The results are taken to be representative of the total lake.

sample size The amount of water, air, waste, and so forth, withdrawn from a larger whole in order to measure the level of some constituent. The term is also used to refer to the number of individual portions removed from the whole. See SAMPLE.

sampling variability Differences among replicate SAMPLES taken to determine the level of some constituent in a larger whole. For example, five separate 500-milliliter portions may be taken at the same location within a lake to determine the level of lead contamination in the water. The differences among the quantities of lead found in the separate portions would be referred to as the sampling variability.

sand filter A device used to remove particles from drinking water prior to distribution to customers. The water is allowed to percolate through a chamber containing sand of various grain sizes, with the finest grain size located on the top. The particles in the water are removed at the surface of the sand and later discarded by reverse flushing.

sandstone aquifer The type of aquifer supplying groundwater to large parts of the United States upper Middle West, Appalachia, and Texas. The water-bearing formation is often contained by shale strata, and the water has high levels of iron and magnesium.

sanitary landfill A method of ground disposal of solid waste in which waste material is spread in relatively thin layers, compacted, and covered with clean earth by the end of each working day, as required by current land disposal regulations for municipal solid waste. The daily COVER minimizes odors, insects, rodents, blowing trash, and smoldering fires that often accompanied the now-banned open dumps.

sanitary sewer A pipe or network of pipes used to transport municipal, commercial, or industrial wastewater (SEWAGE). See SEWER.

sanitary survey An onsite inspection by a trained public health technician to assess environmental conditions in a community, with special emphasis on communicable diseases. The survey could include sewage treatment and disposal facilities, public or private water supplies and distribution, solid waste collection and disposal methods, pest control activities, and public swimming pools.

sanitized copy   A document submitted by a facility to an environmental regulatory agency from which confidential information, such as a trade secret, has been removed.

saprophage   An organism that consumes dead organic matter.

saprophytic   Describing an organism that derives its nutrition from dead organic material. This is contrasted with a parasitic organism which obtains its nutrition at the expense of a living organism.

sarcoma   A malignant tumor, the growth of which begins in connective tissue. See also CARCINOGENESIS.

saturated zone   The zone in the earth's crust, extending from the water table downward, in which pore spaces in the soil or rock are filled with water at greater than atmospheric pressure. Also called the zone of saturation. Compare UNSATURATED ZONE.

saturation current   The movement, by an applied potential, of all ions in a gas. This is a measure of the ion production rate, and can be used as a method of gauging the radioactivity of a substance.

saturation mixing ratio   The maximum water vapor concentration in the atmosphere for a given air temperature. The higher the air temperature, the higher the saturation mixing ratio.

savanna   One of the major types of natural terrestrial communities. A region where the natural flora is dominated by grasses, sedges, and small shrubs.

saxitoxin   The primary toxin produced by dinoflagellate protozoans during blooms known as red tides in marine waters. The genus or protozoan involved in the generation of the red color in the water is *Gonyaulax*. See PARALYTIC SHELLFISH POISONING.

Sax toxicity ratings   A four-point integer scale for chemical toxicity used by N.I. Sax and R. J. Lewis in *Dangerous Properties of Industrial Materials* (Van Nostrand Reinhold, New York; 1988). A 0 rating means nontoxic; 1, slight toxicity; 2, moderate toxicity; 3, severe toxicity. Sax toxicity ratings are used in the HAZARD RANKING SYSTEM for hazardous waste sites.

scale   Deposits of solids (usually salts of calcium) that adhere to the inner surfaces of pipes, boilers, scrubbers, and mist control devices. HARD WATER leaves a deposit (scale) in steam irons, coffee makers, and water heaters.

scaling factor   An adjustment used in cancer risk assessments to compensate for the differences in size and metabolic rate between humans and experimental animals. The various scaling factors applied to dose-response data are based on relative body weights, relative surface areas, and differences in the mass of food and water ingested per body mass per day. See also SURFACE AREA SCALING FACTOR.

scanning   electron   microscope (SEM)   A microscope that employs a beam of electrons as the source of illumination. The specimen to be examined is usually coated with some metal such as gold, and the beam of electrons moves back and forth (scans) over the surface of the specimen. The electrons that are reflected from the surface form an image on a photographic plate. The instrument can also be used to determine the composition of metal ions in a specimen.

scarify   In land restoration activities, to stir the surface of the ground with an implement in preparation for replanting.

scavenging coefficient The exponential constant (Λ) in an EXPONENTIAL DECAY model for the physical removal of particulate from the air by rainfall: $X_t = X_0 e^{-\Lambda t}$, where X_0 is the particulate concentration at time 0, t is the number of time units since the rainfall began, e is the base of the natural logarithm, and X_t is the particulate concentration at time t.

scavenging mechanisms See NATURAL SINK.

scheduled outage A planned shutdown of industrial equipment for periodic inspection and maintenance of machinery.

scheduled wastes A listing of hazardous wastes, ranked by their health and environmental hazards and their volume, for which the United States Environmental Protection Agency was required to set predisposal treatment standards. A waste meeting the treatment standards may be disposed in a landfill or surface impoundment. The list was divided into thirds and the treatment standards were required by the HAZARDOUS AND SOLID WASTE AMENDMENTS to be issued for each third by a certain date, or a restriction on disposal of the waste would take effect. See LAND DISPOSAL BAN and SOFT HAMMER.

schistosomiasis A debilitating disease, common in underdeveloped regions of the world, caused by a small roundworm named *Schistosoma*. The disease is transmitted to humans through contact with water contaminated by fecal material. Infected humans discharge eggs in feces. These eggs hatch in freshwater, producing a small immature form of the parasite that infects snails common in streams, ponds, and lakes. The life cycle of the worm continues in the snail, and a second immature worm reenters the water where it later infects humans who come in contact with the infected water.

Science Advisory Board (SAB) An independent body established by the Administrator of the United States Environmental Protection Agency (EPA) in 1974 and by Congress in 1977. Its purposes are to review the scientific merits of EPA research and the scientific basis for the agency's proposed regulations and standards.

scientific method A systematic method of inquiry that includes the identification of a specific question or problem, the accumulation of the available information relating to that question, the proposal of a tentative answer to the question or problem, the conduct of methodical observations or experiments to test the proposed answer, and the rational interpretation of the results of the observations or experiments.

scientific support coordinator (SSC) Under the NATIONAL CONTINGENCY PLAN for spills or releases of oil or hazardous substances, an official responsible for providing scientific support to the ON-SCENE COORDINATOR or REMEDIAL PROJECT MANAGER. The SSC also serves as the liaison between the federal responders and the scientific community.

scintillation counter An instrument that detects and measures radioactivity by counting flashes of light (scintillations) produced when radiation strikes certain chemicals. A sample that is suspected of containing radioactive substances is placed in a vial and mixed with a solution referred to as a cocktail, a mixture of fluors (compounds that emit light when struck by ionizing radiation). The system is useful in quantifying radiation from weak sources.

scoping An early stage in the environmental impact assessment that is required of all major United States government "actions" by the NATIONAL ENVIRONMENTAL POLICY ACT. Scoping identifies the most important environmental issues raised by the proposed action. All public and private organizations that may be affected and thus should be part of the impact assessment are invited to the scoping meetings or are identified and informed as part of the scoping process. See ENVIRONMENTAL IMPACT STATEMENT.

scram   See SAFETY CONTROL ROD AX MAN.

scree   The stones or debris that accumulates at the base of a hill. Material that has rolled down a hill and come to rest at the bottom.

screening   1. In the testing of chemicals for health effects, any method useful for the preliminary judgment of their toxic potential, which can include short-term biological tests. 2. In public health studies, any method used to indicate the desirability of further testing of individuals, such as a screening questionnaire that indicates alcohol abuse. 3. In water pollution control, the initial separation of large entrained debris from the wastewater as it enters a treatment facility.

scrubber   A device designed to remove pollutant particles or gases from exhaust streams produced by combustion or industrial processes. See below.

scrubber, impingement   An air particulate control device that passes dirty air through a water spray and then removes, the particulate-containing water by deposition (impingement) on a solid surface.

scrubber, spray   An air pollution control device that removes particulates or gases from an airstream by spraying liquid into the air duct and then collecting the pollutant-containing droplets. See SPRAY TOWER.

scrubber, venturi   An air pollution control device that operates by the introduction of a liquid into a narrow throat section (venturi) of an air duct that is carrying a contaminant. The high velocity in the venturi, compared to the low initial liquid velocity, provides efficient contact between the injected scrubbing liquid and the contaminant to be removed.

sea breeze   The sea-to-land surface wind that typically occurs in coastal areas during the day. It is caused by the thermal rising of the air above the land, which warms more readily than the water. See LAND BREEZE.

Secchi depth   A crude measurement of the turbidity (cloudiness) of surface water. The depth at which a Secchi disc, which is about 12 inches in diameter and on which is a black and white pattern, can no longer be seen.

second law of thermodynamics   One of the laws that describe the movement of energy within the environment. Two basic approaches can be used to interpret the second law. One involves the observation that whenever any form of energy is employed to do useful work, the conversion of energy to work is not 100 percent efficient. A portion of the energy source is always lost from the system as heat. Another way to describe the second law relates to the level of organization within a system. Any system tends to become more disorganized. When energy is converted from one form to another, randomness tends to increase. The measure of randomness is ENTROPY.

second third   See LAND DISPOSAL BAN.

secondary air pollutant   A substance formed in the atmosphere by chemical reactions involving PRIMARY AIR POLLUTANTS. These secondary compounds are not released directly by pollutant sources and therefore their control is only accomplished by controls on the compounds (ingredients) from which they form. For example, OZONE is a secondary air pollutant produced by sunlight-driven reactions involving volatile organic compounds and nitrogen oxides and emissions of these precursor materials are controlled to limit ozone formation. See PHOTOCHEMICAL AIR POLLUTION.

secondary association   See SECONDARY (INDIRECT) ASSOCIATION.

secondary clarifier See FINAL CLAR-
IFIER.

secondary consumer An animal that
obtains its nutrition by eating other animals. Secondary consumers are also referred to as carnivores. An animal that
eats only plants is a primary consumer.

secondary containment Any wall,
lining, curbing, or other barrier that contains a spill or leak of the container holding a chemical material.

secondary electron An electron
ejected from an atom or molecule after
a collision with a charged particle or photon of light.

secondary (indirect) association
An apparent link between two variables
that is actually the result of a CONFOUND-
ING VARIABLE. When the confounding
variable is controlled, the link is no longer
seen. For example, a group of workers
exposed to chalk dust is seen to have a
greater risk of lung disease than a group
not exposed to chalk dust, but the exposed group has a much higher portion
of smokers than the unexposed group.
In this example the cigarette smoke is the
most probable cause of the disease increase in the exposed group rather than
the chalk dust, and the association between chalk dust and lung disease is a
secondary one.

secondary irritant A chemical that,
upon overexposure, can cause irritation
to the skin, eyes, or respiratory system at
the site of contact and absorption, but
that exerts an adverse effect on an organ
or organ system elsewhere in the body
that is more pronounced than the contact
irritation. For example, liquid organic solvents can penetrate the unprotected skin
of the hands and forearms and cause
irritation in these areas, but absorption
of the solvent through the skin and entry
into the blood stream can lead to adverse
systemic effects associated with depression of the central nervous system. Compare PRIMARY IRRITANT.

**secondary maximum contaminant
level (SMCL)** The maximum concentration or level of certain water contaminants in public water supplies set by the
United States Environmental Protection
Agency to protect the public welfare. The
secondary levels are written to address
aesthetic considerations such as taste,
odor, and color, rather than health standards. Compare MAXIMUM CONTAMINANT
LEVEL and MAXIMUM CONTAMINANT LEVEL
GOAL.

secondary pollutant A general term
applied to pollutants that are formed by
chemical reaction in the environment.
These reactions form compounds other
than the substances released by the primary sources of pollution. See SECOND-
ARY AIR POLLUTANT, PRIMARY POLLUTANT.

secondary productivity The increase in BIOMASS by CONSUMERS at one
TROPHIC LEVEL per year, equal to the
amount of biomass ingested from lower
trophic levels, less predation and respiratory losses. See also PRIMARY PRODUC-
TIVITY.

secondary recovery The injection of
water into an underground petroleum
deposit to force the remaining oil into
recovery wells. This technique is used to
recover additional oil from old wells following the removal of the oil that can be
easily pumped to the surface (PRIMARY
RECOVERY). See also ENHANCED OIL RE-
COVERY.

secondary settling tank A tank used
to hold wastewater that has been subjected to SECONDARY TREATMENT. Floc or
particles of organic matter formed during
the secondary processes are allowed to
settle from the suspension for subsequent
removal.

secondary standards Allowable
amounts of materials in air or water that
are set to retain environmental qualities
not related to the protection of human
health. Secondary water standards are
set for, among others, taste and color,

and some secondary air standards define concentrations that will not be harmful to plant life. Compare to PRIMARY STANDARDS.

secondary succession　The orderly and predictable changes that occur over time in the plant and animal communities of an area that has been subjected to the removal of naturally occurring plant cover. This type of succession occurs when agricultural fields are taken out of use or when forested areas are subjected to severe fires that destroy all vegetation. In both cases the top soil remains for the regrowth of natural plant communities. Compare PRIMARY SUCCESSION.

secondary treatment　A phase in the treatment of wastewater. This aspect of treatment is usually carried out following the removal of particulate materials from domestic wastewater (see PRIMARY TREATMENT). In this second phase, conditions are established to maximize the functions of bacteria in destroying or mineralizing dissolved or suspended organic materials that are not removed by the primary process. Specific types of units employed in secondary treatment include TRICKLING FILTER, AERATION, ACTIVATED SLUDGE PROCESS, and BIODISC.

Section 404 permit　The wetland dredge and fill permit issued under regulations written to conform to Section 404 of the CLEAN WATER ACT. The permit is actually granted by the United States Army Corps of Engineers; the Section 404 regulations are written by the United States Environmental Protection Agency.

secular equilibrium　The stable relationship established in nature between a RADIOACTIVE element that has a long HALF-LIFE and a decay product that has a much, much shorter half-life. For example, radium-226 has a half-life of about 1600 years. As this element decays and emits radiation, RADON-222 is produced, which has a half-life of about 3.8 days. An equilibrium between the concentrations of these two elements is established such that the activity of each element is equal. The activity is calculated from the equation $A = \lambda N$, where λ is the decay constant for the element and N is the number of atoms of the element present. The decay constant of each element is equal to the natural log of 2 divided by the half-life of the element ($\lambda = [\ln 2]/$half-life). Consequently, an element with a long half-life will demonstrate a low activity, and the opposite will be true for an element with a short half-life. At equilibium, the activity of radium-226 will equal the activity of radon-222 (λN[radium] $= \lambda N$[radon]). Since the λ for radium is much, much smaller than the λ for radon, the concentration of radium will be much, much larger than that of radon when the equilibrium is established.

secular trend　A trend over a relatively long time period. In public health statistics, if a disease is strongly associated with age, as most are, then a long-term rise in the incidence or mortality rate of the disease may be explained by a shift in the age distribution of the population to the older years. Much of the rising secular trend in cancer incidence and mortality rates in the United States during the twentieth century is explained by the aging of the population, with the important exception of cancers associated with tobacco consumption.

secure landfill　A ground location for the deposit of hazardous wastes. The material is placed above natural and synthetic liners that prevent or restrict the leaching of dangerous substances into groundwater. A piping network called a leachate collection system is placed beneath the facility to allow the pumped removal of any liquid that penetrates the bottom or side liners. Access to the location is restricted and wells are used to monitor the leaching of any dangerous materials into the surrounding area. A type of TREATMENT, STORAGE, OR DISPOSAL facility.

sediment   Soil particles, sand, clay, or other substances that settle to the bottom of a body of water. Wind, water, and glacier erosion are natural sources of sediment. Human causes of sediment accumulation are mainly earth-moving activities, such as agricultural or construction operations. The greater the contamination of sediment with organic matter, the greater will be the removal of dissolved oxygen from the water by microorganisms. The term is frequently used to signify the mud at the bottom of a stream or lake or the deposits left behind by receding flood waters.

sediment oxygen demand (SOD) The amount of dissolved oxygen removed from the water covering the sediment in a lake or stream because of microbial activity.

sedimentary   Of geological strata comprising deposits of particulate materials that were deposited in some area by mechanical means. Such materials include gravels, muds, clays, and shales. The term also applies to rocks formed through the consolidation of fine-grained materials. For example, limestone can be formed from accumulations of calcareous shells derived from small marine organisms.

sedimentation   1. The removal of solid materials from a fluid suspension by gravitational settling. Solids suspended in water are removed by allowing the water to stand in a tank, pond, or lagoon for a time sufficient to allow the solids to settle to the bottom. 2. The removal of airborne particles from the natural atmosphere by a gravitational settling or their controlled removal by slowing an exhaust stream through a device sufficiently to allow particles to settle out before the air exits the device.

seeps   Groundwater/surface water connections caused by river or stream erosion into a near-surface aquifer.

selection bias   A source of error in a scientific study that is the result of certain people, animals, water samples, or other group being selected with a greater (or lower) frequency than their occurrence in the sampled population. For example, if 50 of the 60 persons selected for a study of the possible health effects of community air pollution are smokers, but only 50 percent of the population smokes, the study may draw incorrect conclusions when comparing health status of one community to another (if the sample from the other community is properly drawn) because smoking and its associated health effects are overrepresented in one community sample.

selective harvesting   The cutting of mature or diseased trees to encourage the growth of trees of differing ages or species. The removal of trees of only a certain age or species.

self-absorption   A process in which radioactive material itself absorbs the radiation released by RADIOACTIVE DECAY within the material, lowering the external radiation emitted by the material. This phenomenon occurs when the energy of the released radiation is relatively weak.

self-contained breathing apparatus (SCBA)   A respirator equipped with an air supply separate from the ambient air. The device is used in confined spaces (which may not contain adequate oxygen) or if actual or threatened air contamination levels make an AIR-PURIFYING RESPIRATOR otherwise inappropriate.

self-purification   The removal of organic material, plant nutrients, or other pollutants from a lake or stream by the activity of the resident biological community. Biodegradable material added to a body of water will gradually be utilized by the microorganisms in the water, lowering the pollution levels. If excessive amounts of additional pollutants are not added downstream, the water will undergo self-cleansing. This process does not apply to pollution by nonbiodegradable organic compounds or metals.

self-quenched counter tube A device for measuring radioactivity that includes a substance to control the reactions that occur in the device when radiation is encountered. A Geiger counter could contain such a device.

self-scattering The dispersing of radiation by the material emitting radiation. The radioactive substance tends to scatter the radiation as it is released, which tends to increase the measured activity of a sample.

semicontinuous activated sludge test (SCAS test) A test used by the Soap and Detergent Association (a trade and research organization based in New York) to evaluate the BIODEGRADATION of ALKYLBENZENE SULFONATE. The United States Environmental Protection Agency uses a modified SCAS test to establish the biodegradability of carbon compounds. The test involves the addition of ACTIVATED SLUDGE to an aeration tank containing the carbon compound to be tested.

semilogarithmic graph A graph of data with one axis having an arithmetic scale and the other axis having a logarithmic scale. The arithmetic axis could be time units, for example, and the axis with the logarithmic units could show the number of individuals in a population. Exponential growth will result in a straight line when the number of organisms is plotted against time on a semilogarithmic graph.

semistatic test During measurements of the toxicity of chemical substances to aquatic test organisms, a laboratory procedure in which all of the water in a container holding the test organisms is changed by emptying and refilling, frequently on a daily basis. This contrasts with a STATIC TEST, in which the water in the container is not changed for the duration of the procedure, and a FLOW-THROUGH TEST, in which fresh water is constantly flushed through the test container.

senescent Describing plants or specific ecosystems that are nearing the end of their normal life span. For example, a lake that is filling with accumulated aquatic vegetation, dead plant material, and sediments can be described as senescent because it is nearing extinction as a productive lake environment.

sensible heat transfer A removal or addition of heat energy that is accompanied by a change in temperature through CONDUCTION and CONVECTION. Compare LATENT HEAT TRANSFER.

sensitivity 1. The ability of a test to accurately identify true cases, such as the number of carcinogens a screening test identifies in a group of materials containing carcinogens. Expressed as the percent of true cases identified by the test. 2. The relative ability of a laboratory instrument to detect some chemical in a reliable fashion.

sensitization reaction An allergic response to a chemical substance that is preceded by a sensitizing exposure. Once an individual is sensitized to a substance, quite low doses can elicit an adverse response affecting the skin or respiratory system.

septic tank A buried tank used to treat domestic wastes. The sanitary waste from a household is deposited and retained in a covered tank to allow solids to settle to the bottom and to provide an environment for the decomposition of organic components by anaerobic bacteria. The liquid effluent that flows from the tank is allowed to seep into the soil. See LEACHING FIELD.

seral stages The various transitions in the orderly and predictable changes in a biological community from the pioneer stage to the climax stage. See SUCCESSION.

sere A complete sequence of transition stages (seral stages) in the orderly and predictable changes that occur in biological communities over time. *See* SUCCESSION.

settleable solids   Bits of debris, sediment, or other solids that are heavy enough to sink when a liquid waste is allowed to stand in a pond or tank. *See* SETTLING CHAMBER and SETTLING POND.

settling chamber   An enclosed container into which wastewater contaminated with solid materials is placed and allowed to stand. The solid pollutants suspended in the water sink to the bottom of the container for removal.

settling pond   An open lagoon into which wastewater contaminated with solid pollutants is placed and allowed to stand. The solid pollutants suspended in the water sink to the bottom of the lagoon and the liquid is allowed to overflow out of the enclosure.

settling tank   Same as SETTLING CHAMBER.

settling velocity   The rate of downward movement of particles through air or water. This gravitational settling removes particles naturally and is used also in pollution control devices, for example the settling tanks in a sewage treatment facility and the gravity settling chambers to remove larger-diameter (faster-settling) particulate air contaminants. The settling velocity of a particle is often the same as the TERMINAL SETTLING VELOCITY.

7Q10   The period of lowest stream flow during a seven-day interval that is expected to occur once every 10 years. During this time of low flow, the amount of DISSOLVED OXYGEN in the water would be expected to be the lowest encountered under normal conditions. Since such conditions are considered to be the worst natural case, the dissolved oxygen levels during such episodes are used to establish ambient water quality standards for that stream.

Sevin   The brand name for carbaryl, a CARBAMATE insecticide with relatively low acute toxicity to vertebrates.

sewage   Wastewater from homes, businesses, or industries; mainly refers to the water transport of cooking, cleaning, or bathroom waste.

sewage fungus   A thick filamentous growth that develops in water contaminated with sewage. The filamentous material is composed predominately of the bacterium *Sphaerotilus natans*.

sewage lagoon   A shallow pond where natural processes are employed to treat sanitary waste from households or public restrooms. Solid material settles to the bottom and is degraded by anaerobic microbial communities. The enclosure is open to the atmosphere, which permits aerobic mineralization of organic compounds in the upper layers of the water. The decomposition processes are analogous to those in effect in PRIMARY and SECONDARY TREATMENT. The effluent from these ponds is usually allowed to flow into nearby streams without further purification.

sewage treatment plant   Facility designed to receive the wastewater from domestic sources and to remove materials that damage water quality and threaten public health when discharged into receiving streams. The substances removed are classified into four basic areas: grease and fats, solids from human waste and other sources, dissolved pollutants from human waste and decomposition products, and dangerous microorganisms. Most facilities employ a combination of mechanical removal steps and bacterial decomposition to achieve the desired results. Chlorine is often added to discharges from the plants to reduce the danger of spreading disease by the release of pathogenic bacteria.

sewer   The piping system or conduit used to carry runoff water or wastewater. Various types of systems fulfill different functions; for example, a STORM SEWER carries runoff from rainfall, a SANITARY SEWER carries wastewater from a household or public facility, and a COMBINED SEWER transports both runoff and sanitary waste.

shale Rock or mineral deposits that contain solid, waxy hydrocarbons termed kerogen. The hydrocarbons can be extracted and utilized as source material for petroleum products.

shale oil A thick oil recovered from SHALE rock. The liquid is obtained by heating the pulverized rock to high temperatures to vaporize the kerogen layered within the rock, followed by condensation of the vapor. The recovered oil can be modified and refined to yield petroleum products.

shale shaker An oscillating wire mesh that catches oil well drilling cuttings as drilling fluid is passed through it.

Shannon-Weaver index ($\bar{H}$) A SPECIES DIVERSITY expression, equal to

$$-\sum P_i \log P_i$$

where P_i is the number of individuals in species i divided by the total number of individuals in a community, and $\log P_i$ is the (natural or base 2) logarithm of P_i. The sum is taken for all i species in the community. The Shannon-Weaver index incorporates both aspects of SPECIES DIVERSITY, EVENNESS and RICHNESS.

sheet piling Material, often concrete or steel, placed vertically in the ground to contain erosion or the lateral movement of groundwater.

sheetflow A broad, shallow overland flow of stormwater. Also called a sheetflood.

shelterwood cutting The harvesting of all mature trees from a forest area through a systematic removal over one or more decades. This method of cutting is followed to encourage the growth of immature trees where their development is inhibited by the shading of older trees.

shielding Material, such as lead, that is used to reduce the passage of radiation.

Shigella The genus name of a bacterium associated with human dysentery, a disease characterized by severe diarrhea with blood and pus in the feces. The disease is transmitted through the consumption of water, food, or beverages contaminated with fecal material.

shock wave A large-amplitude wave created as an object travels through a medium at a faster rate than the typical wave speed for that medium, for example, faster than the speed of sound in air. In such cases the force of the moving object creates waves that do not travel from the object in all directions, as is the case at slower speeds, but the waves are restricted in movement to the rear and sides of the object. The large amplitude is the result of the great compression at the front of the object and the large degree of rarefaction behind it. A SONIC BOOM is an example of a shock wave.

short-term exposure limit (STEL) An occupational air concentration standard using 15-minute averaging times within an 8-hour workday. A STEL is set for a material that can cause adverse health effects if workers are exposed to higher concentrations for short periods. Materials not requiring a STEL can have short-term peaks without adverse effect, and only their overall 8-hour average must be below a certain limit.

shredding The mechanical division of material into 1- or 2-inch-diameter pieces. Solid waste is often shredded and then compacted to increase its density before disposal in a landfill.

SI units The International System of Units (Le Système International d'Unités) defined by an international gathering convened to establish agreements on the most frequently used units of measurement. The Conference of Weights and Measures in 1960 adopted standard measures based on the meter/kilogram/second units and radiation quantities based on the becquerel, gray, and sievert. The units adopted by the conference represent the currently preferred measures of length, mass, time, radiation, and so on.

Sierra Club   A large organization in the United States that is concerned with environmental issues. Founded in 1892, in 1989 membership numbered about 416,000. Based in San Francisco, California, the club is active in public education, lobbying of legislative and administrative bodies, and, through its affiliated Sierra Club Legal Defense Fund, in the courts.

sievert (Sv)   The SI unit of radiation dose that takes into account both the physical properties of the radiation and the biological properties of the material absorbing the radiation. The unit is equal to the amount of radiation dose actually absorbed by the body (in GRAYS) times a QUALITY FACTOR associated with the energy level and ionizing potential of the particular radiation times modifying factors associated with the tissue being influenced (for example, the distribution of radiation within the tissue being struck). The sievert is the preferred unit of radiation dose for the design of radiation protection devices and for expressing standards limiting human exposure to ionizing radiation. A dose of 0.05 Sv is the maximum annual occupational exposure allowed in the nuclear industry. A dose of 1 Sv in a single incident is thought to increase the probability of leukemia, and a dose of 6 Sv over two days is sufficient to kill 100 percent of the individuals exposed.

sigmoid growth   A population growth pattern that traces out an S-shaped, or sigmoid, curve. A population undergoing sigmoid growth begins at an exponential rate but the rate slows as LIMITING FACTORS are encountered, until a rough

equilibrium level of population is reached, the CARRYING CAPACITY of the ecosystem for this type of population. See EXPONENTIAL GROWTH, R-SELECTED, and K-SELECTED.

SIGMOID GROWTH

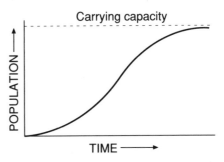

significant biological treatment   A process employed in the SECONDARY TREATMENT of wastewater that provides a biological degradation achieving at least 65 percent removal of 5-day BIOCHEMICAL OXYGEN DEMAND, as a 30-day average.

significant figures   The digits in a number starting with the first nonzero digit to the left of the decimal point (or starting with the first digit to the right of the decimal point if no nonzero digits are to the left of the decimal point) and ending with the rightmost digit. These digits indicate the accuracy of measurements from which the number derives. For example, 78.0 has three significant figures and indicates a known accuracy to the nearest tenth; 0.078 has three significant figures and indicates a known accuracy to the nearest thousandth.

significant new use rule (SNUR)   Under the TOXIC SUBSTANCES CONTROL ACT, the requirement that a producer of a chemical already approved for manufacture notify the United States Environmental Protection Agency if the material will be used in a new and different way. The new use may trigger a regulatory review of the material.

silica   Silicon dioxide. The fibrotic form is crystalline silica, called free silica, or quartz. The quartz content of inhaled dust is positively related to the risk of developing SILICOSIS.

silica gel   A nontoxic, water-absorbing activated silica used as a drying agent for comfort air conditioning and compressed air, in chromatography, and as a catalyst.

silicates   Compounds containing silicon and oxygen, and sometimes also hydrogen or various metals. Crystalline silicates in workplace air are an inhalation hazard. Silicate compounds can occur as deposits on the inside surface of steam and water pipes.

silicosis   A lung disease caused by prolonged exposure to high levels of small-diameter dust with a high crystalline (free) silica content. The disease is characterized by scarring of lung tissue (fibrosis) and a subsequent loss of lung elasticity and gas exchange capacity. Black lung disease is a form of silicosis.

siltation   The settling of a finely divided particulate (previously suspended solids) on the bottom of a lake, stream, or reservoir.

silviculture   A branch of forestry dealing with the cultivation and management of trees in order to produce a crop resource on a continuing basis.

simple asphyxiant   A gas present at a concentration high enough to reduce the oxygen concentration in the air to a level below the minimum required for proper human respiration, about 18 percent by volume. The gases that can act as simple asphyxiants may be nontoxic to humans, but they become dangerous if their presence is at oxygen-excluding concentrations. They are used as purge or blanketing gases in closed tanks and vessels and pose a simple asphyxiation hazard to workers entering these confined spaces for inspection or cleaning. Examples of simple asphyxiant gases include nitrogen, carbon dioxide, helium, ethane, and methane.

sink   See NATURAL SINK.

sinking agents   Chemical additives that, when applied to a floating oil discharge, will cause oil to sink below the surface of the water. The application of sinking agents is generally not permitted by regulations of the United States Environmental Protection Agency governing the treatment of oil discharges.

sinusoidal wave   A wave varying in amplitude with the sine of an independent variable. Sound waves are oscillations around atmospheric pressure and are sinusoidal.

skimmer   See OIL SKIMMER.

skin dose   The amount of radiation absorbed by the skin; the sum of the amount of radiation dose in the air above the skin and that scattered from below by bone and other body parts.

skin sample   Sampled water that is not representative of the cross-sectional flow in a pipe or conduit. A skin sample can result if the tap opens on the inside wall of the pipe and can be avoided by using sampling taps that extend toward the center of the pipe, away from the wall.

skyshine   The radiation released through the roof of a nuclear reactor and reflected back to the ground by the atmosphere.

slag   The nonmetallic glasslike material formed in a blast furnace or smelter when impurities react with a flux. Slag floats as a liquid on top of the process, and after it is cooled and solidified, is used in cement.

slaker   A mechanical device in which dry lime (calcium oxide) or magnesium oxide is powdered and mixed with water to produce alkali consisting of calcium or magnesium hydroxide. Calcium hydroxide is used as a water softener.

slash and burn An agricultural practice involving the rapid destruction of natural forest for limited farming activity. The natural forest is cleared; the residue from the clearing process is burned; and crops are planted for a few years. When the fertility of the soil is depleted, the process is repeated in a new area, and the initial land is abandoned.

slope factor See CANCER POTENCY FACTOR.

slow neutrons Neutrons possessing a kinetic energy of about 100 electron volts or less. This class of neutrons is needed to produce nuclear fission reactions. Compare FAST NEUTRONS.

sludge A general term used to designate a thick suspension of waste products having the consistency of paste or soft mud, such as the particulate waste collected during the treatment of sewage, or the muddy sediment that collects in a boiler used to produce steam. Sludge high in organic matter is produced by municipal sewage treatment plants, food processors, refineries, and paper mills. Sludge low in organic matter is produced by some chemical and power plants.

sludge digestion The biological decomposition of solids collected during the operation of a facility designed to remove organic wastes from domestic or industrial sources. The total volume of solids is reduced by the mineralizing activity of bacteria, and the sludge remaining is rendered less reactive because the easily degraded compounds have been removed.

sludge disposal The removal and discarding of thick watery suspensions of particulate waste matter. Final disposal may involve the removal of excess water and the subsequent burning of the solids, or placing the dewatered material in a landfill.

sludge volume index (SVI) A laboratory test result used to indicate the rate at which sludge is to be returned from the discharge end of an AERATION TANK to the inflow (upstream) end. It is calculated as $SVI = (SV/MLSS) \times 1000$ milligrams/gram, where SV is the sludge volume (solids settled in a 1-liter graduated cylinder after 30 minutes, in milliliters/liter) and MLSS is the level of MIXED LIQUOR SUSPENDED SOLIDS, in milligrams/liter. The SVI has units of milliliters/gram.

slug A short rod containing fissionable material (fuel) for a nuclear reactor.

slurry A watery mixture of insoluble matter such as lime. The mixture is pourable and can be transported by pipe. The form in which some raw material is added to an industrial process. Compare LIQUOR.

slurry wall Material placed vertically in the ground to prevent the lateral movement of groundwater.

small-quantity generator (SQG) A facility that produces small amounts of hazardous waste and is therefore subject to less stringent management requirements by federal regulations. SQGs (often pronounced "squeegees") are currently defined as those producing less than 100 kilograms of hazardous waste per month, with certain exceptions. State regulatory agencies may require SQGs to register.

smog A general term that is currently used to describe (1) any condition in which visibility is reduced because of air pollution, or (2) elevated levels of PHOTOCHEMICAL OXIDANTS in the atmosphere. See PHOTOCHEMICAL AIR POLLUTION.

smoke An AEROSOL produced by the incomplete burning of carbonaceous materials (i.e., wood or fossil fuels). The aerosol consists of a mixture of gases and visible particles.

smoke reader A person trained to quantify the OPACITY (darkness) of emissions from a smokestack. The reader classifies the smoke using a 0–5 scale, from white to deep black. Many pollution control agencies have regulations limiting smoke emissions to a certain number on the opacity scale. See also RINGELMANN CHART.

sodium tripolyphosphate (STPP) A builder, or water softener, contained in many detergents. Water softeners bind calcium and magnesium ions in water, preventing the reaction of the ions with the surfactant (soap) and the loss of cleaning effectiveness. The phosphates in STPP typically are not removed by municipal water treatment plants and are contributors to EUTROPHICATION in the surface waters receiving municipal waste water.

soft hammer The provision in the HAZARDOUS AND SOLID WASTE AMENDMENTS allowing SCHEDULED WASTES to avoid an automatic LAND DISPOSAL BAN if the United States Environmental Protection Agency fails to issue treatment standard regulations for a waste on time. Land disposal is permitted only if the disposal site is equipped with an approved double liner and a leachate collection system and the disposer has attempted to treat the waste in an acceptable manner in the absence of a treatment standard.

soft pesticide A pesticide that is easily biodegradable and is decomposed a short time after distribution into the environment. Compare HARD PESTICIDE.

soft water Water that contains low concentrations of metal ions such as calcium and magnesium. This type of water does not precipitate soaps and detergents. Compare HARD WATER.

softening The removal of metal ions such as calcium and magnesium from water supplies. The converting of HARD WATER to SOFT WATER.

softwood Wood produced from conifers such as pines, spruces, and firs, which dominate the forests of the temperate regions. The term does not relate to the hardness of the wood. Compare HARD WOODS.

soil air Below-ground air in the spaces between soil particles. Decomposition of organic matter in the soil can cause high concentrations of carbon dioxide in the soil air. The carbon dioxide combines with water to form carbonic acid, thereby increasing the acidity of the groundwater.

soil amendment Any material added to soil that enhances plant growth.

soil and sediment adsorption isotherm test A test method that measures the ADSORPTION of a chemical substance to soil or sediment and thus indicates the likely distribution pathways in the environment. If a substance is readily adsorbed, it will increase in concentration in a soil or sediment layer; if it is not adsorbed, it will be free to move through the soil into groundwater or run off into surface water.

Soil Conservation Service (SCS) The federal agency, created by the Soil Conservation Act of 1935, that administers erosion control projects and research. The SCS works through local soil conservation districts.

soil core A sample of soil taken by forcing a cylindrical device into the ground perpendicular to the horizontal. The resulting sample contains a circular section of each layer of sediment.

soil flushing A treatment technique for cleaning soil contaminated with inorganic or organic hazardous waste; the process involves the flooding of the soil with a flushing solution, which may be acidic, basic, or contain surfactants, and the subsequent removal of the leachate via shallow wells or subsurface drains. The recovered leachate is then purified.

soil moisture regime   The changes in the moisture content of soil during a year.

soil profile   The layers of soil revealed when one digs through the earth or removes a core sample. The three basic horizontal layers that may be observed in a soil profile are the A, B, and C horizons. The A horizon, or topsoil, is the main source of plant nutrients. Soluble materials leach from the A horizon down to the subsoil, or B horizon. This is the zone of clay accumulation. The deepest layer, the C horizon, is composed of partially weathered bedrock. See SOIL CORE.

soil sorption coefficient (K_d)   A parameter relating the partitioning of a chemical between soil and water in a soil-water mixture. The coefficient is computed by

$$K_d = \frac{C_s}{C_w}$$

where C_s is the amount of a specific chemical bonded to the soil (micrograms chemical per gram of soil) and C_w is the concentration of the same substance dissolved in the water (micrograms of the chemical per gram of water).

soil structure   The physical properties of different soils. Sand has little structure because sand particles do not tend to stick together, whereas clay has a firm structure because clay particles tend to pack tightly together to form a solid material.

soil texture   A classification of soils based on the size distribution of mineral grains comprising the soil. The relative proportions of silt, sand, clay, and gravel are normally given.

soil thin-layer chromatography   A method used to estimate the potential for leaching a chemical from soil by measuring the mobility of the chemical in soil under controlled conditions.

soil vapor survey   A noninvasive method for the detection of volatile or semivolatile organic contaminants in shallow subsurface soil. The technology is especially useful in the analysis of soils with a high sand content and of shallow groundwater. A probe is driven into the ground, and samples of vapors in the soil are drawn to the surface for analysis on site or in the laboratory.

solar cell   An electrical device that converts sunlight directly into an electric current. Also referred to as a photovoltaic cell, one consists of a thin wafer of silicon containing a small amount of metal that will emit electrons when struck by sunlight.

solar constant   A quantity used to identify the amount of solar radiation striking the upper portion of the atmosphere. Specifically, it is the amount of radiant energy striking a surface positioned at the top of the atmosphere and lying perpendicular to the solar radiation when the earth is an average distance from the sun. The quantity is about 2 calories per square centimeter per minute, or 2 LANGLEYS per minute. The quantity is used in calculations relating to the input of solar energy into the earth's atmosphere.

solar energy   The term is normally applied to the conversion of direct sunlight into usable forms of energy. Passive solar systems are designed to maximize the warming of a residence by the sun, and active solar systems involve the storage of water heated by the sun in tanks with subsequent pumping of the hot water to heat a home or other structure. Sunlight can also be converted to electricity, either directly, using solar cells, or indirectly, by focusing sunlight on a boiler, producing electricity using a steam turbine. The most active use being made of solar energy by individual homeowners is the heating of water. Several experimental projects are underway to generate commercial quantities of electricity employing the energy in sunlight to generate steam. Solar cells are used to power hand-held calculators as well as to provide electricity to operate electrical devices in remote areas.

solar irradiance in water An expression of the amount of light flux from a clear sky, over a certain wavelength range, that is available to cause photochemical transformations (PHOTO-LYSIS) in chemicals present in surface water bodies at shallow depths. Solar irradiance varies with latitude and the time of year. Values are used with the MOLAR ABSORP-TIVITY of the waterborne chemical substance to calculate the HALF-LIFE of the specific chemical in the environment due to photolysis.

sole source aquifer An aquifer designated by a provision of the SAFE DRINK-ING WATER ACT (the Gonzales amendment), as the principal or only source of drinking water for a geographical area. This designation bars the use of federal funds for projects in the RECHARGE ZONE that may lead to a significant hazard to the public health by degrading groundwater quality in the aquifer. See CRITICAL AQUIFER PROTECTION AREA.

solid waste Waste material not discarded into surface waters via water treatment systems or directly into the atmosphere. Under federal regulations the term can include waste in solid or liquid form, as well as gaseous material. MUNIC-IPAL SOLID WASTE and all HAZARDOUS WASTE is by definition solid waste.

solid waste management unit (SMU) The property on which hazardous waste management occurs, such as a surface impoundment, landfill, incinerator, waste pile, or tank and the adjacent land used for storage, transfer, or preliminary treatment of the waste.

solidification The addition of agents to convert liquid or semiliquid hazardous waste to a solid before burial to reduce the leaching of the waste material and the possible migration of the waste or its

constituents from the facility. Usually accompanied by stabilization. See STABILI-ZATION, WASTE.

solubility The relative capacity of a substance to serve as a solute. Sugar has a high solubility in water, whereas gold has a low solubility in water.

solubility product constant (K_s) The product of the molar concentrations of the ions in solution that result from a solid chemical compound partly dissolving in water. Each compound has its own equilibrium dissolved concentration, and the solubility product is a constant for that compound. For example, K_s for an equilibrium between solid ferric hydroxide (FeOH) and dissolved Fe^{-3} and OH^- ions is 1×10^{-38}, which means that the product of the molar concentrations of Fe^{-3} and OH^- is 1×10^{-38}. Usually expressed as pK_s, which is the negative logarithm of K_s. The pK_s for ferric hydroxide is 38.

solum The top two soil layers, composed of the topsoil (A-horizon) and the subsoil (B-horizon, or layer of leached material deposition). The solum excludes the parent material layer (C-horizon). See also SOIL PROFILE.

solute The substance that is dissolved in a solution.

solution A homogeneous mixture of a solute in a solvent. For example, when sugar (the solute) is dissolved in water (the solvent), the molecules that comprise the sugar crystal are separated from one another and dispersed throughout the liquid medium.

solution mining The removal of a mineral deposit that is soluble in water. Water is injected into the geological strata containing the mineral and the dissolved material is recovered by wells. The method is applied to mine salt (sodium chloride) and potash (potassium chloride), among others.

solvent   The dissolving medium, or liquid portion, of a solution. Water is frequently referred to as the universal solvent. The term also applies to organic materials (e.g., benzene, acetone, or gasoline) used to clean (dissolve) oils or grease from machinery, fabrics, or other surfaces, or to extract hydrocarbons from some source material. Many organic solvents are flammable and/or toxic. See also SOLVENT REFINING.

solvent recovery   A method to minimize hazardous waste by recovering process solvents for reuse. Common techniques are distillation or absorption of the solvent from a solvent-containing mixture.

solvent refining   A process used to remove sulfur and other contaminants from coal. Powdered coal is mixed with an organic solvent, such as anthracene, which puts about 95 percent of the carbon compounds in the coal into solution. The coal recovered from the solution contains less than 1 percent of the sulfur and ash-forming material present in the original coal.

somatic cells   All the cells of the body except the GERM CELLS, which produce gametes.

sone   A unit used to express the perceived loudness of a sound. The unit is equal to the loudness of a 1000-hertz tone with a sound pressure of 40 decibels. If listeners judge that a sound is three times as loud, for example, as this 1000-hertz tone, the sound has a loudness of 3 sones.

sonic boom   The intense sound waves created by objects traveling faster than the speed of sound (about 1100 feet per second or 750 miles per hour). The shock wave pressure may be as large as 2000–3000 newtons per square meter, compared to normal speech sound levels of 0.01 newton per square meter, and is strong enough to shake houses. Concern over sonic booms restricts some faster-than-sound flight paths. For example, supersonic flight by the Concorde is restricted to areas over open water.

sorbent   A material that ABSORBS or ADSORBS solids, liquids, gases, or vapors, such as the material in workplace respirators that selectively removes gases or vapors as air passes through it. Sorbents must be matched to the type of substance(s) to be removed.

sorbing agent   A material placed on an oil spill to ABSORB or ADSORB the oil; the oil and the material can then be removed together.

sorption   The physical or chemical linkage of substances, either by ABSORPTION or by ADSORPTION.

sorting   The separation of solid municipal waste into recyclable categories of material: paper, metals, glass, for example. Sorting can be performed by households or at centralized facilities.

sound absorption coefficient   See SABIN ABSORPTION COEFFICIENT.

sound intensity   The average sound power passing through a unit area perpendicular to the direction that the sound is traveling. Common units for sound intensity are watts per square meter.

sound intensity level   The expression of SOUND INTENSITY in decibel units. The sound intensity level (L_I), in decibels, is computed as

$$L_I = 10 \log \frac{I}{I_0}$$

where I is the measured sound intensity and I_0 is the reference intensity $(1 \times 10^{-12}$ watts per square meter).

sound level   See SOUND PRESSURE LEVEL.

sound power   The sound energy emitted by a source per unit time, usually expressed in watts. Sound power causes SOUND PRESSURE.

sound power level   The sound energy emitted by a sound source per unit time, and expressed in decibels. SOUND POWER *(W)*, in watts, is converted to sound power level *(L_W)*, in decibels, by

$$L_W = 10 \log \frac{W}{W_0}$$

where W_0 is the reference power $(1 \times 10^{-12}$ watts).

sound pressure   The periodic fluctuation above and below atmospheric pressure created by an oscillating body. The pressure differences are often expressed in newtons per square meter.

sound pressure level   The expression of SOUND PRESSURE in DECIBEL units. Because the average of the sound pressure fluctuations above and below atmospheric pressure (equal positive and negative values) would be zero, the measured pressures are squared, summed, and averaged. The square root of the average, termed the root mean square, is then converted to decibels by $L_p = 20\log_{10}(p/p_0)$, where L_p is the sound pressure level, p is the root-mean-square sound pressure, and p_o is the reference sound pressure (commonly 2×10^{-5} newtons per square meter). Noise meters perform these conversions internally and display sound pressure level in decibels.

sound wave   A sinusoidal variation around atmospheric pressure caused by a vibrating body.

sour gas   Natural gas that contains a high level of HYDROGEN SULFIDE, which has a very foul odor and poses a severe health hazard.

source material   Any material that is not a SPECIAL NUCLEAR MATERIAL but which contains at least 0.05 percent uranium, thorium, or any combination of the two.

source measurement   The noise level produced by a tool, machine, or process. See AMBIENT NOISE.

source reduction   A lowering of the amount of hazardous or nonhazardous waste requiring treatment or disposal through a reduction in the amount of waste material produced.

source/release assessment   An estimate of the types, quantities, and locations of the release of harmful substances or energy into the indoor or outdoor environment. The results are used in a RISK ASSESSMENT.

source separation   The separation, by members of a household, of municipal waste into those items that can be recycled and those that cannot.

span drift   A gradual change in the instrument response to a SPAN GAS. This source of measurement error is reduced by frequent calibration.

span gas   A gas with a known concentration of a chemical substance introduced to an analytical instrument as a calibration step. The initial concentration of a span gas is typically about 80 percent of the operating range of an instrument. Other, more dilute gas concentrations are then used to create a CALIBRATION CURVE.

special nuclear material   Elements capable of undergoing nuclear fission: includes plutonium-239, uranium-233, uranium-235, or uranium enriched with uranium-233 or uranium-235. See also SOURCE MATERIAL.

Special Review   A declaration of the United States Environmental Protection Agency issued under its implementation of the FEDERAL INSECTICIDE, FUNGICIDE, AND RODENTICIDE ACT that the registration (use permit) of a pesticide is in question because data indicate the agent to be acutely or chronically toxic. The declaration does not imply cancellation or suspension of the chemical; the Agency can still weigh the risks and benefits of the pesticide and allow a continuation of the registration. Formerly called a rebuttable presumption against registration (RPAR).

speciation The formation of two or more new genetically distinct groups of organisms following a division within a single group or SPECIES. A group of organisms capable of interbreeding is segregated into two or more populations that gradually develop barriers to reproduction. These barriers can be extrinsic, that is, the separated populations may reproduce at different locations or at different times of the year, or intrinsic, that is, some genetic barrier that makes attempts to crossbreed unsuccessful. When the reproductive isolation is maintained long enough, the separated groups may develop into separate, distinct, and identifiable species.

species The members of a group of organisms that can successfully interbreed with each other under natural conditions. The members of the group generally resemble each other and occupy a specific geographic region. The term is also applied to taxonomic classifications devised by man into which individual specimens are placed. These taxonomic groupings usually, but not always, correspond to the natural biological groups.

species composition The different types, and abundance of each type, of organisms inhabiting a specified locality.

species density The total numbers of individuals of a specific species found in a specific area of a habitat for a specified time period. Such calculations are of value in that they provide information on the magnitude of the population of some species occupying a given area. The formula is

$$D = \frac{n/a}{t}$$

where D is the density, n is the number of individuals, a is the area studied, and t is the time period during which the study was conducted.

species diversity A measurement that incorporates both the number of different species, or individual types of organisms, that inhabit a given location and the number of individuals of each type present. Generally, undisturbed locations have a higher species diversity than that found in similar habitats that have undergone extensive environmental alteration. See SHANNON-WEAVER INDEX.

species frequency The percent of sampling areas in which a representative of a specific species is represented. The frequency is independent of the number of individuals of each species located in each sampling area.

species richness index A mathematical expression that indicates the number of species in a community relative to the total number of individuals in that community. The number of organisms in each species is not considered in the index.

specific activity The total radioactivity of a specified substance per gram of that compound or element.

specific collection area (SCA) The ratio of the total COLLECTING SURFACE area of an ELECTROSTATIC PRECIPITATOR to the VOLUMETRIC FLOW RATE of the airstream.

specific conductance A measure of the ability of a solution to conduct an electric current, mainly dependent on the concentrations and types of ions in the solution, expressed in units of conductance/length, such as micromhos per centimeter. Used in groundwater monitoring as an indication of the presence of ions of chemical substances that may have been released by a leaking landfill or other waste disposal facility. A higher specific conductance in water drawn from DOWNGRADIENT WELLS when compared to UPGRADIENT WELLS indicates possible contamination from the facility.

specific gravity   A comparison of the density of a substance (frequently a liquid) to the density of a reference substance (normally water). Specific gravity (SG) is given by

$$SG = \frac{\rho}{\rho_w}$$

where ρ is the density (weight per unit volume) of the unknown substance and ρ_w is the density of water. The parameter has no units and is frequently used to determine the concentration of a SOLUTION.

specific heat   The ratio of the amount of heat (calories) required to raise the temperature of one gram of a substance one degree Celsius to the amount of heat required to raise the temperature of one gram of water one degree Celsius (which is one calorie). Every substance has a characteristic specific heat. For example, the value is 1.000 for water and 0.092 for copper, meaning that it takes more heat to raise the temperature of one gram of water than it does to produce the same temperature increase in one gram of copper.

specific molal volume   The volume occupied by one mole of a liquid. The volume can be calculated by dividing the molecular weight of the substance by the liquid density. Common units are cubic centimeters per gram-mole.

specific rate   In public health statistics, a RATE that is limited to a certain age/sex/ethnic group/year category, such as the 1987 death rate for black 70-year-old females. See CRUDE RATE.

specific weight   The weight (force) per unit volume of a material, expressed as $\gamma = \rho g$, where ρ is the material density and g is the acceleration of gravity.

specific yield   The volume of water available per unit volume of aquifer, if

drawn by gravity. Specific yield is expressed as a percent. For example, if 0.2 cubic meter of water will drain from 1 cubic meter of aquifer sand, the specific yield is 20 percent.

specifications standards   Environmental or workplace health and safety standards that stipulate in detail what actions must be taken to be in compliance, such as how a control device will be configured, where labels will be placed on a container, and what size print they will have, and so on. Compare PERFORMANCE STANDARDS.

spectrometer   An instrument used to identify and measure the wavelength of electromagnetic radiation emitted by elements when heated to high temperatures. Since each chemical element emits a unique spectrum, the technique can be used to detect small quantities of substances. The elemental composition of stars can be deduced from studying the spectrum of radiation emitted by the star.

spectrophotometry   An analytical method that uses the intensity of radiation absorbed at certain wavelengths to detect the presence of specific chemical elements.

spent fuel   Fuel, from a nuclear power reactor, that can no longer efficiently sustain a continuous fission reaction. The rods containing the nuclear fuel used to power a nuclear power plant become depleted of fission fuel (i.e., certain forms of uranium), and waste products from the fission reactions accumulate within the fuel elements. As a result the efficiency of the nuclear reaction decreases and the rods are said to be spent. The rods are usually removed after three years of use. The used rods are very radioactive and constitute a significant disposal or recycling problem.

spent fuel pool   A storage facility where fuel rods that have been removed from a nuclear reactor are held underwater for cooling. The used rods are held in this manner for several years before disposal.

Spill Cleanup Inventory (SKIM) A listing of public and private response and support equipment available if needed by the NATIONAL RESPONSE CENTER in the event of an accidental release of oil or hazardous substances into the environment. The inventory is maintained by the United States Coast Guard.

Spill Prevention Control and Countermeasure Plan (SPCC Plan) A written description of prevention, detection, and containment measures a facility has or will use to avoid or minimize oil spills into water. Regulations written by the United States Environmental Protection Agency under the authority of the CLEAN WATER ACT require certain "nontransportation-related onshore and offshore facilities" to have a SPCC plan. Transportation-related facilities that have the potential to spill oil in harmful quantities are regulated by the United States Department of Transportation.

spirometer An instrument for determining lung flow and volume capacities. The lung function test results are used to detect lung disease or adverse effects on the respiratory system from the inhalation of occupational or community air contaminants.

spirometry The testing of lung function through the use of a SPIROMETER.

split samples Divided environmental samples sent to two (or more) analytical laboratories for comparison of the results.

spoil 1. The refuse or rubble that accumulates when soil, rock or sand is removed to allow access to mineral deposits. 2. The material that is removed from a channel when it is dredged.

spongy parenchyma Chlorophyll-containing cells in a leaf that lie under the palisade parenchyma. If excessive air contaminant exposure to vegetation damages the spongy parenchyma or palisade cells to the point of tissue collapse, the leaf will be discolored and the dam-

aged areas may fall away, leaving holes in the leaf.

spray chamber A device that removes certain organic compounds from an airstream by condensation. A cooling material, usually water, is sprayed into a chamber, and the condensed organics exit with the water. In addition to the removal of the condensable contaminants, the condensation greatly reduces the volume of the waste exhaust. Also called a contact condenser.

spray tower A device used for the removal of gases or particles from an exhaust gas. The dirty gas stream typically is directed through the bottom of a tower and flows past a finely divided spray that removes the pollutants. The cleaned gas exits the top of the tower and the water drains to the bottom. See also PACKED TOWER, TRAY TOWER.

spring turnover The mixing of water in a lake during the spring of the year. This process takes place most frequently in lakes located in temperate zones where the winter temperatures are low enough during the winter to result in freezing of the lake surface. As the ice melts in the early spring, the water temperature at the surface gradually warms to that of the underlying water (about 4°C). Since the water temperature is then uniform from top to bottom, the force exerted by the winds is sufficient to promote mixing of the surface and deeper waters. Compare FALL TURNOVER.

spurious counts A term used to describe the recording of false radiation readings caused by malfunctions of, interference with, or improper use of radiation detection devices.

stability See ATMOSPHERIC STABILITY and STABILITY CLASS, ATMOSPHERIC.

stability class, atmospheric A classification of atmospheric turbulence, or air contaminant dispersive capability, developed by F. Pasquill and F. A. Gifford. Atmospheric turbulence is categorized by the letters A through F, with A being the most dispersive and F the least. The most important variables determining the stability class at a given time are wind speed and the amount of sunshine or solar insolation.

stability index See LANGELIER INDEX.

stabilization A broad expression used to denote a process that is intended to lessen the damage that a pollutant or discharge causes in the environment. For example, the stabilization of sewage involves allowing microorganisms to degrade those components that can be decomposed. See STABILIZATION, WASTE and SOLIDIFICATION.

stabilization/solidification (S/S)
See STABILIZATION, WASTE and SOLIDIFICATION.

stabilization, waste A hazardous waste treatment process that decreases the mobility or solubility of waste constituents by means other than SOLIDIFICATION. Stabilization techniques include chemical precipitation or pH alteration to limit solubility and mixing the waste with sorbents such as fly ash to remove free liquids.

stabilized grade The grade (slope) of a water channel at which no erosion or deposition will occur.

stable 1. Describing an element or isotope that does not undergo radioactive decay. A stable material may be the end of a radioactive series. 2. Describing air that will not readily disperse pollutants.

stable isotope A form of an element that is not radioactive; that is, an isotope that does not undergo radioactive decay.

stack downwash The movement of a smokestack plume toward the ground, instead of, more commonly, rising. Stack downwash (also called stack-tip downwash) can occur if the exit velocity of the

stack gas is significantly less than the wind speed at the top of the stack. Certain AIR QUALITY DISPERSION MODELS have a stack downwash option, which will simulate this phenomenon. Model results under the stack downwash assumption, all other variables being equal, will predict higher ground-level concentrations.

stack flue The innermost channel in a smokestack through which exhaust gas is carried.

stack gas The gaseous product of a process (usually combustion) that exits through a stack, or flue.

stack sampling The collection of representative portions of the gases and particulate matter that is being discharged through a smokestack or duct. This type of sampling allows for a direct estimation of the amount and types of air pollutants being released.

stagnation Persistent atmospheric conditions with limited vertical and horizontal air motion, resulting in an increase in the concentration of air contaminants. See TEMPERATURE INVERSION.

standard addition technique In analytical chemistry, a method in which a known amount of a standard solution is added to the sample being measured. It is used if a constituent in the sample other than the one being measured will interfere with the response of the analytical instrument, thus giving incorrect results because the pure calibration standards are not subject to this interference. The standard addition technique corrects for the interference.

standard air Air under defined standard temperature and pressure conditions. The United States Environmental Protection Agency uses 25°C and one atmosphere as STANDARD CONDITIONS. Many occupational health calculations use 21°C and one atmosphere, and most other applications use 0°C and one atmosphere.

standard air density The density of dry air at the chosen STANDARD CONDITIONS. At 0°C and one atmosphere of pressure, the density of dry air is 1.293 kilograms per cubic meter.

standard conditions Arbitrary reference conditions established for comparing reactions that involve gases. The density and other properties of gases depend on the temperature and pressure under which the gas exists. Standard conditions represents a way that gases can be compared when under different circumstances. Standard conditions are frequently set at 0°C and one atmosphere, but environmental and occupational calculations assume different air temperatures. See STANDARD AIR.

standard cubic feet per minute (SCFM) Common volumetric flow rate units for air. The STANDARD CONDITIONS used will determine the density and therefore the standard flow rate expressed.

standard deviation (σ) The square root of the VARIANCE of a data set, an indication of the "spread" of the data set around its mean, or average. The variance for a large data set is calculated as follows:

$$\sigma = \sqrt{\frac{\Sigma(x_i - \bar{x})^2}{n}}$$

where x_i is an individual observation, $\bar{x}$ is the mean, and n is the number of observations. For smaller data sets (less than 50) the sample standard deviation (s) is calculated by replacing n with $n-1$ in the equation.

Standard Industrial Classification (SIC) A numbering system used by the federal government to group industrial facilities by category. The full SIC number has four digits. The first two digits (range 01–99) indicate the major industrial category, and the last two digits classify the company further. For ex-

ample, major category codes 20–39 identify manufacturing facilities; code 28 within this range is the major category number for chemicals and allied products, and SIC number 2869 is the classification for industrial organic chemicals, not elsewhere classified. Environmental regulations are often applied to facilities within certain SIC codes.

Standard Methods A short form for "Standard Methods for the Examination of Water and Wastewater" which is prepared and published jointly by the American Public Health Association, American Water Works Association, and Water Pollution Control Federation. The book serves as the primary reference for analytical methods employed in investigations and monitoring of water purification, sewage treatment and disposal, water pollution, sanitary quality, and other functions. A new edition is published every five years.

standard population A group used to compare the mortality or morbidity rates of two different populations. The distribution of some characteristic within the standard population as it is related to mortality or morbidity (especially age) is used as the standard, and any differences in the distribution of the characteristic in the two groups is statistically taken into account in the comparison. See AGE ADJUSTMENT, DIRECT METHOD and AGE ADJUSTMENT, INDIRECT METHOD.

standard mortality ratio (SMR) The ratio of the occurrence of a specific cause of death in a given test population to the mortality rate from the same cause in a standard population. The numerical value is based on 100 incidents in the standard population; therefore, a ratio of 500 means that the test population had a mortality rate five times that of a standard population due to the same specific cause.

standards, environmental Allowable conditions or actions that will protect against unwanted effects on human health or welfare, wildlife, or natural processes. Such standards can be defined as ambient chemical concentrations, chemical or radioactive material emission or effluent rates, emission or effluent chemical concentrations, sound levels, or radioactivity levels, as well as harder-to-measure taste, odor, and appearance criteria. See AMBIENT STANDARD, EMISSION STANDARD, and EFFLUENT STANDARD.

standing crop   The quantity of plant biomass in a given area. For example, the amount of plant material per acre of forest or swamp, or per cubic meter of water. Usually expressed as mass (dry) per unit area, or energy content per unit area.

standing stock   See STANDING CROP.

standing, legal   In environmental law, the legal position required to validly seek judicial review of the actions of a government agency. Standing is gained by a plaintiff experiencing an adverse impact (legal or bodily injury) to himself, called injury in fact. Once thought to be a barrier to third party (concerned citizen) intervention in environmental agency decisions, the definition of standing has been much relaxed by court decisions, giving environmental interest groups ready access to the courts. Also, most major environmental statutes have specific provisions for private citizen legal action that do not require injury in fact to a plaintiff. See CITIZEN SUIT PROVISION.

stannosis   The presence of tin oxide dust in the lungs, resulting from occupational exposure. Stannosis is detected by a chest x-ray. None of the fewer than 200 reported cases has been accompanied by the chronic lung damage, such as fibrosis, which is associated with dusts containing free silica or asbestos.

State and Local Air Monitoring System (SLAMS)   Nonfederal air quality monitoring stations meeting siting and quality assurance requirements established by the United States Environ-

mental Protection Agency. Some stations are used to track compliance with air quality standards, some to measure the air in areas of expected high concentrations for certain pollutants, and others to determine the background concentration for air contaminants.

State and Territorial Air Pollution Program Administrators (STAPPA)   A group of nonfederal air pollution control agency representatives responsible for implementing the CLEAN AIR ACT. The group shares information on air pollution issues and regulatory compliance. Affiliated with the Association of Local Air Pollution Control Officials.

State Emergency Response Commission (SERC)   A group in each state appointed by the governor in accordance with provisions of the Emergency Planning and Community Right-to-Know Act, which is TITLE III of the SUPERFUND AMENDMENTS AND REAUTHORIZATION ACT OF 1986. The commission is responsible for forming and working with LOCAL EMERGENCY PLANNING COMMITTEES (at the county level) to collect information on chemicals stored and released by affected facilities. The SERC also reviews emergency plans written by the Local Emergency Planning Committees and responds to public requests for the inventory and chemical release information it compiles.

state implementation plan (SIP)   The written agreement between an individual state and the United States Environmental Protection Agency describing how the state will comply with the provisions of the CLEAN AIR ACT, especially the attainment and maintenance of the NATIONAL AMBIENT AIR QUALITY STANDARDS.

state variables   The components that make up a system in mathematical models used to describe the environment. The state variables have certain states, or conditions, at a given time. For example, in a simple BOX MODEL, the amounts of phosphorous in aquatic vegetation, sediments, and water at a given time could be chosen as the state variables.

static pressure The pressure exerted by a fluid at rest. Static pressure in air is due to the weight of the atmosphere and is usually equivalent to BAROMETRIC PRESSURE. See HYDROSTATIC PRESSURE.

static reserve index A resource use model that assumes resource use rates will remain constant. Under this assumption, the depletion time for the resource is calculated by dividing the current RESERVES by the use rate.

static test A laboratory test in which the water is not changed and there is no flow in or out of the test vessel when the toxicity of chemical substances to aquatic test organisms is being measured. See SEMISTATIC TEST and FLOW THROUGH TEST.

stationary growth phase A particular growth phase in the cultivation of bacteria in which there is no increase in the number of cells over time. Any cell division that takes place is balanced by cell death. This phase of growth follows the EXPONENTIAL GROWTH phase.

stationary source A fixed (nonmoving) source of air emissions, such as an oil refinery or power plant. See MOBILE SOURCE, AREA SOURCE.

statistical tests of significance Mathematical methods of stating the probability that two data sets are not from the same population; in other words, there is an actual difference in some characteristic between the two groups. These tests are commonly used for cases in which one group has been exposed to a treatment (or pollutant) and the other has not. After the treatment, some characteristic of the two groups is measured. A difference between the treated and untreated groups may be due to SAMPLING VARIABILITY or actually due to the exposure. A statistical test can determine the probability that the difference is not due to sampling variability.

statistically significant Describing a difference between two groups, one of which functions as the control and is not subjected to manipulation in the laboratory or environment and one which is subjected to some alteration in the physical, chemical, or biological surroundings (termed experimental or test group). For example, an experimental group of animals may be subjected to increased noise levels while the control group is held under the same conditions except for the noise. If subsequent measurements of the ability of the animals in both groups to respond to sound demonstrate a difference between the two groups, STATISTICAL TESTS OF SIGNIFICANCE can be applied to determine if the observed differences are due to the exposure to the sound or to natural sampling variability.

statutory law Written law passed by state or federal legislatures, such as the CLEAN WATER ACT. The common law is not written by legislative bodies, but is based on precedent. Environmental law is mainly concerned with statutory law and the regulations issued from it.

stay time The time that personnel may remain in a restricted area before accumulating the MAXIMUM PERMISSIBLE DOSE of radiation.

steady state In a system with a flow-through of material or energy, the equilibrium condition in which flow in equals the flow out.

steady-state or apparent plateau In testing chemical substances for their BIOCONCENTRATION potential in fish, the situation in which the amount of chemical substance taken into the test fish from the water is equal to the amount being eliminated from the test fish. If the substance bioconcentrates in the fish, the steady-state condition will follow an uptake phase in which the absorbance rate exceeds the elimination rate.

steam stripping   The removal of volatile compounds from wastewater by forcing steam through the liquid. The higher wastewater temperature increases the evaporation rate of the volatile contaminants.

Stefan-Boltzmann law   The equation describing the rate of radiation emitted by an object. Expressed as $H = Ae\sigma T^4$, where A is the surface area, e is an EMISSIVITY factor (between 0 and 1), σ is the Stefan-Boltzmann constant, and T is the ABSOLUTE TEMPERATURE of the object.

steno-   A prefix meaning narrow, used in ecology to describe tolerance ranges for organisms. For example, a fish species may be described as stenothermal, meaning it can grow and reproduce well within only a narrow range of water temperatures. See EURY-.

sterile   1. A condition in which a quantity of water, soil, or other substance does not contain viable organisms such as viruses and bacteria. The term, however, is not synonymous with clean.   2. Animals or humans that are not capable of reproduction because of the absence of gametes.

sterilization   1. The process of killing, inactivating, or removing microorganisms (bacteria, viruses, or fungi) from a quantity of water, soil, or other liquid or solid material. The normal methods of sterilization involve agents such as heat, chemicals, or radiation.   2. The process of rendering an animal incapable of reproduction.

sticky tape sampler   A device designed to collect airborne particulate matter by surface adhesion.

stippling   The spotty injury to a leaf surface caused by excessive exposure to air contaminants. Chronic exposure to OZONE, for example, causes stippling in broad-leaved vegetation.

stochastic   Describing an event or process that involves random chance or probability.

stock solution   A concentrated solution of a chemical used as a reagent in some laboratory procedure. For example, a laboratory test may require 100 milliliters (ml) of a 0.01 percent solution of sodium chloride. To avoid the requirement of weighing 0.01 gram (g) of salt and adding that amount to 100 ml of distilled water every time that one is to perform the test, a stock solution of sodium chloride containing 1 g salt per 100 ml could be prepared. When the test is to be performed, 1 ml of the concentrated solution could be added to 99 ml of distilled water, and the test performed in a much more timely fashion.

stocks   In ecological cycles and models, the amounts of a material in a certain medium or reservoir; for example, the stock of the carbon dioxide in the atmosphere, which equals about 2.7×10^{15} kilograms.

stoichiometric ratio, conditions   In air pollution control, the air-fuel mixture containing just enough air to completely oxidize the fuel. Under these conditions all air oxygen would be consumed and the STACK GAS would contain no oxygen. Combustion of fuels actually needs more oxygen than the amount indicated by the theoretical stoichiometric ratio, and EXCESS AIR is introduced to the combustion chamber.

stoker   Mechanical equipment (or person, in the past) for feeding waste materials or coal into an incinerator or furnace.

Stokes diameter   An equivalent diameter for a (usually airborne) particle, defined as the diameter of a perfect sphere with the same SETTLING VELOCITY and density as the particle. For example, a nonspherical particle that is 6 microns across and with density of 2 grams per cubic centimeter may have a Stokes diameter of 5 microns, meaning that the nonspherical particle has the same settling velocity as a 5-micron sphere with a density of 2 grams per cubic centimeter. See AERODYNAMIC DIAMETER.

Stokes' law The relationship that defines, for low velocities, the frictional force resisting the movement of a spherical body through a fluid (air or water) as 6π times the product of the radius of the sphere, the velocity of the sphere, and the VISCOSITY of the fluid. Stokes' law is used to determine the rate at which an object will fall through air or water, which is important in the design of pollution control systems. Expressed as $F = 6\pi r \eta c$, where F equals the force resisting the sphere radius r at speed c in a fluid of viscosity η.

stomata Small openings in the surface of plant leaves that allow gas transfer. Air pollutants can cause direct internal damage to vegetation by entering the stomata.

stopping power The rate of energy loss by IONIZING RADIATION per unit thickness or per unit mass of radiation shielding.

storage In reference to groundwater reservoirs, the water naturally retained in an aquifer plus any water artificially recharged to the aquifer by pumping.

Storage and Retrieval of Aerometric Data (SAROAD) A database of the United States Environmental Protection Agency containing air quality measurements from federal, state, and local monitoring stations. Each station has a unique SAROAD number.

storage facility In hazardous waste management, a facility that stores HAZARDOUS WASTE before treatment or disposal on the site or before transport to another treatment or disposal facility. Industries or businesses that generate hazardous waste usually are not classified as storage facilities. Regulations for hazard-

ous waste generators state that, with certain exceptions, hazardous waste may accumulate on-site for at most 90 days before the generator is subject to the rigorous permit requirements of a storage facility.

storm sewer A network of pipes and conduits buried underground that drain rain water from city streets.

stormwater runoff Rain water that runs off land surfaces directly into drainage facilities, rivers, lakes, streams, and other waters.

straggling The variation in energy content of particles released in nuclear events. Although they may all be of the same energy level upon release, over time they may vary in their range of energy level because of interactions within the material being traversed.

straight-chain hydrocarbons Compounds of carbon and hydrogen in which multiple carbon atoms are bonded to each other in a straight line. Other hydrocarbons can be branched, such as when carbons are bonded at right angles to the primary chain. Some hydrocarbons can have two or more ISOMERS, one straight-chain and the other(s) branched. Straight-chain hydrocarbons are ALIPHATIC compounds. Compare AROMATIC.

strata Distinct horizontal layers in geological deposits. Each layer may differ from adjacent layers in terms of texture, grain size, chemical composition, or other geological criteria. The term is also applied to layering of other material such as the atmosphere.

stratification The layering of a body of water caused by temperature or salinity differences among the layers. Also, the layering of materials in sedimentary rock.

stratopause The boundary in the atmosphere between the STRATOSPHERE and the next highest layer, the mesosphere.

stratosphere The second layer of the atmosphere above the earth. The air surrounding the earth consists of distinct zones, distinguished by the temperature gradient within each layer. The stratosphere begins at about 7 miles of altitude and extends up to about 30 miles. The air temperature in this layer generally increases as altitude increases. The OZONE LAYER of the atmosphere is located within the stratosphere. See ATMOSPHERE.

streamline A line parallel to the nonturbulent flow of a fluid, indicating the path that the fluid takes around objects, such as airflow around a building.

stressed waters A portion of an aquatic environment with poor SPECIES DIVERSITY due to human actions. If a facility applying for a water permit will discharge into an aquatic system that is stressed by the actions of others, then it will not be held responsible for the existing poor conditions but must demonstrate to the environmental agency issuing the permit that further degradation will not occur as a result of its effluent.

strip mining See SURFACE MINING.

stripper well A low-output oil well that is only marginally profitable to operate; a well producing less than 10 barrels of oil per day.

stripping Methods for the removal of unwanted dissolved gases from water. Stripping techniques involve increasing the surface area of the water to be stripped and maintaining the atmospheric PARTIAL PRESSURE of the gas(es) to be removed at a low level relative to the partial pressure of the gas dissolved in the water. Oxygen, ammonia, hydrogen sulfide, volatile organic compounds, and carbon dioxide are commonly stripped from water. Also called air stripping.

strontium-90 (^{90}Sr) A radioactive isotope released by nuclear weapons testing. With a HALF-LIFE of about 28 years, it can BIOACCUMULATE in the bones and replace calcium in the human body.

structure-activity relationship (SAR) The connection between the molecular structure of a chemical and its toxicity. Although only poorly predictive, an untested chemical similar in composition and spatial arrangement to a material known to be a human toxicant would be suspected to exhibit similar toxic effects based on a SAR analysis.

stuff and burn The practice of introducing waste to an incinerator at a rate greater than the waste combustion rate in the incinerator.

subbituminous A grade of coal that has a heat content higher than that of lignite but lower than bituminous. See COAL.

subchronic exposure In toxicology, doses that extend to approximately one-tenth of the lifetime of an organism. For humans, 5–10 years; for rats, 2–4 months.

subcutaneous Beneath the skin; a route of exposure of chemical substances to test organisms.

subduction zone According to the theory of PLATE TECTONICS, a region at the boundary of two of the plates that form the crust of the earth where one of the plates is forced downward into the mantle. The deep ocean trenches off the Pacific coasts of Mexico and South America are examples of these regions. The zones are also referred to as convergent plate boundaries.

sublimation The conversion of a substance in the solid phase directly to the gas phase without an intervening liquid phase, as in the changing of snow directly into water vapor without melting.

submerged plants Aquatic vegetation that has roots, stems, and leaves. The plants are rooted in the bottom of a water course but the leaves remain submerged below the surface of the water.

submicron   A distance less than 1 micron, or less than one millionth of a meter. Often used to describe particle diameters.

subsidence   The sinking of the land. The process can be natural, as in the sinking of the land relative to sea level in coastal areas, or man-made as in the collapse of the ground because of the removal of water, coal, or mineral deposits from the underlying strata.

subsidence inversion   A temperature inversion. The sinking air in the center of a high-pressure weather system compresses the air below it and raises the temperature of this upper-level air above that of the air below it. Inversion conditions suppress vertical motion of air contaminants and thus allow their air concentrations to increase. The high-pressure systems associated with subsidence inversions can remain in an area for several days or more.

substituted ring compound   A chemical compound consisting of one or more closed rings of carbon atoms and their attached hydrogen atoms, but with one or more of the hydrogen atoms replaced by another chemical substance. For example, polychlorinated biphenyl is composed of two benzene rings with a varying number of hydrogens replaced by chlorine atoms.

substrate   1. The solid surface on which an organism moves about or attaches, such as the sediments, rock, or sand on the bottom of a water body. 2. In biochemical reactions, the chemical material on which enzymes act. Enzyme action is typically specific to a particular substrate. The combination of enzyme and substrate is called the enzyme-substrate complex.

Subtitle C   The section of the federal RESOURCE CONSERVATION AND RECOVERY

ACT that applies to hazardous waste management.

Subtitle D   The section of the federal RESOURCE CONSERVATION AND RECOVERY ACT in which standards for the construction and operation of municipal solid waste disposal facilities are outlined. The standards apply primarily to the management of nonhazardous waste.

succession   The orderly and predictable changes in the biological communities within a specific locality over time from an initial, or pioneer community to a final stage, or climax. The climax community is stable and prevented from further change by climate, or other environmental conditions. The term is also refered to as ecological succession.

suction lysimeter   A sampling device for the collection of groundwater from the unsaturated zone; a sample is drawn by applying a negative pressure to a porous ceramic cup embedded in the soil layer.

sulfates   In the atmosphere, small-diameter aerosols composed of oxidized SULFUR DIOXIDES in the form of sulfate salts (e.g., ammonium sulfate) or sulfuric acid. Sulfates are an inhalation hazard and an important contributor to ACID DEPOSITION. In the soil, inorganic molecules (SO$_4^{-3}$) that serve as plant nutrients when present in appropriate concentrations. In the salt marsh, precursors for the microbial production of HYDROGEN SULFIDE.

sulfide ores   Natural metal deposits chemically combined with sulfur. Copper, zinc, mercury, and lead are commonly found in nature as sulfides. The removal of sulfur from the ore in smelters, called roasting, is an important (localized) source of atmospheric SULFUR DIOXIDE.

sulfur dioxide (SO$_2$)   A compound of sulfur containing one sulfur atom and two oxygen atoms. This compound is a major air pollutant produced by the burning of sulfur-containing compounds.

sulfur oxides (SO$_x$) A mixture of compounds of sulfur containing an indeterminate number of oxygen atoms. When sulfur-containing compounds are burned, a variety of compounds containing sulfur and oxygen can be produced (SO$_2$, SO$_3$). When the exact composition is not known, the notation SO$_x$ is used to signify the unknown mixture. Sulfur oxides are a major source of air pollution.

sulfuric acid A strong, highly corrosive acid. It is a colorless, oily liquid that is a strong dehydrating and oxidizing agent. The acid is a widely used industrial chemical and is formed in the atmosphere from emission of SULFUR DIOXIDE. See ACID RAIN.

sunset provision The stipulation that a regulation or an administrative agency will automatically lose effect or cease operations on a specific date unless positive measures are taken to justify its continuance.

Superfund See HAZARDOUS SUBSTANCES SUPERFUND.

Superfund Amendments and Reauthorization Act (SARA) 1986 amendments to the COMPREHENSIVE ENVIRONMENTAL RESPONSE, COMPENSATION, AND LIABILITY ACT (the Superfund law) that included provisions that increased the size of the HAZARDOUS SUBSTANCES SUPERFUND, required new cleanup standards, and started the SUPERFUND INNOVATIVE TECHNOLOGY EVALUATION (SITE) program. TITLE III of SARA, the Emergency Planning/Community Right-to-Know Act of 1986 requires that, for certain chemicals, facilities make public annually the amounts they routinely (or accidentally) release into the air, water, or ground. Title III also requires, for certain chemicals, that facilities make public annually the amounts stored and their locations within the facility. The chemical storage provision also is accompanied by the creation of STATE EMERGENCY RESPONSE COMMISSIONS and LOCAL EMERGENCY PLANNING COMMITTEES, which are to compile this information and plan for the possible accidental release of the stored chemicals.

Superfund Comprehensive Accomplishments Plan (SCAP) The fiscal year budget of the United States Environmental Protection Agency for technical support at SUPERFUND SITES. The document is available at all regional EPA offices.

Superfund Innovative Technology Evaluation program (SITE) A cooperative arrangement between private companies and the United States Environmental Protection Agency to demonstrate, evaluate, and encourage the commercial development of improved technologies for the permanent cleanup of SUPERFUND SITES.

Superfund site A hazardous waste landfill on the NATIONAL PRIORITIES LIST being cleaned up by the responsible parties or using proceeds from the HAZARDOUS SUBSTANCES SUPERFUND.

supernatant The clear fluid that is removed from the top of tanks or ponds used to allow solids to settle from suspension. It is also referred to as overflow.

supersonic Faster than sound. See SHOCK WAVE and SONIC BOOM.

supplied-air respirator A respirator delivering clean air to the wearer from an external source through a hose. In contrast, an AIR-PURIFYING RESPIRATOR cleans the air by filtering or chemical sorption as the wearer inhales.

surface area scaling factor In cancer risk assessments, a factor used to convert animal dose rates to equivalent human dose rates based on their body surface areas. This method relies on the proportionality between body surface area and the basal metabolic rate. The dose rate units are typically milligrams of a chemical per square meter of surface area per day. Of the several adjustment methods available, surface area scaling is the method most frequently used by the United States Environmental Protection Agency. See SCALING FACTOR.

surface casing The well pipe inserted as a lining nearest to the surface of the ground to protect the well from near-surface sources of contamination.

surface collecting agents Chemical additives spread on oil spills in an aquatic environment to control the thickness of the oil layer.

surface compaction Increasing soil density by applying force at the surface; the process is used in the installation of clay LINERS. In contrast, solid waste undergoes COMPACTION into bales before the waste is buried.

surface impoundment See IMPOUNDMENT.

surface mining The process of removing mineral deposits that are found close enough to the surface so that the construction of tunnels is not necessary. The soil and strata that cover the deposit are removed to gain access to the mineral deposit. The primary environmental concerns related to this technique are the disposition of spoils removed to gain access to the deposit and the scoring of the landscape that remains following the complete removal of the deposit. Water pollution is also a concern because runoff from the mining area is frequently rich in sediments and minerals. Also called strip mining. Compare UNDERGROUND MINING.

Surface Mining Control and Reclamation Act (SMCRA) A federal law passed in 1977 requiring all surface coal mining operators to meet detailed performance standards, including the restoration of the surface-mined land to its original condition. The Act also imposed a fee on each ton of coal removed, to be used for the reclamation of land subject to unregulated or poorly regulated strip mining in the past.

surface water Water that occupies rivers, streams, lakes, reservoirs, and wetlands. Also, water that falls to the ground as rain or snow and does not evaporate or percolate into the ground.

surfactant An agent that is used to decrease the surface tension of water, useful for removing or dispersing oils or oily residues. Most detergents are surfactants. The term is derived from *surface active agent.*

surrogate standard An organic compound used in gas CHROMATOGRAPHY as a standard for accuracy and precision estimates. The compound is chosen to be of similar chemical makeup to the actual organic compounds being measured and therefore will behave similarly in the analytical procedure. A known amount of the surrogate is introduced along with the sample containing other organic compounds as an internal control of methods and instrumentation.

survival curve A graph obtained by plotting the fraction of organisms surviving an increasing dose of some dangerous agent such as radiation or toxic chemical.

survivorship curve A graph of the number of individuals born in the same year surviving at the beginning of a series of succeeding time periods. For example, if in year one 100 individuals are born, at the beginning of year two perhaps 90 are living, at the start of year three 87 survive, and so on.

suspended particulate matter A sample drawn from natural water or from a wastewater stream consists of a mixture of both dissolved and suspended matter. Those solid materials that are retained on a filter prescribed by the specific technique being followed are referred to as particulate matter. The suspended particulate matter can by subdivided into two fractions: volatile and fixed. The volatile particulates are those that are lost when the filter is heated to about 550°C, and the fixed particulates are those that are not lost upon heating to 550°C. The volatile substances are generally considered to be of biological origin, and the fixed solids are considered to be minerals. See PARTICULATE MATTER; PARTICULATE MATTER, 10-MICRON DIAMETER.

suspended solids Same as SUSPENDED PARTICULATE MATTER.

suspension The dispersion of small particles of a solid or liquid in a liquid or gas.

sustainable development Describes efforts to guide economic growth, especially in less-developed countries, in an environmentally sound manner, with emphasis on natural resource conservation.

sustained nuclear reaction A continuous series of fission events. Nuclear fission involves the splitting of the nucleus of an atom such as uranium-235. As a result of the fission process, heat energy, radiation, large fission products (which are actually atoms of other elements produced by the fragmentation of the uranium atom), and neutrons are released. The process will continue with individual uranium atoms undergoing sequential fission reactions if certain conditions are met: the energy level of neutrons that are released must be lowered or moderated, and some of the moderated neutrons must hit another atom of uranium-235. When these two conditions are satisfied, the chain reaction will perpetuate itself.

sustained-yield harvesting The removing of a renewable resource, such as trees, at a rate that allows sufficient regrowth to maintain a continuous supply of trees for cutting in the future.

swamp A tract of land that is saturated with water and that is covered intermittently with standing water. The area is usually overgrown with thick vegetation dominated by shrubs and trees.

sweetening The removal of odorous sulfur contaminants from petroleum products.

swill A thick liquid waste material consisting of food scraps and water.

symbiosis An association between two different organisms so that both members of the association profit from the relationship. See also MUTUALISM.

symbiotic Describing a relationship between either bacteria and animals or plants, an animal and plant, animals of different species, or plants of different species such that both members of the pair benefit from the association. The association is so strong that neither member of the pair can exist or carry out certain activities alone. For example, the presence of bacteria of the genus *Rhizobium* residing within root structures of legumes (plants that produce seeds in a pod) results in the fixation of atmospheric nitrogen for use as a nutrient. Neither the bacterium nor the plant can carry out the function alone.

sympatric Describing two or more species occupying the same geographical area.

synapse The point at which a nerve impulse travels from one nerve cell to another. This transfer requires the release and activity of a variety of chemicals. Some toxic agents, such as insecticides, have an effect on organisms by suppressing the functioning of these chemicals.

synecology The study of the interactions among different POPULATIONS within an ecosystem. Compare AUTECOLOGY.

synergistic effect An effect resulting from two or more agents acting in such a way that the total effect is greater than the predicted sum of the individual agents acting alone. For example, the two air pollutants SULFUR DIOXIDE and PARTICULATE MATTER have a greater adverse effect on human health than would be expected from the sum of their individual toxicities.

synfuel   A gas or liquid hydrocarbon fuel produced from coal or shale. The use of coal to produce these alternative fuels presents a variety of environmental problems, ranging from damage done at the mine to air pollution related to the release of sulfur oxides, hydrocarbons, and fly ash. The problems related to the use of shale center around the high demand for water during processing and the substantial amount of solid waste that results.

synoptic scale   Large-scale weather patterns, as used on a typical weather map, with horizontal units of several hundred to several thousand kilometers and time units of one day to one week. This scale of analysis combines atmospheric data from hundreds of weather stations to produce patterns such as low-pressure systems and fronts.

synthetic fuel   Same as SYNFUEL.

synthetic natural gas (SNG)   A gaseous hydrocarbon fuel produced by the processing of coal, primarily composed of methane. See SYNFUEL.

synthetic seawater   An artificial product of the approximate ionic composition of seawater.

systemic   In toxicology, affecting the whole body or portions of the body other than the site of entry of a chemical substance. For example, a material may contact the skin and cause a localized irritant effect at the site of contact. If the material also penetrates the skin and gains access to the blood, it can also have systemic effect.

systems audit   A comprehensive examination of an environmental sampling and analysis project, including sampling techniques, calibration methods, and an appraisal of the QUALITY ASSURANCE plan.

systems ecology   The use of mathematical analyses to study the operations, factors, and processes that influence the association of organisms and their surroundings.

T

tacking   The binding of mulch fibers by mixing them with an adhesive chemical compound during land restoration projects.

tagged molecule   An atom of a radioactive element used within some molecule for the purpose of studying the behavior of that molecule. For example, carbon dioxide (CO_2) containing a radioactive isotope of carbon (^{14}C) might be used to study carbon dioxide fixation by plants through determining the inclusion of the radioactive form of carbon into the carbohydrates produced by the plant during photosynthesis.

taiga   See BOREAL FOREST.

tailings   The waste material remaining after metal is extracted from ore.

talc   A soft mineral, usually light tan or white in color, composed of magnesium silicate. The main ingredient in talcum powder. Chronic exposure to this agent by workers in the rubber and cosmetics industries results in pulmonary fibrosis.

tar balls   Nonvolatile hydrocarbon clumps remaining in water after the volatile fractions have evaporated from crude oil that has been discharged or spilled into the marine environment. When washed ashore, these residues, which range from marble-size to beachball-size, spoil beaches.

tar sands   Sandy deposits containing bitumen, a viscous, petroleum-like material that has a high sulfur content. Bitumen can be thermally removed after surface mining of the sands and upgraded to a synthetic crude oil. Large tar sand deposits are in Alberta, Canada. The smaller amounts in the United States are almost all in Utah.

tare   The determination of the empty weight of a container or vessel in order to allow for the future quantification of its contents by weighing the full or partially full container and computing the difference.

target organ   The body organ (e.g., the liver) or organ system (respiratory, nervous) that is most likely to be adversely affected by overexposure to a chemical or physical agent.

target theory   A concept used to explain the interaction of ionizing radiation and biological specimens in which the ionization that is produced in the cell by the radiation damages a specific target or location within that cell. One or more ionizing events may be required to take place within the cell before a specific physiological condition is observed.

Taylor Grazing Act   A 1934 federal statute that, with the FEDERAL LAND POLICY AND MANAGEMENT ACT, governs the management and preservation of federal public land.

technologically enhanced natural radioactivity (TENR)   A naturally occurring source of radioactive elements to which humans are exposed as a result of some human activity, such as underground mining or the drilling of wells.

technology-based   Describing emission or effluent limitations that are not defined in terms of allowable releases that achieve a desirably low ambient pollutant concentration but instead are based on the pollutant control efficiency that is achievable using current technology.

technology-forcing   Describing standards or levels of control called for in environmental statutes or regulations for which existing technologies are inadequate and therefore require technical advancements to achieve.

temperate deciduous forest   A geographic region characterized by distinct seasons, moderate temperatures, and rainfall from 30 to 60 inches per year. These forests are found in eastern North America; eastern Australia; western, central, and eastern Europe; and parts of China and Japan. Typical deciduous trees in the North American deciduous forest are oak, hickory, maple, ash, and beech.

temperature   A measure of the average energy of the molecular motion in a body or substance at a certain point.

temperature inversion   In the atmosphere, the condition in which air temperature increases with increasing altitude over a certain altitude range. The inversion layer can be at ground level or aloft. The condition results in a layer of warmer air above cooler air, a circumstance that inhibits atmospheric mixing and dispersion of pollutants. The two types are RADIATION INVERSION and SUBSIDENCE INVERSION.

temporary hardness   Water hardness that can be reduced or removed by heating the water. Heating drives off carbon dioxide, shifting the carbonate buffer system equilibrium so that carbonate ions combine with dissolved calcium or magnesium ions, form solids, and precipitate. This lowers the calcium/magnesium ion water concentration, lowering the hardness. Also called carbonate hardness.

temporary threshold shift (TTS)

The short-lived reduction in hearing ability for certain frequencies caused by noise overexposure. Exhibited by an increase in the HEARING THRESHOLD LEVEL. Recovery is usually complete within 16 to 24 hours. See PERMANENT THRESHOLD SHIFT.

ten percent rule

In ecology, the maxim stating that about 10 percent of the energy available at one trophic level is passed upward and stored in the bodies of organisms at the next-higher trophic level. The relatively small transfer of energy is due in part to respiration requirements of the energy consumers. Also, parts of ingested organisms are excreted, and not all available organisms are harvested. See ECOLOGICAL PYRAMID.

teratogen

A chemical substance or physical agent that significantly increases the risk of malformation or adverse development of a fetus.

terminal settling velocity

For a particle falling in a nonturbulent fluid (liquid or gas), the maximum possible velocity reached when the drag, or frictional resistance, on the particle equals the gravitational force on the particle. The measure is used in the design of chambers in which particles are removed from air or from water by gravitational settling. The horizontal flow rate through the chamber must allow time for the particles to reach the bottom of the settling chamber. See STOKES' LAW.

terracing

A series of levels on a hillside, one above another. Hillside farming on terraces greatly reduces water erosion of soil.

terrestrial radiation

The INFRARED RADIATION emitted by the surface of the earth and the atmosphere. Solar radiation is absorbed in the atmosphere, by surface waters, and by the ground; the energy is reradiated as heat (longer-wavelength infrared radiation) and, after atmospheric absorption and transfers, eventually lost to space. See WIEN'S LAW and GREENHOUSE EFFECT.

tertiary treatment

Treatment methods for wastewater beyond the conventional methods of PRIMARY TREATMENT (physical) and SECONDARY TREATMENT (biological). Facilities may apply one or more of the tertiary treatment methods, which remove inorganic nitrogen and phosphorus or dissolved organic compounds. Methods range from chemical techniques, such as the addition of lime to remove phosphates, to biological techniques, such as the application of wastewater to land to allow the growth of plants to remove plant nutrients.

tetrachlorodibenzofuran (TCDF or TCDBF)

A polyhalogenated hydrocarbon found as a contaminant in commercial preparations of POLYCHLORINATED BIPHENYLS (PCBs). This contaminant may be responsible for some of the adverse effects of PCBs. The molecule is very similar in chemical structure to TETRACHLORODIBENZO-PARA-DIOXIN (TCDD) and is frequently referred to as DIBENZO-DIOXIN. The toxicity, potency, and biological effects are very similar to those of TCDD.

tetrachlorodibenzo - *para* - dioxin (TCDD)

An aromatic halogenated hydrocarbon that is one of the most toxic compounds known. It is produced during the synthesis of precursors used in the manufacture of TRICHLOROPHENOXYACETIC ACID (2,4,5-T). The compound is toxic to liver and kidney function and has been shown to induce a variety of tumors in animal models. Adverse effects on the immune system of mammals have also been noted. CHLORACNE is the most common symptom resulting from human exposure. TCDD has been involved in a number of well-publicized environmental cases, the most famous of which is contamination of the herbicide mixture known as AGENT ORANGE used as a defoliant during the Vietnam War. Also referred to as dioxin and dibenzo-*para*-dioxin.

TETRACHLORODIBENZO-PARA-DIOXIN

tetraethyl lead An ANTIKNOCK ADDITIVE in gasoline that was the single largest source of lead emissions into the atmosphere until the phaseout of the additive in the United States completed in 1989.

theoretical oxygen demand (TOD) The amount of oxygen that is calculated to be required for complete decomposition of a substance. The calculation is based on the empirical formula of specific substances.

therm A unit of heat equal to 100,000 BRITISH THERMAL UNITS (BTUs).

thermal inversion See TEMPERATURE INVERSION.

thermal NO_x NITRIC OXIDE formed by the heating of combustion air to the point at which atmospheric nitrogen and atmospheric oxygen combine. It accounts for a significant percentage of the nitrogen oxide emissions from human activities. Compare FUEL NO_x.

thermal plume The hot water discharged from a power generating facility or other industrial plant. When the water at elevated temperature enters a receiving stream, it is not immediately dispersed and mixed with the cooler waters of the lake or river. The warmer water moves as a single mass (plume) downstream from the discharge point until it cools and gradually mixes with that of the receiving stream. See THERMAL POLLUTION.

thermal pollution The addition of excessive waste heat to a water body, usually by the discharge of cooling water from an electric power plant. The shift to a warmer aquatic environment can cause a change in SPECIES COMPOSITION and lower the DISSOLVED OXYGEN content of the water.

thermal power plant A facility that generates electricity by using some energy source to convert water to steam. The steam is then used to turn turbine generators.

thermal stratification The formation or condition of well-defined water temperature zones with depth in a lake or pond. See HYPOLIMNION, EPILIMNION, THERMOCLINE, FALL TURNOVER.

thermal treatment, of hazardous waste Any treatment of hazardous waste that involves exposure of the material to elevated temperatures in an effort to change the characteristics of the waste, for example, incineration, PYROLYSIS, and microwave discharge.

thermal turbulence Randomly fluctuating air motion caused by ground-level air being heated at the surface and rising past and through the upper air. The erratically moving eddies thus produced are typically larger than those produced by MECHANICAL TURBULENCE and are more effective in the dilution of air pollutants by mixing cleaner air with contaminated air. See EDDY DIFFUSION.

thermocline The sharp boundary between the EPILIMNION and the HYPOLIMNION in certain lakes and ponds in the temperate zone during the summer months. Characterized by a rapid change in water temperature over a short distance, from the warmer epilimnion to the cooler hypolimnion. Little mixing occurs across the thermocline.

thermodynamics The study of the involvement of heat energy in chemical or physical reactions and the conversion of energy from one form to another. See FIRST LAW OF THERMODYNAMICS and SECOND LAW OF THERMODYNAMICS.

thermonuclear   A nuclear reaction that requires extremely high temperature (10^7–10^8 K) as the activation energy to initiate the fusion process.

thickener   A settling pond or tank where the concentration of solids is increased by allowing settling and the removal of clarified liquid. The solids that are pumped from the bottom of the pond or tank are much thicker than the incoming fluid.

Thiobacillus   An aquatic or terrestrial genus of bacteria that is capable of oxidizing elemental sulfur, sulfide ions, thiosulfates, and other forms of inorganic sulfur to derive the energy needed in metabolism. Bacteria belonging to this genus fix carbon dioxide (are AUTOTROPHIC) and produce sulfuric acid as an end product. They can increase the acidity of soil or water to levels that result in the destruction of natural environments. Bacteria belonging to this genus are termed CHEMOAUTOTROPHS.

third third   See LAND DISPOSAL BAN.

thorium   A naturally occurring element with an atomic number of 90 and an atomic weight of about 232. The element can be converted to uranium-233, a fissionable isotope of uranium, by exposure to neutron irradiation.

threatened species   A specific plant or animal species whose population level in some sections of its natural range is very low; however, the breeding stocks are not sufficiently low throughout the entire range to place the continued existence of the species in jeopardy. Compare ENDANGERED SPECIES.

three-way catalyst   A device that controls all three of the top automobile emissions. An oxidizing catalyst converts hydrocarbons to carbon dioxide and water

and oxidizes carbon monoxide to carbon dioxide; a reducing catalyst converts nitrogen oxides to atmospheric nitrogen.

threshold dose   The smallest physical or chemical exposure that results in an observable adverse biological effect.

threshold effect   The adverse effect caused by the THRESHOLD DOSE for a particular physical or chemical agent and route of exposure; the effect often varies by individual.

threshold hypothesis   The concept that holds that no damage is done to a cell or organism when that unit is exposed to chemical agents or radiation below certain concentrations or intensities.

threshold level   See THRESHOLD DOSE.

threshold limit value (TLV)   The airborne concentration of a gas or particle to which most workers can be exposed on a daily basis for a working lifetime without adverse effects. The TLVs are set by the AMERICAN CONFERENCE OF GOVERNMENTAL INDUSTRIAL HYGIENISTS. See PERMISSIBLE EXPOSURE LEVEL.

threshold planning quantity (TPQ)   For a chemical designated as an EXTREMELY HAZARDOUS SUBSTANCE by TITLE III of the SUPERFUND AMENDMENTS AND REAUTHORIZATION ACT, the inventory amount at a facility that results in a requirement for the material to be reported by the facility and be included in the emergency response plan for chemical spills or releases.

threshold shift   A change in the ability of an individual to hear minimum sounds at a given frequency. The change can be either permanent or temporary. See TEMPORARY THRESHOLD SHIFT and PERMANENT THRESHOLD SHIFT.

tidal energy   The mechanical energy associated with the rising and falling of water level during the movement of the tides. See TIDAL POWER.

tidal power   The conversion of the mechanical energy associated with the rising and falling of water level caused by the tides into electric energy. Rising and falling waters are forced into large turbines, which then turn and activate generators.

tidal volume   1. The volume of air in one normal breath, about 0.5 liter in adult humans. 2. The volume of water entering and leaving a bay or salt marsh as the water level fluctuates because of the tides.

Tier I, Tier II reports   As required by the EMERGENCY PLANNING AND COMMUNITY RIGHT-TO-KNOW ACT, chemical inventory forms that must be submitted to the STATE EMERGENCY RESPONSE COMMISSION (SERC), the LOCAL EMERGENCY PLANNING COMMITTEE (LEPC), and the local fire department. Tier I reports are annual public disclosures of the estimated ranges and the maximum amounts of certain hazardous chemicals stored at a facility, the average daily amounts, and the general locations of the stored materials. Tier II forms give more detailed information on the same chemicals covered by Tier I, such as the storage conditions of the chemicals and their specific location within the facility. Tier II reports may be submitted in place of Tier I forms voluntarily, and they must be submitted upon request from the SERC, LEPC, or local fire department.

tiering   The preparation of an ENVIRONMENTAL IMPACT STATEMENT (EIS) for a broad action by a federal agency (called a program EIS) and the subsequent preparation of short, detailed statements for site-specific projects.

tilth   1. The general physical condition of soil as it relates to agricultural use. 2. Land used for agriculture, as opposed to pasture or forest.

time constant   The time required for a physical quantity to rise from zero to $1 - 1/e$, or 63.2 percent of its final steady-state value, when it varies with time (t) as $1 - e^{-kt}$, where $e = 2.71828$. Also, the time required for a physical unit to fall to $1/e$ (36.8 percent) of its initial value when it varies with time as e^{-kt}.

time-weighted average (TWA)   A method employed to calculate the exposure of workers to airborne materials, which considers the duration of exposure to various airborne concentrations, usually during an 8-hour day. Expressed as

$$\text{TWA} = \frac{\sum\limits_{i=1}^{N} C_i T_i}{N}$$

where C_i is the air concentration measured during time period i, T_i is the length of time period i, and N is the number of time periods. For example, the TWA for an exposure to a chemical level of 20 parts per million (ppm) for 2 hours and 100 ppm for 6 hours is 80 ppm $(2 \times 20 + 6 \times 100$, divided by 8).

tipping   At a solid waste disposal facility, the dumping of the contents of a waste truck, often by hydraulic lifting of one end of the load.

tipping fee   The per truck or per ton monetary charge to dispose of solid waste at a sanitary landfill.

Title III   The section of the SUPERFUND AMENDMENTS AND REAUTHORIZATION ACT that deals with emergency planning for chemical accidents, which includes the public release of information about chemicals used, stored, or accidentally released in a community. Additionally, the Act requires facilities to make periodic public reports of the amounts of routine or accidental releases of any of over 300 toxic chemicals to the air, water, or land. Also known as the Emergency Planning and Community Right-to-Know Act.

tolerance limits   See LIMITS OF TOLERANCE.

tolerances, pesticide An extensive listing (prepared by the United States Environmental Protection Agency and published in the Code of Federal Regulations) of the amounts of specific pesticides that are allowable in or on raw agricultural products as a result of pesticide application prior to harvest or slaughter. The amounts are expressed in parts by weight of each specific chemical per one million parts by weight of the commodity (ppm).

tons per day (TPD) A mass per time unit often used to describe municipal solid waste generation rates. When describing disposal at a municipal solid waste facility, the annual tonnage divided by 260 days (52 five-day weeks).

total carbon (TC) A measure of the amount of carbon-containing compounds in water. The measure includes both organic and inorganic forms of carbon as well as compounds that are soluble and insoluble. The typical laboratory analysis involves the conversion of all forms of carbon to carbon dioxide and the subsequent measurement of the carbon dioxide produced. The parameter represents an estimate of the strength of wastewater and the potential damage that an effluent can cause in a receiving stream as a result of the removal of DISSOLVED OXYGEN from the water. The measurement of total carbon requires less sample, is more rapid, and yields more reproducible results than the measurement of either CHEMICAL OXYGEN DEMAND or BIOCHEMICAL OXYGEN DEMAND. See TOTAL ORGANIC CARBON.

total dissolved solids (TDS) A measure of the amount of material dissolved in water (mostly inorganic salts). The inorganic salts are measured by filtering a water sample to remove any suspended particulate material, evaporating the water, and weighing the solids that remain. An important use of the measure involves the examination of drinking water. Water that has a high content of inorganic material frequently has taste problems and/

or water hardness problems. As an example, water that contains an excessive amount of dissolved salt (sodium chloride, a dissolved solid that is frequently encountered) is not suitable for drinking.

total inorganic carbon (TIC) The total amount of inorganic salts of carbonates and bicarbonates present in water without regard as to whether the salts are in suspended particulate form or dissolved. Water that contains an excessive amount of these salts is considered to be hard water. The dissolved materials interfere with the functioning of soaps and detergents and can form adherent scale in boilers, pipes, and steam equipment.

total maximum daily load (TMDL) The maximum quantity of a particular water pollutant that can be discharged into a water body without violating a water quality standard. The amount of pollutant is set by the United States Environmental Protection Agency when it determines that existing, technology-based EFFLUENT STANDARDS on the water pollution sources in the area will not achieve one or more ambient water quality standards. The process results in the allocation of the TMDL to the various sources in the area.

total organic carbon (TOC) A measure of the amount of organic materials suspended or dissolved in water. The measure is very similar to the assay of the total CARBON content; however, samples are acidified prior to analysis to remove the inorganic salts of CARBONATES and BICARBONATES. The assay of total organic carbon represents an estimation of the strength of wastewater and the potential damage that an effluent can cause in a receiving stream as a result of the removal of DISSOLVED OXYGEN from the water. The measurement of total organic carbon requires less sample, is more rapid, and yields more reproducible results than the measurement of either CHEMICAL OXYGEN DEMAND or BIOCHEMICAL OXYGEN DEMAND. As a pollution indicator, this method is more reliable than the assay of total carbon when the waste-water contains high amounts of total inorganic carbon. See TOTAL CARBON.

total reduced sulfur (TRS)   The sum of the nonoxidized sulfur compounds emitted to the air from a facility. Sulfur emissions from a kraft paper mill might include hydrogen sulfide, methyl mercaptan, dimethyl sulfide, and dimethyl disulfide. TRS from a sulfur recovery process in an oil refinery might include hydrogen sulfide, carbonyl sulfide, and carbon disulfide.

total solids (TS)   A measure of the amount of material that is either dissolved or suspended in a water sample, obtained by allowing a known volume to evaporate and then weighing the remaining residue. Total solids equals the sum of the measurements of TOTAL DISSOLVED SOLIDS and TOTAL SUSPENDED SOLIDS.

total suspended particulate (TSP)   The measured concentration of airborne particulate matter, often expressed as micrograms of particulate per cubic meter of sampled air.

total suspended solids (TSS)   A measure of the amount of particulate matter that is suspended in a water sample. The measure is obtained by filtering a water sample of known volume. The particulate material retained on the filter is then dried and weighed. Compare TOTAL DISSOLVED SOLIDS.

total trihalomethanes (TTHMs)   The sum of the concentrations of individual members of a family of halogenated derivatives of methane in drinking water. The concentrations of the following are employed to compute the sum in milligrams per liter: chloroform, dibromochloromethane, bromodichloromethane, and bromoform. See TRIHALOMETHANES.

toxaphene   A commercial chlorinated hydrocarbon insecticide. Preparations consist of a mixture of up to 170 different chlorobornanes and chlorobornenes. The compounds that constitute the mixtures vary widely in biological activity. This insecticide has a lower persistence in the environment when compared to other chlorinated hydrocarbon pesticides and has therefore been used extensively. Chronic exposure to toxaphene in laboratory animals has resulted in degenerative changes in liver and kidney tissue, and some tendency to promote tumor development has also been noted.

toxemia   A pathological condition in a person or animal caused by the presence of a toxic substance in the body.

toxic   Describing a material that can cause acute or chronic damage to biological tissue following physical contact or absorption.

Toxic Substances Control Act (TSCA)   The 1976 federal law that authorizes the United States Environmental Protection Agency to prohibit the manufacture, sale, or use of any new or existing chemical substance if it determines that the material poses an unreasonable risk to human health or the environment. The main focus of the law is on the review of new chemicals or proposed new uses of existing materials, which starts with a PREMANUFACTURE NOTICE (PMN) sent by the chemical producer to the agency. The PMN may include the results of toxicity testing of the chemical and other pertinent information required for the "unreasonable risk" determination.

toxic waste   Superfluous chemical compounds or mixtures that have the potential to cause adverse effects on human health or the environment. See HAZARDOUS SUBSTANCE, HAZARDOUS WASTE.

toxicant   Any chemical that has the potential of causing acute or chronic adverse effects in animals, plants, or humans.

toxicity   The ability of a chemical substance to cause acute or chronic adverse health effects in animals, plants, or humans.

toxicity characteristic leaching procedure (TCLP)   A test that measures the mobility of organic and inorganic chemical contaminants in wastes. The test, designed by the United States Environmental Protection Agency, produces an estimate of the potential for LEACHATE formation by a waste if it is placed in the ground. If the TCLP is applied to a solid waste sample and the extract leached from the waste or the solid waste sample itself contains concentrations of specified materials exceeding allowable levels, the waste is defined as a HAZARDOUS WASTE, meeting the toxicity characteristic. The TCLP replaced the EXTRACTION PROCEDURE TOXICITY TEST. See CONSTITUENT CONCENTRATIONS IN WASTE EXTRACT TABLE and CONSTITUENT CONCENTRATIONS IN WASTE TABLE).

toxicology   The study of chemical agents that cause diminished health and death in organisms, including humans. The study involves the chemistry, recognition, identification, measurement, distribution, and metabolism of hazardous substances to which organisms are exposed. The science also includes the prediction of potential adverse effects of chemicals on organisms, including humans.

Toxicology Information on Line (TOXLINE)   A database, operated by the National Library of Medicine in Bethesda, Maryland, containing information on adverse effects of chemical compounds. The National Library of Medicine is part of the National Institutes of Health, Public Health Service, Department of Health and Human Services.

toxin   A substance produced by plants or microorganisms that has the ability to cause adverse health effects in animals or humans.

trace elements   Elements essential to plant or animal life but required in only

small amounts, such as the trace amounts of manganese, zinc, iron, molybdenum, cobalt, and copper.

trace metals   Metals present in low concentrations in air, water, soil, or food chains.

tracer   A stable, easily detected substance or a RADIOISOTOPE added to a material to follow the location of the substance in an organism or in the environment or to detect any physical or chemical changes it undergoes. Some applications of tracers are to study atmospheric dispersion of pollutants, to follow the uptake of fertilizers by plants, and in the study of the metabolism and excretion of compounds introduced to the human body.

Tragedy of the Commons   The title of an essay by Garrett Hardin originally published in *Science* in 1968. Hardin likened the human effect on the planetary environment to overgrazing a commons; in other words, individual decisions based on incremental personal benefit (adding another animal to the pasture, driving an automobile) will, when added together, ruin the common environment (overgrazing, pollution). His chief concern in the essay was world overpopulation, but the analogy has been widely applied to pollution of common resources such as air and water.

trammel net   A net used in fish sampling; the net contains a layer of large-mesh netting on either side of a smaller-mesh gill net.

transfer station   In solid waste management, a facility at which waste collected by smaller trucks is reloaded into larger trucks for more efficient transport to a distant disposal site.

transformation   The conversion of a normal animal or human cell to a tumor cell.

transformed cell Cells that have been converted from normal cells to tumor cells. The term is frequently used in association with cells grown in tissue culture in the laboratory and converted by exposure to some chemical or radiation.

translocation A disorder of cellular chromosomes caused by breakage of a chromosome, followed by a combining of the broken fragment with the wrong chromosome as the cell attempts to repair the damage caused by the initial breakage. Such damage can be caused by exposure to an excessive amount of x-rays.

translocation factor In models of the distribution of chemical materials in the environment, a factor, different and specific for each chemical, used to estimate the daily movement of a chemical from soil to plants, from animal feed to animal tissues, or from animal feed to milk.

transmission electron microscopy (TEM) A type of microscopy in which the illumination is provided by a beam of electrons passed through electron-transparent items. Cells, even bacteria, must be sliced very thinly and treated with specific reagents to allow the observation of the interior of the cell. Compare SCANNING ELECTRON MICROSCOPY.

transmissivity (T) The rate at which water can flow through a unit width of an aquifer under a unit hydraulic gradient. Calculated as $T = KD$, where K is the hydraulic conductivity and D is the aquifer thickness. Transmissivity has units of area/time.

transmutation The process of radioactive decay during which one element is converted to another element through the loss of radiation of some type.

transparency The portion of light that passes through water without distortion or absorption. A measure of the turbidity of water or other liquid.

transpiration The direct transfer of water as a gas from plant leaves to the atmosphere. Transpiration combined with evaporation from the soil is called EVAPOTRANSPIRATION.

transuranic waste (TRU waste) A type of HIGH-LEVEL WASTE containing uranium-233 and its daughter products and radioactive elements with atomic numbers greater than 92 (the atomic number of uranium) produced during the operation of nuclear reactors and from nuclear weapons fabrication. The alpha activity of TRU waste is greater than 100 nanocuries per gram.

transuranium Those elements with an atomic number greater than that of uranium, which is 92.

trash The fraction of solid waste that is not food waste; includes paper, glass, wood, and aluminum cans. Some components are biodegradable (such as paper and wood), and others potentially can be recycled (such as aluminum cans and glass). Also called rubbish.

tray tower An air pollution control device that removes pollutants from an exhaust by forcing the gas up through a tower containing trays above which the scrubbing liquid is introduced. Holes in the trays increase the gas-liquid contact. See also PACKED TOWER, SPRAY TOWER.

treatment, storage, or disposal (TSD) Describing a facility where hazardous waste is treated, stored, and/or disposed as defined by the RESOURCE CONSERVATION AND RECOVERY ACT. Such facilities include landfills, surface impoundments, waste piles, and incinerators among others.

treatment, storage, or disposal facility (TSDF) See TREATMENT, STORAGE, OR DISPOSAL.

tree line The dividing point, caused by changing latitudes or altitudes, between areas with environmental conditions supporting trees and those that do not. The tree line in North America defined by latitude runs across northern Canada. Tree lines also exist at lower latitudes because of the altitudes in such places as in the Rocky Mountains.

tremolite A fibrous mineral of the asbestos group consisting of hydrated silicates of calcium and magnesium. A small amount of this mineral is marketed as asbestos.

trench method, landfill A technique for the placement of municipal solid waste in a landfill. The waste is spread and compacted in 2-foot layers within trenches 100–400 feet long, 15–25 feet wide, and 3–6 feet deep. The cover material for a full trench is obtained from the excavation of an adjacent trench. See also AREA METHOD, LANDFILL.

trial burn An extensive, controlled test of the DESTRUCTION AND REMOVAL EFFICIENCY of an incinerator. The results are used to establish the operating conditions specified in the permit for a facility.

trichloroethylene A two-carbon aliphatic compound (C_2HCl_3) belonging to the general class of chemicals referred to as halogenated or chlorinated hydrocarbons. It is used as a solvent and industrial chemical. The compound has been shown to cause liver and kidney damage and to cause tumors in some animal models.

trichlorophenoxyacetic acid (2,4,5-T) A chlorophenoxy herbicide used to control broadleaf weeds and woody plants. An ingredient, along with dichlorophenoxyacetic acid (2,4-D) in AGENT ORANGE, the toxic defoliant used during the Vietnam War. Synthesis of the chemical results in a slight contamination with TETRACHLORODIBENZO-PARA-DIOXIN.

trickling filter A wastewater treatment apparatus used to remove soluble or colloidal organic compounds. The filter consists of a large open tank filled with small stones. Clarified wastewater

(from which the particulate material has been removed) is sprayed over the surface of the stone bed and allowed to trickle through. Microbial communities that coat the stones mineralize the organic compounds by aerobic metabolic processes as the water trickles downward. See SECONDARY TREATMENT.

trihalomethanes (THMs) A group of low-molecular-weight, halogenated hydrocarbons including chloroform, bromodichloromethane, dibromochloromethane, and bromoform. The group includes suspect human carcinogens. Small amounts of THMs have been detected in raw water collected from surface sources used as a public water supply, and concentrations have been shown to be increased during the chlorination phase of the water purification process. The most marked increase during chlorination has been recorded in water containing suspended particles and/or humic substances.

trip blank A sample container to which a clean sample matrix (water, soil, etc.) is added and then is carried to and from the location where environmental samples are collected. The trip blank is analyzed along with the actual samples and serves as a check for sample or sample container contamination during transit.

tritium A man-made isotope of hydrogen containing one proton and two neutrons in the nucleus.

tromeling The removal of small-diameter dense solids using a TROMMEL SCREEN. Shredded solid waste is placed in a rotating screened drum and the smaller particles fall through the screen.

trommel screen A cylindrical screen with a shaft through the center, around which the screen rotates. The apparatus is used for waste separation. Smaller-sized, denser materials, such as glass fragments and grit, in waste that is placed on the rotating screen will fall out while the larger components remain behind.

trophic level   A feeding level within a FOOD WEB. The nature of the food of an organism determines its trophic level, numbered by the levels up from PRODUCERS; for example, the second trophic level is occupied by organisms feeding on organisms in the first trophic level, which feed directly on the producers.

trophic structure   The food chain within a specific biological community.

tropical rain forest   The richest natural terrestrial environment in terms of the amount of biomass and the number of species present. These forests occupy the equatorial regions of South and Central America, Central Africa, and Southeast Asia. Such forests include up to five times the number of species of trees as found in temperate regions and an unrivaled number of species of associated plants and animals. Paradoxically, the soil of the typical rain forest is very thin and low in plant nutrients. The temperatures usually remain between 70° and 95°F year round, and rainfall typically varies between 50 and 200 inches per year. The destruction and disturbance of tropical rain forests are of environmental concern because of the loss of many species (some yet to be described), the conversion of the land to a semidesert condition, disruption of the climactic patterns of the region, and a worsening of the global balance in carbon dioxide production and utilization, among other factors.

tropopause   The boundary in the atmosphere between the layer next to the surface of the earth (troposphere) and the next highest layer (stratosphere). See ATMOSPHERE.

troposphere   The layer of the atmosphere closest to the surface of the earth extending up to about 9 to 16 km. See ATMOSPHERE.

tumor   See NEOPLASM, BENIGN NEOPLASM, and MALIGNANT NEOPLASM.

tumorigenicity   The ability of cells from a tissue culture to grow and produce tumors when inoculated into a host organism.

tundra   A biome located in the northern regions of the continents of North America, Europe, and Asia, characterized by the absence of trees and the presence of permafrost (permanently frozen subsoil). The natural vegetation consists of small shrubs, mosses, and lichens.

turbidity   A measure of the amount of suspended matter in water or some other fluid as determined by the relative light transmission of the suspension. The typical scales used are percent transmission or transmittance, which varies from 0% to 100% of the light passing through the sample, and optical density or absorbance, which is a logarithmic scale varying from 2 to 0, in which 2 is the most turbid and 0 is the least turbid. About 0.3 on the absorbance scale corresponds to 50% transmittance.

turbine   A device with blades connected to a central shaft that converts a fluid's energy into rotational mechanical energy. In steam and gas turbines, hot gases expand through and rotate the turbine blades. Water and wind turbines capture and convert the energy of falling water or blowing wind into rotation. The mechanical energy from turbines is used in pumps and electric generators.

turbulence   See MECHANICAL TURBULENCE and THERMAL TURBULENCE.

turbulent flow   Fluid flow exhibiting random fluctuations in speed and direction. Compare LAMINAR FLOW; see also REYNOLDS NUMBER.

turnaround   The period during which an industrial facility or generating station is shut down for planned maintenance.

turnover   1. In ecology, the rate of PRODUCTIVITY divided by the STANDING CROP, or biomass. Expressed as

$$T = \frac{P}{B}$$

where P is productivity (in units of mass per area-time) and B is biomass (in units of mass per area). Turnover has units of 1/time.   2. See FALL TURNOVER.

turnover rate   The rate (for example, in milligrams per cubic meter of water per day) at which some material is metabolized or decomposed in the environment.

turnover time   1. In ecology, the average time *(t)* required for the biomass *(B)* in an ecosystem to replace itself, the inverse of TURNOVER, in time units. Expressed as

$$t = \frac{B}{P}$$

where P is the productivity (in units of mass per area-time) and B has units of mass per area.   2. In microbiology, the time required to metabolize a specific substance in a body of water or in soil.

type A/type B packaging   Containers designed to meet federal regulations for the transport of radioactive materials. The containers must maintain their integrity without leakage of radioactive material or loss of shielding under certain instances of heat, cold, free drop, and other factors. Type B packaging must meet higher standards in terms of radiation shielding and structural integrity.

type A/type B quantities   Types and maximum amounts of radioactive materials that can be transported in type A and type B packaging, respectively.

Type I error   When judging the results of a scientific study, the rejection of the NULL HYPOTHESIS when in fact the null hypothesis is true. See TYPE II ERROR.

Type II error   When judging the results of a scientific study, the acceptance of the NULL HYPOTHESIS when in fact the null hypothesis is false. See TYPE I ERROR.

U

ultimate analysis   A chemical analysis that measures the concentrations of certain individual elements or compounds in a material. When performed for combustible waste, the concentrations of carbon, oxygen, sulfur, nitrogen, hydrogen, and noncombustibles are typically determined. Compare PROXIMATE ANALYSIS.

ultrasonic   Describing acoustic waves with frequencies above 20 kHz and therefore not audible by the human ear.

ultraviolet (UV)   That portion of the electromagnetic spectrum that extends from the violet range of visible light (wavelength equal to about 400 nanometers) to the x-rays (wavelength equal to about 10 nanometers). Radiation in this range does not produce ions when it interacts with matter and is therefore termed nonionizing. That radiation falling between the wavelengths of x-rays and 160 nanometers is referred to as extreme or vacuum ultraviolet and is of little biological interest since it is not transmitted through air. The remaining portion of the spectrum (160 to 400 nanometers) is divided into A, B, and C ranges for the purpose of differentiating the biological effects of UV light at different wavelengths. Ultraviolet radiation within the range of 130 to 200 nanometers is responsible for the generation of OZONE within the stratosphere. See ULTRAVIOLET RADIATION–ACTINIC RANGE, ULTRAVIOLET RADIATION–A RANGE, –B RANGE, –C RANGE.

ultraviolet photometry   An analytical method that employs the selective molecular absorption of particular ultraviolet wavelengths to identify and measure the concentrations of certain chemicals in air or water. The method is also called ultraviolet spectrophotometry.

ultraviolet radiation–actinic range   That range of the spectrum of ultraviolet radiation that has the most pronounced biological effect, such as mutations and skin cancer; the wavelengths extend from 240 to 320 nanometers (all of the UV-B and part of the UV-C range).

ultraviolet radiation–A range (UV-A)   That part of the spectrum of ultraviolet radiation that encompasses the wavelengths from 320 to 400 nanometers. This portion of the spectrum is also referred to as near-ultraviolet since the range is nearest to that of visible light (wavelengths longer than 400). Radiation within this range is transmitted through air and somewhat through glass. A portion of the radiation within this class will cause tanning of the skin, especially when the radiation is of high intensity. Most (95 to 99.9 percent) of the ultraviolet light used in tanning salons is within the A range.

ultraviolet radiation–B range (UV-B)   That part of the spectrum of ultraviolet radiation that encompasses the wavelengths from 290 to 320 nanometers (middle-ultraviolet). Radiation within this range is not well transmitted through glass. UV-B is the primary cause of sunburn and of the cancer-causing properties of ultraviolet light.

ultraviolet radiation–C range (UV-C)   That part of the spectrum of ultraviolet radiation that encompasses the wavelengths from 160 to 290 nanometers. This portion of the spectrum is also referred to as far-ultraviolet since the range is most distant from the wavelengths of visible light. The germicidal effect of UV light is strongest within this region, with a wavelength of 260 nanometers being the most effective at the killing of microorganisms.

ultraviolet spectrophotometry   See ULTRAVIOLET PHOTOMETRY.

ultraviolet–visible absorption spectrum (UV-VIS)   The absorption pattern by certain chemical compounds of electromagnetic energy at wavelengths within those of visible and ultraviolet radiation. Various molecules absorb radiation at specific wavelengths and consequently can be analyzed in both a qualitative and quantitative manner. The technique is especially useful in the analysis of organic molecules that contain double bonds between carbon atoms.

uncertainty factor (UF)   An adjustment applied to experimental toxicity data to set acceptable human dose levels. The UF is intended to account for the uncertainties introduced by the extrapolation of animal data to humans, the variation in susceptibility within the human population, the use of acute exposure data to predict safe chronic exposure levels, and/ or the use of data from an oral route of exposure to set inhalation standards. United States Environmental Protection Agency guidelines recommend the following uncertainty factors: The higher the UF, the lower the acceptable human dose.

Factor	Applied to
10	Valid data on human exposures
100	Valid chronic animal studies
1000	Animal studies with less than chronic exposure
1–10	An additional factor when using a LOWEST OBSERVED ADVERSE EFFECT LEVEL instead of a NO OBSERVED ADVERSE EFFECT LEVEL

unconfined aquifer   An aquifer with no low-permeability zones between the zone of saturation and the surface; an aquifer with a water table. Also called a water table aquifer.

underflow   The slurry of concentrated solids or sludge that is removed from the bottom of a settling tank, clarifier, or thickener.

Underground Injection Control (UIC)   A program required in each state by a provision of the SAFE DRINKING WATER ACT for the regulation of injection wells, including a permit system. An applicant must demonstrate that the well has no reasonable chance of adversely affecting the quality of an underground source of drinking water before a permit is issued.

underground mining A technique of removing coal and mineral ores from the earth that involves cutting tunnels below the surface to gain access to the deposit(s). The environmental problems that can be associated with underground mining include the runoff from the SPOIL brought to the surface and ACID MINE DRAINAGE. Compare SURFACE MINING.

underground source of drinking water (USDW) An aquifer that supplies or potentially can supply a public water system.

underground storage tank (UST) Under the RESOURCE CONSERVATION AND RECOVERY ACT regulations, a tank with at least 10 percent of its volume beneath the ground, including attached pipes. Underground storage tanks must meet certain performance standards, have spill and overfill controls, and be monitored regularly for leaks.

unit density A density of one gram per cubic centimeter or one gram per milliliter; the density of water at 4°C.

unit risk The lifetime risk of cancer per unit dose of exposure. See UNIT RISK ESTIMATE and CANCER POTENCY FACTOR.

unit risk estimate The CANCER POTENCY FACTOR expression for air exposures; expressed as an inverse concentration, usually (micrograms per cubic meter)$^{-1}$. The actual or estimated air concentration of a chemical, in micrograms per cubic meter, multiplied by the unit risk estimate, gives the lifetime risk of contracting cancer as a result of inhaling the chemical.

$$\text{Lifetime risk} \atop \text{of cancer} = \frac{1}{\frac{\mu g}{m^3}} \times \frac{\mu g}{m^3}$$

unit	air
risk	concentration
units	units

The exposure is assumed to be 24 hours per day for 70 years. See CANCER POTENCY FACTOR.

United Nations Environment Program (UNEP) An agency of the United Nations, headquartered in Nairobi, Kenya, responsible for coordinating intergovernment efforts to monitor and protect the environment.

United Nations/North America number (UN/NA number) A four-digit number used internationally to identify a hazardous material; for example, UN/NA 1203 is gasoline.

United States Code (U.S.C.) A multivolume compilation of all federal statutes. A new edition is published every six years, with supplements issued during the intervening period.

United States Code Annotated (U.S.C.A.) A multivolume collection of the entire UNITED STATES CODE plus state and federal court decisions applicable to specific Code sections and other reference material.

United States Code Supplement (U.S.C. Supp.) A cumulative supplement to the UNITED STATES CODE published during the six-year interval between releases of new editions of the complete code.

United States Geological Survey (USGS) An agency within the United States Department of the Interior based in Reston, Virginia; responsible for assembling technical geographical and geological information about public lands and waterways for use by other government agencies.

universal gas constant *(R)* The constant of proportionality in the EQUATION OF STATE and the IDEAL GAS LAW for one gram-mole of a gas, $PV = RT$; where P is the pressure, V is the volume, and T is the ABSOLUTE TEMPERATURE. The constant is available is many different units, to match the units for P, V, and T. A common form is 82.06 atm-cm^3 gram-mole^{-1} K^{-1}, where atm is the pressure in ATMOSPHERES and K is the Kelvin temperature.

Universal Transverse Mercator coordinates (UTM coordinates)   A map coordinate system covering the world from 80° north to 80° south with 60 north-south zones, each covering 6° of longitude and divided into 8° latitude sections. The zones overlap 0.5° on each side. Each zone has an individual origin, and the coordinates are read in meters east and meters north of the origin. The coordinates often are used to define emission locations in air pollution dispersion modeling.

unsaturated zone   The upper layers of soil in which pore spaces or rock are filled with air or water at less than atmospheric pressure; the area of the ground from the surface down to the water table. The region is also called the zone of aeration or vadose zone. Compare SATURATED ZONE.

unstable   Describes elements or compounds that react easily or spontaneously to form other elements or compounds. For example, ozone (O$_3$) is an extremely unstable gas because it reacts readily with many materials. Likewise, radioactive materials form new elements as they undergo radioactive decay: uranium-238 is converted to thorium as a result of the release of radiation.

upgradient well   A groundwater monitoring well, such as those required at facilities that treat, store, or dispose of hazardous waste using surface impoundments or landfills, that allows sampling and analysis of groundwater that is upstream from the facility, before the groundwater is possibly affected by any escaping contaminants. The results of the analyses are used for comparison to the results of groundwater sampled from DOWNGRADIENT WELLS.

upper-bound risk level   A risk level derived from a cancer risk assessment. The upper bound is derived by using conservative (on the side of increased risk) assumptions in the risk assessment and the upper CONFIDENCE LIMIT of the statistically estimated risk value. The level of risk is reported as the 95 percent upper-bound risk, meaning that there is only a 5 percent chance that the true risk is greater than the upper-bound risk.

upper explosive limit (UEL)   The highest concentration of a substance in air that will burn or explode when ignited. At concentrations higher than this level, the amount of oxygen in the mixture is not sufficient to support combustion. Compare LOWER EXPLOSIVE LIMIT.

upper flammable limit (UFL)   See UPPER EXPLOSIVE LIMIT.

upper respiratory tract (URT)   The mouth, nasal passages, pharynx, and larynx.

upset   An exceptional incident, beyond the reasonable control of a company holding a water discharge or air emission permit, in which there is an accidental release of pollutants greatly in excess of the discharges or emissions allowed by the permit.

uptake   The incorporation of a pollutant into an exposed organism.

upwelling   The appearance of water from the deep ocean at the surface. This usually occurs along the coasts of continents (such as the coast of Peru along the west coast of South America) where the prevailing winds tend to push the surface waters away from the land area, allowing waters from the deep ocean to rise to the surface. The deep waters carry a significant input of plant nutrients to the surface, causing an elevated level of primary production and abundant fish populations.

uranium (U) A rare heavy metal with an atomic number of 92. The most common atomic weight in naturally occurring uranium is 238. Other ISOTOPES range from 234 to 239. The element was little used until about 1942 when it was discovered that it could be employed as a FISSION fuel for the construction of nuclear reactors and weapons. Uranium is found in natural deposits combined with other elements, principally oxygen.

uranium enrichment A process that results in an increase in the amount of the fissionable isotope of uranium (URANIUM-235) in a given mass of uranium. As it is recovered from natural deposits, uranium consists of about 0.7 percent of the fissionable isotope (^{235}U) and 99.3 percent of the nonfissionable isotope URANIUM-238 (^{238}U). The relative abundance of the fissionable uranium-235 must be increased to about 3 percent before the uranium can be used as a fuel in nuclear reactors. See GRAHAM'S LAW.

uranium fuel cycle The mining, refining, fabrication, transport, and recycling of uranium along with the disposal of wastes from both the purification processes and the reprocessing of used fuel rods.

uranium tailings The residue that remains from the mining of uranium ores and the purification of uranium from those ores. This residue presents a disposal problem because it releases radioactive elements such as radon gas.

uranium-235 (U-235 or ^{235}U) A fissionable isotope of the element uranium found in naturally occurring uranium deposits at a concentration of about 0.7 percent of the total uranium present in the geological deposit.

uranium-238 (U-238 or ^{238}U) A nonfissionable isotope of the element uranium. U-238 contains three more neutrons than U-235 and is the most commonly occurring form of the element (99.3 percent).

urban heat island The elevated release and retention of energy characteristic of an urban area in which large quantities of fossil fuels are burned relative to the surrounding countryside and that contains asphalt, brick, concrete, and other materials that absorb and retain higher amounts of heat energy than does vegetative cover. This phenomenon is illustrated by the higher average annual temperatures measured in urban areas compared to adjacent rural areas.

urban plume The downwind impact zone of air contaminants or products of atmospheric reactions involving compounds released within the confines of an urban area.

Users' Network for Applied Modeling of Air Pollution (UNAMAP) A collection of over 30 air quality dispersion models in the form of computer programs available from the United States Environmental Protection Agency.

V

vadose zone The area of the ground below the surface and above the region occupied by groundwater. See UNSATURATED ZONE.

valence The number of electrons an atom contributes or receives when a chemical bond is formed or when ions are formed from the elements. Atoms contributing electrons in reactions or during ionization, such as metals, have positive valences. Nonmetals, such as chlorine, have negative valences; that is, they receive electrons during bonding.

van Dorn sampler See EKMAN WATER BOTTLE.

vapor The gaseous form of a material that is normally found in the solid or liquid state at room temperatures. Vaporization occurs when particles in the solid or liquid gain enough energy to escape the material, especially with an increase in temperature. The term is frequently employed to refer to water in the gas phase.

vapor density   The density of a pure gas or vapor compared with the density of hydrogen or to the density of air. When used in chemical hazard analysis, the comparison is with air and the ratio is based on an assumed air density of 1.0; therefore, lighter-than-air gases have a vapor density less than 1.0, and those with values greater than 1.0 are heavier than air. Note that a gas or vapor released into the environment will mix with the surrounding air, causing the vapor density of the mixture to approach 1.0 in many cases, and behavior of the vapor from a spilled liquid may not be accurately predicted using a vapor density based on the pure material.

vapor incinerator   See AFTERBURNER.

vapor pressure   The pressure exerted by a gas or vapor. These pressures are experimentally determined by establishing an equilibrium between the gas and liquid phases of the substance in a closed vessel at a specific temperature. The higher the vapor pressure, the greater the tendency of the liquid to evaporate. Common units are mm Hg and POUNDS PER SQUARE INCH (ABSOLUTE).

vaporize   To change into a VAPOR; to evaporate.

variance   1. A statistical parameter that measures the spread of a data set about the mean, or average. See STANDARD DEVIATION. 2. A waiver of certain environmental regulatory requirements for a facility, process, or unit.

vascular bundle   Tissue in a plant responsible for transport of materials from the roots up and from the leaves down.

vector   An organism, such as a mosquito, flea, or tick, that carries a pathogenic microorganism from one host to another.

vector-borne   Describing a disease transmitted by a VECTOR.

velocity (v)   The distance moved in a given direction per unit time, such as meters per second (m s^{-1}).

velocity head (v$_H$)   The kinetic energy in a hydraulic system. Velocity head is given by

$$v_H = \frac{v^2}{2g}$$

where v is the fluid flow velocity and g is the acceleration due to gravity. Expressed in length units. The sum of the ELEVATION HEAD, PRESSURE HEAD, and velocity head equals the total energy of a hydraulic system. See HEAD, TOTAL.

velocity pressure (VP)   For air flowing in a duct, the air pressure attributable to the impaction of the moving air molecules. Expressed in the same way as velocity head,

$$VP = \frac{v^2}{2g}$$

where VP is the velocity pressure, v is air velocity, and g is gravitational acceleration. The total pressure in a duct is the sum of the STATIC PRESSURE and the velocity pressure.

velometer   A device for measuring air velocity, often used in studies and analysis of workplace airflow and in the design and operation of local exhaust ventilation systems. The portable instrument indicates air velocity using a spring-loaded vane (plate) that is deflected by the moving air. Also called a swinging vane anemometer.

ventilation   In exposure studies, air intake to the lungs.

ventilation rate (respiration) The volumetric breathing rate: for adult males, about 23 cubic meters/day; adult females, 21 cubic meters/day; and children, 15 cubic meters/day.

venturi effect The increase in the velocity of a fluid stream as it passes through a constriction in a channel, pipe, or duct. Calculated by the CONTINUITY EQUATION, $Q = VA$, where Q is the volumetric flow rate, A is the area of flow, and V is the fluid velocity. Because Q does not change, if A gets smaller then V must increase.

venturi scrubber An air pollution control device that removes particulate matter or acid mists from gas streams by injecting a scrubbing fluid, usually water, into a narrow section of duct called the venturi throat. The high gas velocity in the venturi atomizes the water into very small droplets with relatively low velocities compared to the particulate or· mist in the gas stream, achieving high collection efficiencies.

vertical dispersion coefficient See DISPERSION COEFFICIENT.

vinyl chloride A gaseous organic compound composed of carbon, hydrogen, and chlorine; vinyl chloride monomers (individual units) are used to make the polymer (a long chain), POLYVINYL CHLORIDE. Occupational exposure to vinyl chloride is linked to a rare liver cancer, angiosarcoma. Workplace exposure to vinyl chloride is regulated by the OCCUPATIONAL SAFETY AND HEALTH ACT. The U.S. Environmental Protection Agency has set a community air emission standard for vinyl chloride under the NATIONAL EMISSION STANDARD FOR HAZARDOUS AIR POLLUTANTS provision of the CLEAN AIR ACT.

virtually safe dose (VSD) A dose or exposure level for a carcinogen corresponding to an individual lifetime cancer risk considered to be essentially zero; sometimes suggested as a lifetime risk of one in one million.

viscosity (η) A measure of the resistance of a fluid to flow. For liquids, viscosity increases with decreasing temperature. For gases, viscosity increases with increasing temperature. Expressed as mass per length-time (e.g., kilograms per meter-second). A common viscosity unit is the poise. One poise equals 1.0 gram per centimeter-second. Also called dynamic viscosity. Compare KINEMATIC VISCOSITY.

visible range That part of the electromagnetic spectrum that can be seen by the human eye. The wavelengths range from about 400 nanometers (violet light) to about 710 nanometers (red light). Also called visible light.

visibility In the atmosphere, the distance to which an observer can distinguish objects from their background. The determinants of visibility include the characteristics of the target object (shape, size, color, pattern), the angle and intensity of sunlight, the observer's eyesight, and the extent of light absorption and scattering caused by air contaminants. See KOSCHMIEDER RELATIONSHIP, AIR QUALITY RELATED VALUE.

vitrification A process forming a highly stable noncrystalline material; glassification. Proposed as a treatment for radioactive waste, which would then resist high temperatures and leachate formation indefinitely.

volatile Describes a substance that evaporates or vaporizes rapidly at room temperatures.

volatile hydrocarbons Organic compounds composed of carbon and hydrogen that evaporate rapidly at room temperatures, for example gasoline, methanol, and benzene.

volatile organic carbon (VOC) A measure of the amount of particulate material in a water sample that is lost upon heating. The measure is obtained by passing a given quantity of water through a glass fiber filter and then drying and weighing the solids retained on the filter. The preweighed filter is then heated to about 500–600°C and a second weight is obtained. The amount lost during the heating process is termed VOC.

volatile organic compounds (VOCs) A category of volatile organic compounds with relatively high vapor pressures, a major category of air contaminants. Most VOCs are carbon-hydrogen compounds (hydrocarbons), but they also may be aldehydes, ketones, chlorinated hydrocarbons, and others. Thousands of individual compounds exist, including the unburnt hydrocarbon compounds emitted from automobiles or industrial processes and the organic solvents lost to evaporation from household, commercial, or industrial cleaning and painting operations, and other activities. Some VOCs participate in the atmospheric reactions that lead to PHOTOCHEMICAL AIR POLLUTION, and excessive exposure to certain individual compounds is associated with skin irritation, central nervous system depression, and/or an increased risk of cancer.

volatile organic sampling train (VOST) The air-sampling apparatus specified by the United States Environmental Protection Agency for the collection of organic material that is not completely destroyed in a waste incinerator. Analysis of the collected organics is used to determine incinerator efficiency. See also DESTRUCTION AND REMOVAL EFFICIENCY, PRODUCTS OF INCOMPLETE COMBUSTION, and PRINCIPAL ORGANIC HAZARDOUS CONSTITUENTS.

volatility A measure of the tendency of a solvent or other material to evaporate at normal temperatures.

volatilization The process of evaporation.

volumetric flow rate For a liquid or gas, the volume moving past a point per unit time. For example, a smokestack exhaust of 10 cubic meters per second

would result if gas exited a stack 2 meters square at 5 meters per second. Actual flow rate (Q) is expressed as $Q = AV$, where A is the cross-sectional area of the pipe, duct, or stack and V is the velocity of the liquid or gas.

volumetric water content That portion of the volume of a soil sample that is occupied by water, expressed as percent by volume.

W

waste exchange A clearinghouse that matches disposers of waste materials to facilities that can use the waste as a fuel or raw material.

waste immobilization See IMMOBILIZATION.

Waste Isolation Pilot Plant (WIPP)
An underground facility near Carlsbad, New Mexico, that is to be used for the disposal of radioactive wastes arising from nuclear weapons activities. Wastes are to be buried 2000 + feet below the surface in salt beds. Opening has been delayed due to safety and environmental concerns.

waste separation The extraction of recyclable components (metals, glass, newspaper) from municipal solid waste. This can be accomplished by households, in which case it is called source separation, or a central facility.

waste stabilization See STABILIZATION/SOLIDIFICATION.

wasteload allocation (WLA) A system designed to limit the total discharge of pollutant materials into a receiving stream. Each POINT SOURCE is allowed to release a specific fraction of the total amount of pollutant materials that can be expected to be assimilated by the stream. Pollution from nonpoint sources comprises the stream's LOAD ALLOCATION.

wastewater Water discharged from homes, businesses, and industries that contains dissolved, suspended, and particulate inorganic or organic material. The term is also used as a synonym for sewage. Also called domestic wastewater.

water-air ratio (K_w) A parameter expressing the partitioning of a substance present in a dilute solution between the water and the overlying air. The ratio is computed by

$$K_w = \frac{C_W}{C_A}$$

where C_W is the chemical's water concentration (micrograms of the chemical per cubic centimeter of water) and C_A is the chemical's air concentration (micrograms per cubic centimeter).

water balance A measure of the amount of water entering and the amount leaving a system.

waterborne Of or related to something that is carried by water, for example, a disease transmitted by water contaminated by a disease-causing microorganism.

water column A hypothetical cylinder of water from the surface to the bottom of a stream, lake, or ocean within which the physical and/or chemical properties can be measured.

water-cooled reactor A NUCLEAR REACTOR that employs water to cool the reactor core. A nuclear reactor is a device designed to promote the FISSION of an appropriate fuel (such as uranium-235) in a controlled manner. The heat produced during the fission event must be removed from the device to prevent an excessive buildup. Water is usually used as the heat transfer agent. Other coolants used in nuclear reactors of other designs

are liquid sodium and inert gases. See LIGHT-WATER REACTOR.

water dilution volume (WDV) The volume of water required to dilute radioactive waste to a concentration meeting drinking water standards. Typically expressed in cubic meters of water per metric ton of radioactive waste.

water hyacinth A floating freshwater plant belonging to the genus *Eichhornia*. The plant was introduced into the United States in the late nineteenth century and has become a prolific nuisance weed that clogs waterways in the southern part of the country.

water quality criteria The aqueous concentration limits for pollutants in water that is to be used for specific purposes. The criteria are set for individual pollutants and are based on different water uses, such as a public water supply, an aquatic habitat, an industrial supply, or for recreation.

water quality limited segment A portion of a stream where the condition of the water does not meet water quality standards and/or where standards are not expected to be achieved after EFFLUENT LIMITATIONS on all POINT SOURCES are applied. Therefore, controls beyond the TECHNOLOGY-BASED discharge limits will be required for the stream segment to meet the ambient standards.

water reactive Describing any substance that reacts spontaneously with water to release a flammable or toxic gas, such as sodium metal.

water softener An apparatus designed to remove divalent metal ions (the most important of these are calcium, magnesium, and iron) from water, often replacing the divalent or trivalent ions with the monovalent sodium ion. See ION EXCHANGE.

water-soluble Of a material that dissolves in water.

water table The uppermost level of the below-ground, geological formation that is saturated with water. Water pressure in the pores of the soil or rock is equal to atmospheric pressure.

water treatment The processing of source water (well water or surface water) for distribution in a public drinking water system.

watershed That area of land that drains into a lake or stream.

waterwall incinerator An energy recovery system used in some municipal waste incinerators. The combustion chamber of the incinerator is lined with steel tubes containing circulating water. The heat from the combustion boils the water, and the steam can be sold or used to turn turbines in an electric generator.

watt (W) The SI unit of power equal to one JOULE per second. Expressed as $1 \text{ W} = 1 \text{ J s}^{-1}$.

weak acid A compound that releases small amounts of hydrogen ions when dissolved in water. As a result, low concentrations of hydronium ions are formed relative to the volume of water employed. An example of a weak acid is acetic acid, which is used as vinegar. See ACID.

weathering The breakdown of rock through a combination of chemical, physical, geological, and biological processes. The ultimate outcome is the generation of soil.

weighted average For a series of recorded observations, the sum of the products of the frequency of certain values and the value of the observation, divided by the total number of observations. For example, for one measurement of 5 grams, three measurements of 7 grams, and two measurements of 2 grams, the weighted average is $[1(5) + 3(7) + 2(2)]/6 = 5$ grams. See TIME-WEIGHTED AVERAGE.

weight fraction An expression of concentration for materials in solutions or mixtures. The weight fraction of a certain material is its weight divided by the total weight of the solution or mixture. For example, in a mixture of A, B, and C,

$$A\text{'s weight fraction} = \frac{\text{weight of A}}{\text{weight of A} + B + C}.$$

weighting networks A frequency-specific adjustment made by a sound level meter to the measured decibel levels to account for the increased sensitivity of the human ear to higher-frequency sounds and its lower sensitivity to lower frequencies. Three networks are available, called A, B, and C. The A-weighted scale is by far the most frequently used because it best matches the sound frequency sensitivity of the human ear at the sound levels most commonly encountered. See FLETCHER-MUNSON CONTOURS and DECIBELS, A-WEIGHTING NETWORK.

weir An underwater dam or barrier in a channel or ditch placed to limit or control water flow; water flows over the top of the weir.

well plug Any watertight or gastight seal installed in a well to prevent the flow of fluids or gases.

well stimulation Cleaning, enlarging, or increasing the pore space of a well used for the injection of fluids into subsurface geological strata.

wet adiabatic lapse rate The rate of temperature decrease as a parcel of air saturated with water rises and the pressure decreases, given by

$$\gamma_s = -\frac{dT}{dz}$$

where dT is the temperature change, dz is the change in altitude, and γ_s is the saturated (wet) ADIABATIC LAPSE RATE. Because moisture is condensing in the rising parcel of air and releasing latent heat, the temperature drop with increasing altitude is less than the (dry) adiabatic lapse rate, or about 0.6°C per 100 meters. The rate assumes that there is no exchange of heat between the parcel and the surrounding air by conduction or mixing.

wet deposition The introduction of acidic material to the ground or to surface waters by sulfuric and nitric acids dissolved in rainfall or snow. Compare DRY DEPOSITION.

wet scrubbing A process that removes particles, gases, or vapors from an exhaust gas by passing the exhaust through a shower of water or water that contains an agent to react with the material to be removed. See SCRUBBER, IMPINGEMENT; SCRUBBER, SPRAY; SCRUBBER, VENTURI.

wet stack A stack that is capable of handling moisture that condenses from the exhaust gas after it exits a SCRUBBER.

wet test meter A laboratory instrument used mainly to calibrate the volume of airflow in other instruments. It consists of a set of compartments of equal volume that can rotate inside an outer casing. The compartments are partially submerged in water or other liquid. Air enters the meter and forces the liquid out of a compartment, which causes the compartments to rotate. As the open end of each compartment rotates past the top of the meter, its air flows out of the meter. The measurement of the liquid level in the meter indicates the amount of air released by each compartment, and a counter determines the number of compartments that were filled with air. This information is used to compute the airflow.

wetlands A land area that is covered or saturated with water for a large portion of the year.

wet-bulb temperature The temperature reading from a thermometer with a wetted wick surrounding its bulb. The

evaporative loss of LATENT HEAT from the wick lowers the temperature reading. Used with the DRY-BULB TEMPERATURE and a table to compute RELATIVE HUMIDITY.

whey The clear fluid that separates from the solid curd when milk is allowed to coagulate, or sour. The curd contains most of the protein solids from the milk (casein), and the whey contains most of the small, soluble compounds found in milk. Whey represents a waste liquid produced in the manufacture of some cheeses, and has a high BIOCHEMICAL OXYGEN DEMAND.

white damp Carbon monoxide remaining in an underground mine after a mine fire or explosion. See DAMP.

white goods Refrigerators, stoves, clothes washers and dryers, and other appliances contained in municipal solid waste.

whole-body dose The exposure of the entire human body to radiation, for example BACKGROUND RADIATION. This type of exposure is also of importance when a radioactive isotope is inhaled or ingested and then uniformly distributed throughout the body.

Wien's law The physical law relating the peak wavelength of electromagnetic radiation emitted by a radiating body to the surface temperature of that body. The higher the temperature, the shorter the peak wavelength of the energy emitted. For example, the sun emits an energy spectrum with a peak wavelength of about 480 nanometers, which is within the visible part of the electromagnetic spectrum; since the surface temperature of the earth is much lower than that of the sun, the energy emission spectrum of the earth has a much longer peak wavelength, about 10,000 nanometers, which is in the infrared, or radiant heat, range. It can be expressed as $\lambda T = $ constant, where λ is the peak wavelength in nanometers, T is the absolute temperature of the body in kelvins, and the constant is equal to 2.9×10^6 nanometers-K.

Wilderness Act   See WILDERNESS AREA.

wilderness area   Land where the effects of man are not apparent. Large tracts of land that are set aside and allowed to develop without the intervention of man. Such activities as the construction of roads, development of recreational facilities, removal of trees, or hunting are prohibited. In some cases, even the fighting of fires started by natural means is limited. The 1964 Wilderness Act allows the U.S. government to set aside sections within the national forests, national parks, and national wildlife refuges as wilderness areas. There are about 450 areas within the United States totaling 90 million acres, two-thirds of which is in Alaska.

Wilderness Society, The   An American environmental organization concerned with protecting wildlife habitat and wildlife refuges as well as the preservation of public lands. In 1989, membership numbered 300,000.

window (infrared)   The wavelength range of the infrared emissions from the surface of the earth, between about 8 and 12 micrometers, that is poorly absorbed in the atmosphere by water vapor and carbon dioxide. This allows a portion of the heat radiated by the earth to escape directly to space.

wind profile power law   The expression used to estimate the (higher) wind speed at the top of a smokestack using a measure of wind speed at ground level; the law is applied in air quality dispersion modeling. It is expressed as

$$\frac{u_2}{u_1} = \left(\frac{z_2}{z_1}\right)^p$$

where u_2 is the wind speed at the higher altitude z_2, u_1 is the wind speed at the lower altitude z_1, and the exponent p varies with atmospheric turbulence.

wind rose   A diagram depicting the strength and direction of the winds as measured over time at a specific station.

windrow   A long, narrow compost pile. In large-scale operations, the design allows convenient access by machines, which turn (mix) the material periodically.

Winkler method   A standard procedure for measuring the level of DISSOLVED OXYGEN in water. This laboratory analysis, also called the iodometric method, is a reliable titrametric procedure. A solution of divalent manganese and a strong alkali are added to a water sample. The oxygen dissolved in the water oxidizes the manganous hydroxide to hydroxides of higher valences. In the presence of iodide ions and acidification, the oxidation of the metal is reversed with the release of iodine in amounts equivalent to the amount of oxygen dissolved in the original water sample. The released iodine is titrated with a standard solution of thiosulfate using starch as the indicator.

wipe test   A sampling method used to determine the presence of hazardous or radioactive substances that can be removed from surfaces. For the detection of radioactive contamination, for example, a small piece of absorbent material (filter paper) is wiped across a table, sink, or other surface, and the absorbent material is then assayed for radioactivity.

withdrawal, of water   The removal of water from a surface or underground source without regard to any quality change that occurs by its withdrawal and use; if it is degraded such that its reuse is limited, the water is said to be consumed. See CONSUMPTION, OF WATER.

withdrawal, public land   In the management of public lands, the temporary or permanent suspension of all or some of the laws that allow public land to be used for certain purposes, such as mineral extraction. Withdrawal followed by

WIND ROSE

N

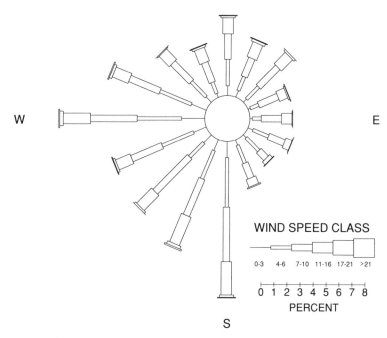

W

E

WIND SPEED CLASS

0-3 4-6 7-10 11-16 17-21 >21

0 1 2 3 4 5 6 7 8
PERCENT

S

RESERVATION for a single use, such as a wilderness area, has been a method of setting aside public lands for environmental preservation.

work A force acting over a distance. The work W done by a constant force F on a body that undergoes a displacement s is expressed as $W = (F \cos \Theta)s$, where Θ is the angle between the force and the displacement. The SI work unit is the JOULE, or 1 newton-meter. The work unit in the U.S. system is the FOOT-POUND.

working face The location within a solid waste disposal site where waste is discharged and compacted prior to burial with cover material.

working level (WL) A measure of the radioactivity of gases released by the decay of radon. The unit is used as an expression of radon exposure level for individuals. One WL is the combined radioactivity of radon and its short-lived daughter products in one liter of air that

emit a total alpha energy of 1.3×10^5 million electron volts upon complete decay to ^{210}Pb. United States Environmental Protection Agency guidelines for residential radon levels recommend taking actions to lower exposure if indoor WL measurements are greater than about 0.02 WL.

working-level month (WLM) A unit for individual exposure to radioactivity resulting from the inhalation of air contaminated with a sufficient amount of the decay products of radon to equal one WORKING LEVEL of radioactivity for 170 working hours (one month of working in a contaminated environment).

World Health Organization (WHO)

A multinational agency dedicated to improving and maintaining public health; founded 1948. The organization works through regional offices with national governments to support programs in food and water sanitation, epidemiological surveillance, communicable disease control, and many other areas. Headquarters are in Geneva, Switzerland.

X

xenobiotics   A general term for chemical materials not normally present in the environment, such as pesticides.

xenon (Xe)   A heavy, relatively inert gas found in very minute quantities in the atmosphere. The gas in employed in flash bulbs used for photographic purposes, and the gas may be produced as a fission product in nuclear reactors.

xeric   Describing an organism that requires little moisture or a habitat containing little moisture. Dry.

xerophyte   A plant, such as cactus, adapted to a dry environment. Compare HYDROPHYTE, MESOPHYTE.

x-ray   A form of electromagnetic radiation having a wavelength shorter than that of ultraviolet light and usually longer than that of gamma rays. This form of radiation is termed ionizing since it has excellent penetrating ability and produces ions within material as it interacts with the atoms of the material through which it passes. Excessive exposure to x-rays has resulted in mutations, abnormal fetal development and cancer development.

x-ray diffraction   An analytical method that involves the exposing of samples to x-rays. The radiation is reflected from the sample in response to the structure of the crystals that compose the sample. Consequently, the method is useful in the identification of elements in materials and coatings. Because of the structure of the asbestos crystal, the method is especially useful in detecting this element in old construction material.

x-ray fluorescence (XRF)   A technique that can be used to visualize the presence of x-rays. Certain substances will absorb the electromagnetic energy of x-rays and then emit that energy as an altered form of electromagnetic radiation, usually as visible light. It is the glowing of a material when struck by x-rays. The technology is employed to produce x-ray images on video screens (fluoroscopy) that allow for instant viewing or for the viewing of a system such as the gastrointestinal tract in motion rather than the still photographic images produced on standard x-ray films. The technology is also employed to lessen the medical exposure of patients to x-rays. Intensifying screens made of fluorescent materials are used to produce x-ray images using radiation at a much lower dose than that required by standard film techniques.

Y

years of life lost   The expected shortening of a lifetime due to exposure to a dangerous material or radiation. Calculated from the expected lifetime for a nonexposed person minus the expected lifetime of a person exposed to the dangerous agent.

yellow cake   Uranium oxide (U_3O_8) that results from the refining of uranium ore. The purified material contains 99.3 percent uranium-238 and 0.7 percent uranium-235. The term yellow cake is applied because of the color and texture of the material.

Z

Z tables   Listings of workplace exposure limits for toxic and hazardous substances published by the Occupational Safety and Health Administration. The tables are found in the CODE OF FEDERAL REGULATIONS, Subpart Z, Title 29, Part 1910.

zero discharge The goal, in the preamble to the CLEAN WATER ACT, of zero pollutants in water discharges.

zero-infinity dilemma In a risk analysis that ranks environmental hazards by the product of their magnitude and probability, the policy problem posed by a catastrophic harmful scenario that has a very low probability of occurrence. For example, the introduction of an extremely potent human carcinogen as an additive to aspirin might cause tens of thousands of excess cancers over several decades, but the likelihood of such a chemical evading the normal toxicity screenings for additives is very low.

zero-order reaction A chemical reaction in which the rate of reaction is independent of a reactant's concentration or the concentration of any other chemicals present. Compare FIRST-ORDER REACTION.

zero population growth (ZPG) A condition in which a population in a given location neither increases nor decreases over time. The increases due to births and immigration are balanced with decreases caused by deaths and emigration.

zero pressure A complete vacuum; the zero reference point on the ABSOLUTE PRESSURE scale. Note that zero GAUGE PRESSURE equals ATMOSPHERIC PRESSURE.

zero tolerance A requirement that no amount of pesticide may remain on an agricultural commodity when shipped. Tolerance levels are established under the authority of the FEDERAL INSECTICIDE, FUNGICIDE, AND RODENTICIDE ACT and the FOOD, DRUG, AND COSMETICS ACT.

zone of aeration See UNSATURATED ZONE.

zone of engineering control The area occupied by a hazardous waste treatment, storage, or disposal facility which the owner or operator can readily decontaminate if a leak is detected, thus preventing hazardous waste or its constituents from entering groundwater or surface water.

zone of initial dilution (ZID) That area within a lake or stream where the discharge from an outfall first mixes with the receiving water.

zone of saturation See SATURATED ZONE.

zoonotic Describing a pathogen that normally infects wild animals but which can infect humans if the carrier and man have contact. For example, Lyme disease is caused by a pathogenic bacterium normally carried by ticks that infect deer. The bacterium can be transmitted to humans by the bite of deer ticks, given a close association of deer and man in some locations.

zooplankton The small, often microscopic, animals in the aquatic environments that possess little or no means of propulsion. Consequently, animals belonging to this class drift along with the currents.

APPENDIX

Contents

Acronyms (and other abbreviations)

AA	atomic absorption spectrophotometer
ABS	alkylbenzene sulfonate
A/C	air-to-cloth ratio
ACFM	actual cubic feet per minute
ACGIH	American Conference of Governmental Industrial Hygienists
ACLs	alternate concentration limits
ACM	asbestos-containing material
ADI	acceptable daily intake
AFR	air/fuel ratio
AI test	adsorption isotherm test
AL	action level
ALARA	as low as reasonably achievable
amu	atomic mass unit
ANOVA	analysis of variance
ANPR	advance notice of proposed rulemaking
APA	Administrative Procedure Act
APCA	Air Pollution Control Association
API	American Petroleum Institute
APR	air-purifying respirator
APWA	American Public Works Association
AQCR	air quality control region
AQMA	air quality maintenance area
AQRV	air quality related value
AQS	air quality standard
AQS	ambient quality standard
ARAR	applicable or relevant and appropriate requirement
As	arsenic
ASCE	American Society of Civil Engineers
ASTM	American Society for Testing and Materials
ATP	adenosine triphosphate
ATSDR	Agency for Toxic Substances and Disease Registry
AWMA	Air and Waste Management Association
AWWA	American Water Works Association
BACT	best available control technology
BADT	best available demonstrated technology
BaP	benzo(a)pyrene
BART	best available retrofit technology
BAT	best available technology economically achievable
BCF	bioconcentration factor
BCT	best conventional control technology
BADT	best available demonstrated technology
BDAT	best demonstrated available technology
BEI	Biological Exposure Index

BEIR report	Biological Effects of Ionizing Radiation report
BEJ	best engineering judgment
BHC	benzene hexachloride
BLM	Bureau of Land Management
BMP	best management practices
BNA	Bureau of National Affairs
BOD	biochemical oxygen demand
BPJ	best professional judgment
BPT	best practicable control technology
BTU	British thermal unit
BWR	boiling-water reactor
^{14}C	carbon-14
CAA	Clean Air Act
CAER	Community Awareness and Emergency Response
CAG	Carcinogen Assessment Group
CAI	carcinogenic activity indicator
CAIR	comprehensive assessment information rule
CANDU	Canadian deuterium-uranium reactor
CAPA	critical aquifer protection area
CARB	California Air Resources Board
CASAC	Clean Air Scientific Advisory Committee
CAS number	Chemical Abstracts Service Registry number
CBA	cost-benefit analysis
CCW table	constituent concentrations in wastes table
CCWE table	constituent concentrations in waste extract table
CDBF	chlorinated dibenzofurans
CDC	Centers for Disease Control
CEA	cost-effectiveness analysis
CEC	cation exchange capacity
CEM	continuous emission monitoring
CEQ	Council on Environmental Quality
CERCLA	Comprehensive Environmental Response, Compensation, and Liability Act
CERCLIS	Comprehensive Environmental Response, Compensation, and Liability Information System
CFCs	chlorofluorocarbons
CFR	Code of Federal Regulations
CGL policy	comprehensive general liability policy
CHEMTREC	Chemical Transportation Emergency Center
CHIPs	Chemical Hazard Information Profiles
CHRIS	Chemical Hazard Response Information System
Ci	Curie
CI pipe	cast-iron pipe
CITES	Convention on International Trade in Endangered Species of Wild Fauna and Flora
CMA	Chemical Manufacturers Association
CO	carbon monoxide
COD	chemical oxygen demand
Coh	coefficient of haze
COHb	carboxyhemoglobin
CPF	cancer potency factor
CPF	carcinogenic potency factor

CPSC	Consumer Product Safety Commission
^{137}Cs	cesium-137
CTGs	Control Techniques Guidelines
CWA	Clean Water Act
CWP	coal workers' pneumoconiosis
CZMA	Coastal Zone Management Act
2,4-D	dichlorophenoxyacetic acid
DAL	defect action level
dB	decibel
dBA	decibels, A-weighting network
DDD	dichlorodiphenyldichloroethane
DDE	dichlorodiphenyldichloroethene
DDT	dichlorodiphenyltrichloroethane
DEIS	draft environmental impact statement
DES	diethylstilbestrol
DMR	discharge monitoring report
DNA	deoxyribonucleic acid
DO	dissolved oxygen
DOC	dissolved organic carbon
DOE	Department of Energy
DOM	dissolved organic matter
DRE	destruction and removal efficiency
ECCS	emergency core cooling system
ECD	electron capture detector
ECx	experimental concentration–percent
EDB	ethylene dibromide
EDF	Environmental Defense Fund
EGR	exhaust gas recirculation
EHS	extremely hazardous substance
EIL policy	environmental impairment liability policy
EIQ	emission inventory questionnaire
EIS	environmental impact statement
EOX	extractable organic halogens
EPA	Environmental Protection Agency
EPCRA	Emergency Planning and Community Right-to-Know Act
EPRI	Electric Power Research Institute
EP toxicity test	extraction procedure toxicity test
ERA	expedited removal action
ERCs	emission reduction credits
ERDA	Energy Research and Development Administration
ERT	environmental response team
ESA	Endangered Species Act
ESP	electrostatic precipitator
ETS	environmental tobacco smoke
eV	electron volt
EW	equivalent weight
f/cc	fibers per cubic centimeter
FD	forced draft
FDA	Food and Drug Administration
FDFs	fundamentally different factors

FEPCA	Federal Environmental Pesticide Control Act
FERC	Federal Energy Regulatory Commission
FEV_1	forced expiratory volume
FGD	flue gas desulfurization
FID	flame ionization detector
FIFRA	Federal Insecticide, Fungicide, and Rodenticide Act
FLPMA	Federal Land Policy and Management Act
F/M ratio	food-to-microorganism ratio
FOE	Friends of the Earth
FONSI	finding of no significant impact
FPC	Federal Power Commission
FPC	fish protein concentrate
FTP	Federal Test Procedure
FVC	forced vital capacity
FWPCA	Federal Water Pollution Control Act
g	gram
GAC	granular activated carbon
GAO	General Accounting Office
G/C	gas-to-cloth ratio
GC	gas chromatography
GC/MS	gas chromatography/mass spectrometry
GCR	gas-cooled reactor
GEMS	Global Environment Monitoring System
GEMS	Graphical Exposure Modeling System
GEP stack height	good engineering practice stack height
GLC	ground-level concentration
GLP standards	Good Laboratory Practice Standards
GMW	gram molecular weight
GPCD	gallons per capita per day
GPO	Government Printing Office
GRAS	generally recognized as safe
GW	gigawatt
Gy	gray
HACS	Hazard Assessment Computer System
HAZOP	hazard and operability study
HCB	hexachlorobenzene
HDD	halogenated dibenzo-p-dioxin
HDF	halogenated dibenzofuran
HDPE	high-density polyethylene
HEPA filter	high-efficiency particulate air filter
HIT	Hazard Information Transmission
hi-vol	high-volume air sampler
HLW	high-level waste
HMTA	Hazardous Materials Transportation Act
HOCs	halogenated organic compounds
HRS	Hazard Ranking System
HSI	heat stress index
HSL	hazardous substance list
HSWA	Hazardous and Solid Waste Amendments
HTGR	high-temperature gas reactor
HVL	half-value layer

HWM facility	hazardous waste management facility
HWR	heavy-water reactor
HYVs	high-yielding varieties
IAEA	International Atomic Energy Agency
IARC	International Agency for Research on Cancer
ICRP	International Commission on Radiological Protection
ICS	intermittent control system
ID	induced draft
IDL	instrument detection limit
IDLH	immediately dangerous to life and health
I & I	infiltration and inflow
I & M	inspection and maintenance
INFOTERRA	International Environmental Information System
[I]/[O]	indoor/outdoor concentration ratio
IPM	integrated pest management
IRIS	Integrated Risk Information System
IRLG	Interagency Regulatory Liaison Group
IRPTC	International Register of Potentially Toxic Chemicals
ISC model	Industrial Source Complex model
ISR	indirect source review
IUCN	International Union for Conservation of Nature and Natural Resources
JTU	Jackson turbidity unit
K	carrying capacity
K	Kelvin
K_a	acid dissociation constant
K_b	base dissociation constant
K_d	soil sorption coefficient
K_{oc}	organic carbon partition coefficient
K_{ow}	octanol-water partition coefficient
K_s	solubility product constant
K_w	water-air ratio
kcal	kilocalorie
kg	kilogram
kW	kilowatt
kWh	kilowatt-hour
L_{dn}	day-night sound level
LA	load allocation
LAER	lowest achievable emission rate
LAS	linear alkyl sulfonate
LC_{50}	lethal concentration—50 percent
LC_{50}	median lethal concentration
LC_{LO}	lethal concentration, low
LD	lethal dose
LD_{50}	lethal dose—50 percent
LD_{LO}	lethal dose, low
LEL	lower explosive limit
LEPC	local emergency planning committee
LET	linear energy transfer

LFL	lower flammable limit
LIDAR	light detection and ranging
LLW	low-level waste
LMFBR	liquid-metal fast breeder reactor
ln	natural logarithm
LNG	liquefied natural gas
LOAEL	lowest-observed-adverse-effect level
LOCA	loss-of-coolant accident
LOEL	lowest-observed-effect level
LPG	liquefied petroleum gas
LSI	listing site inspection
LSS	life-span study
LUST trust fund	leaking underground storage tank trust fund
LWR	light-water reactor
m^3	cubic meter
MAC	maximum allowable concentration
MACT	maximum achievable control technology
MATC	maximum acceptable toxicant concentration
MCL	maximum contaminant level
MCLG	maximum contaminant level goal
MDL	method detection limit
MEDLINE	Medical Literature Analysis and Retrieval System
MEI	most exposed individual
meq/l	milliequivalents per liter
mg	milligram
mg/l	milligrams per liter
MHz	megahertz
MIR	maximum individual risk
ml	milliliter
MLD	median lethal dose
MLSS	mixed liquor suspended solids
MLVSS	mixed liquor volatile suspended solids
MMMFs	man-made mineral fibers
MOE	margin of exposure
MOS	margin of safety
MPC	maximum permissible concentration
MPD	maximum permissible dose
MPN	most probable number
MPRSA	Marine Protection, Research, and Sanctuaries Act
mrem	millirem
MS	mass spectrometer
MSDS	material safety data sheet
MSR	mammalian selectivity ratio
MSW	municipal solid waste
MTD	maximum tolerated dose
MTTP	maximum total trihalomethane potential
μ	micron
μg	microgram
μg/m^3	micrograms per cubic meter
μl	microliter
μm	micrometer
MWTA	Medical Waste Tracking Act

NAA	nonattainment area
NAAQS	national ambient air quality standards
NAMS	National Air Monitoring System
NASQAN	National Air Stream Quality Accounting Network
NATICH	National Air Toxics Information Clearinghouse
NBAR	nonbinding preliminary allocation of responsibility
NCI	National Cancer Institute
NCP	National Contingency Plan
NCP	National Oil and Hazardous Substances Pollution Contingency Plan
NCP	net community productivity
NCRPM	National Council on Radiation Protection and Measurements
NDIR	nondispersive infrared analysis
NDT	nondestructive testing
NEPA	National Environmental Policy Act
NESHAP	National Emission Standards for Hazardous Air Pollutants
NFPA	National Fire Protection Association
NIEHS	National Institute for Environmental Health Sciences
NIHL	noise-induced hearing loss
NIMBY	not-in-my-backyard syndrome
NIOSH	National Institute for Occupational Safety and Health
NMHC	nonmethane hydrocarbons
NMOC	nonmethane organic compounds
NO	nitric oxide
NO_2	nitrogen dioxide
NO_x	nitrogen oxides
NOAA	National Oceanic and Atmospheric Administration
NOAEL	no-observed-adverse-effect level
NOEL	no-observed-effect level
NORM	naturally occurring radioactive material
NOW	nonhazardous oil field waste
NPDES	National Pollutant Discharge Elimination System
NPDWR	National Primary Drinking Water Regulations
NPL	National Priorities List
NPP	net primary productivity
NRC	National Research Council
NRC	National Response Center
NRC	Nuclear Regulatory Commission
NRDC	Natural Resources Defense Council
NRT	National Response Team
NSDWR	National Secondary Drinking Water Regulations
NSF	National Science Foundation
NSF	National Strike Force
NSPS	new source performance standards
NSR	new source review
NTIS	National Technical Information Service
NTP	National Toxicology Program
NTU	nephelometric turbidity unit
NWPA	Nuclear Waste Policy Act
OHMTADS	Oil and Hazardous Materials Technical Assistance Data System
OMB	Office of Management and Budget

OSC	on-scene coordinator
OSHA	Occupational Safety and Health Administration
OSHAct	Occupational Safety and Health Act
OSWER	Office of Solid Waste and Emergency Response
OTEC	ocean thermal energy conversion
PAH	polycyclic aromatic hydrocarbons
PAH	polynuclear aromatic hydrocarbons
PAIR	preliminary assessment information rule
PAN	peroxyacetyl nitrate
PAN	peroxyacylnitrates
PAR	population at risk
PA/SI	preliminary assessment and site inspection
Pb	lead
PCBs	polychlorinated biphenyls
PCDFs	polychlorinated dibenzofurans
PCP	pentachlorophenol
PCV	positive crankcase ventilation
PE	population equivalent
PEL	permissible exposure limit
PF	protective factor
PFLT	paint filter liquids test
PICs	products of incomplete combustion
PID	photoionization detector
PM_{10}	particulate matter, 10-micron diameter
PMN	premanufacturing notice
POC	point of compliance
POC	purgeable organic carbon
POHCs	principal organic hazardous constituents
POM	particulate organic matter
POM	polynuclear organic matter
POTW	publicly owned treatment works
POU/POE	point-of-use/point-of-entry
POX	purgeable organic halogens
ppb	parts per billion
ppm	parts per million
PRA	probabilistic risk assessment
PRP	potentially responsible party
PSD	prevention of significant deterioration
PSES	pretreatment standards for existing sources
PSI	Pollutant Standards Index
psia	pounds per square inch (absolute)
PSNS	pretreatment standards for new sources
PSP	paralytic shellfish poisoning
PTS	permanent threshold shift
^{239}Pu	plutonium-239
PVC	polyvinyl chloride
PWR	pressurized-water reactor
PYAR	person-years-at-risk
q	quad
Q	quint

QA/QC	quality assurance/quality control
QF	quality factor
R	roentgen
R_0	net reproductive rate
RA	Regional Administrator
RACM	reasonably available control measures
RACT	reasonably available control technology
rad	radiation absorbed dose
Raw	airway resistance
RBC	rotating biological contactor
RBE	relative biological effectiveness
RCRA	Resource Conservation and Recovery Act
RDA	recommended daily allowance
RDF	refuse-derived fuel
RDI	reference daily intake
rem	roentgen equivalent man
RfD	reference dose
RFP	reasonable further progress
RFT	respirator fit test
RGR	mean relative growth rate
RIA	Regulatory Impact Analysis
RI/FS	remedial investigation/feasibility study
RMS sound pressure	root-mean-square sound pressure
ROD	record of decision
RPAR	rebuttable presumption against registration
RPM	remedial project manager
RQ	reportable quantity
RRC	Regional Response Center
RRT	Regional Response Team
RSD	risk-specific dose
RTECS	Registry of Toxic Effects of Chemical Substances
RTP	Research Triangle Park
RVP	Reid vapor pressure
SAB	Science Advisory Board
SAR	structure-activity relationship
SARA	Superfund Amendments and Reauthorization Act
SAROAD	Storage and Retrieval of Aerometric Data
SCA	specific collection area
SCAP	Superfund Compliance Accomplishments Plan
SCAS test	semicontinuous activated sludge test
SCBA	self-contained breathing apparatus
SCFM	standard cubic feet per minute
SCRAM	Safety Control Rod Ax Man
SCS	Soil Conservation Service
SDWA	Safe Drinking Water Act
SEM	scanning electron microscope
SERC	State Emergency Response Commission
SIC	Standard Industrial Classification
SIP	state implementation plan
SITE	Superfund Innovative Technology Evaluation program
SKIM	Spill Cleanup Inventory

SLAMS	State and Local Air Monitoring System
SMCL	secondary maximum contaminant level
SMCRA	Surface Mining Control and Reclamation Act
SMR	standardized mortality ratio
SMU	solid waste management unit
SNG	synthetic natural gas
SNUR	significant new use rule
SO_2	sulfur dioxide
SO_x	sulfur oxides
SOD	sediment oxygen demand
SPCC Plan	Spill Prevention Control and Countermeasure Plan
SQG	small-quantity generator
^{90}Sr	strontium-90
S/S	stabilization/solidification
SSC	scientific support coordinator
STAPPA	State and Territorial Air Pollution Program Administrators
STEL	short-term exposure limit
STPP	sodium tripolyphosphate
Sv	sievert
SVI	sludge volume index
2,4,5-T	trichlorophenoxyacetic acid
TC	total carbon
TCDBF	tetrachlorodibenzofuran
TCDD	tetrachlorodibenzo-*para*-dioxin
TCDF	tetrachlorodibenzofuran
TCLP	toxicity characteristic leaching procedure
TDS	total dissolved solids
TEM	transmission electron microscopy
TENR	technologically enhanced natural radioactivity
THMs	trihalomethanes
TIC	total inorganic carbon
TLV	threshold limit value
TMDL	total maximum daily load
TOC	total organic carbon
TOD	theoretical oxygen demand
TOXLINE	Toxicology Information on Line
TPD	tons per day
TPQ	threshold planning quantity
TRS	total reduced sulfur
TRU waste	transuranic waste
TS	total solids
TSCA	Toxic Substances Control Act
TSD	treatment, storage, or disposal
TSDF	treatment, storage, or disposal facility
TSP	total suspended particulate
TSS	total suspended solids
TTHMs	total trihalomethanes
TTS	temporary threshold shift
TWA	time-weighted average
^{235}U	uranium-235
^{238}U	uranium-238

UEL	upper explosive limit
UF	uncertainty factor
UFL	upper flammable limit
UIC	Underground Injection Control
UNAMAP	Users' Network for Applied Modeling of Air Pollution
UNEP	United Nations Environment Program
UN/NA number	United Nations/North America number
URT	upper respiratory tract
USDW	underground source of drinking water
U.S.C.	United States Code
U.S.C.A.	United States Code Annotated
U.S.C. Supp.	United States Code Supplement
USGS	United States Geological Survey
UST	underground storage tank
UTM coordinates	Universal Transverse Mercator coordinates
UV	ultraviolet
UV-A	ultraviolet radiation–A range
UV-B	ultraviolet radiation–B range
UV-C	ultraviolet radiation–C range
UV-VIS	ultraviolet-visible absorption spectrum
VOC	volatile organic carbon
VOCs	volatile organic compounds
VOST	volatile organic sampling train
VP	velocity pressure
VSD	virtually safe dose
W	watt
WDV	water dilution volume
WHO	World Health Organization
WIPP	Waste Isolation Pilot Plant
WL	working level
WLA	wasteload allocation
WLM	working-level month
XRF	x-ray fluorescence
ZID	zone of initial dilution
ZPG	zero population growth

Unit Prefixes

tera (T)	1×10^{12}	trillion
giga (G)	1×10^{9}	billion
mega (M)	1×10^{6}	million
kilo (k)	1×10^{3}	thousand
hecto (h)	1×10^{2}	hundred
deka (da)	1×10^{1}	ten
deci (d)	1×10^{-1}	one-tenth
centi (c)	1×10^{-2}	one-hundredth
milli (m)	1×10^{-3}	one-thousandth
micro (μ)	1×10^{-6}	one-millionth

| nano (n) | 1×10^{-9} | one-billionth |
| pico (p) | 1×10^{-12} | one-trillionth |

Approximate Unit Equivalents

1. Mass
   1 kilogram = 1000 grams = 2.205 pounds
   1 metric ton = 1000 kilograms
   1 pound = 453.6 grams = 7000 grains = 16 ounces = 0.4536 kilogram
   1 ton = 2000 pounds = 907.2 kilograms

2. Length
   1 meter = 100 centimeters = 3.281 feet = 39.37 inches
   1 kilometer = 1000 meters = 0.6214 mile = 3281 feet
   1 mile = 5280 feet = 1760 yards = 1609 meters
   1 inch = 2.54 centimeters
   1 micron = 1×10^{-6} meter
   1 nanometer = 1×10^{-9} meter
   1 Angstrom = 1×10^{-10} meter

3. Area
   1 square meter = 1×10^4 square centimeters
   $\qquad$ = 10.764 square feet
   $\qquad$ = 1550 square inches
   $\qquad$ = 1.196 square yards
   1 square foot = 929.03 square centimeters
   $\qquad$ = 0.0929 square meter
   $\qquad$ = 144 square inches
   $\qquad$ = 0.1111 square yard
   1 hectare = 10,000 square meters = 2.47 acres
   1 acre = 0.405 hectare = 43,560 square feet
   1 square kilometer = 0.386 square mile
   1 square mile = 2.59 square kilometers = 640 acres

4. Volume
   1 cubic centimeter = 0.001 liter = 0.061 cubic inch
   1 liter = 1000 cubic centimeters
   $\qquad$ = 1.057 quarts (U.S.)
   $\qquad$ = 61.02 cubic inches
   $\qquad$ = 0.0353 cubic foot
   1 cubic meter = 1000 liters = 1 stere
   $\qquad$ = 1×10^6 cubic centimeters
   $\qquad$ = 35.31 cubic feet
   $\qquad$ = 264.2 gallons (U.S.)
   1 gallon (U.S.) = 4 quarts (U.S.) = 3.785 liters
   1 quart (U.S.) = 2 pints (U.S.) = 0.946 liter
   1 pint (U.S.) = 16 ounces = 0.473 liter
   1 cubic mile = 4.17 cubic kilometers
   1 acre-foot = 43,560 cubic feet = 1234 cubic meters
   1 barrel of petroleum = 42 gallons (U.S.)

5. Energy
   1 joule = 1 kilogram-meter^2/second^2 = 1 newton-meter = 1 watt-second
   $\qquad$ = 0.239 calorie = 9.48×10^{-4} British thermal unit
   $\qquad$ = 0.7376 foot-pound
   $\qquad$ = 2.78×10^{-7} kilowatt-hour
   $\qquad$ = 1×10^7 ergs = 6.24×10^{18} electron volts
   1 calorie = 4.184 joules
   1 Calorie (food) = 1 kilocalorie = 1000 calories
   1 British thermal unit (Btu) = 1055 joules = 252 calories
   1 kilowatt hour = 3.6×10^6 joules = 3,413 Btu
   1 quad = 1×10^{15} Btu = 1.05×10^{18} joules
   Energy source equivalents (approximate)
   1 gallon of gasoline = 1.4×10^8 joules (J) = 1.3×10^5 Btu
   1 42-gallon barrel of petroleum = 5.8×10^9 J = 5.5×10^6 Btu
   1000 cubic feet (MCF) of natural gas = 1.1×10^9 J = 1×10^6 Btu
   1 ton of bituminous coal (varies) = 2.7×10^{10} J = 2.6×10^7 Btu

6. Power
   1 watt = 1 joule/second = 3.413 British thermal units/hour
   1 kilowatt = 1000 watts = 1.34 horsepower
   1 horsepower = 550 foot-pounds/second = 0.746 kilowatt

7. Force
   1 newton = 1 kilogram-meter/second^2 = 1×10^5 dynes

8. Pressure
   1 pascal = 1 newton/square meter
   $\qquad$ = 1.45×10^{-4} pound per square inch
   $\qquad$ = 9.87×10^{-6} atmosphere
   1 atmosphere = 101,325 pascals = 101.33 kilopascals
   $\qquad$ = 14.7 pounds per square inch
   $\qquad$ = 760 millimeters (mm) Hg = 29.92 inches Hg
   $\qquad$ = 406.8 inches of water = 33.9 feet of water
   $\qquad$ = 1013 millibar = 1.013 bar
   $\qquad$ = 760 torr
   1 pound per square inch = 0.068 atmosphere = 6893 pascals
   $\qquad$ = 51.7 mm Hg = 27.68 inches of water
   1 torr = 133.32 newtons/square meter = 1/760 atmosphere
   1 bar = 1×10^5 pascals = 0.9869 atmosphere

9. Speed
   1 meter/second = 3.281 feet/second = 2.237 miles per hour
   1 mile/hour = 0.447 meter/second = 1.609 kilometers/hour
   $\qquad$ = 1.467 feet/second = 88 feet/minute

10. Viscosity (dynamic)
   1 poise = 100 centipoise = 0.1 kilogram/(meter-second)
   $\qquad$ = 1 dyne-second/square centimeter
   1 centipoise = 1×10^{-3} kilogram/(meter-second)
   $\qquad$ = 0.01 gram/(centimeter-second)
   $\qquad$ = 3.6 kilogram/(meter-hour)
   $\qquad$ = 2.42 pounds/(foot-hour)

11. Radioactivity and radiation dose
    1 curie $= 3.7 \times 10^{10}$ nuclear disintegrations per second
    1 becquerel $= 1$ nuclear disintegration per second
    1 roentgen $=$ an exposure to x-rays or gamma rays causing an electric charge of
        2.58×10^{-4} coulomb per kilogram of dry air
    1 rad $= 100$ ergs of absorbed radiation per gram of absorbing medium
        $= 0.01$ joule per kilogram of medium
    1 gray $= 100$ rad $= 1$ joule per kilogram
    1 rem $=$ an equivalent radiation dose
        $=$ rads times a quality factor   (QF)
        for gamma, beta, and x-rays   QF $= 1$
        fast neutrons and protons   QF $= 10$
        alpha particles   QF $= 20$
    1 sievert $= 100$ rem

12. Temperature
    degrees Celsius $= 5/9 \times$ (degrees Fahrenheit $- 32$)
    degrees Fahrenheit $= (9/5 \times$ degrees Celsius) $+ 32$
    degrees Kelvin $=$ degrees Celsius $+ 273.15$

Concentrations

Concentration is an expression of how much of a material is in a given amount of another material or environmental medium (air, water, soil, food). Environmental contaminant concentrations are usually described as the mass of a chemical in a given mass (or volume) of a medium or as the volume of a material in a given volume of a medium.

Concentrations of contaminants in water are expressed as the mass of a contaminant per given volume (typically one liter) of water or as the mass of a contaminant per given mass of water. An example of a mass/volume concentration is the number of milligrams of chemical X in one liter of water. Now, one liter of water (at 4°C) has a mass of 1000 grams. Therefore the mass of chemical X, in milligrams/liter, is equivalent to the mass of chemical X in milligrams/1000 grams, a mass/mass expression. There are one million milligrams in 1000 grams, or one milligram is one-millionth of 1000 grams. Therefore a concentration of one milligram per liter is equal to one part per million. It follows that one microgram per liter is equal to one part per billion, and so on. The following units are used interchangeably in expressions of water pollution concentrations.

mass/volume	*mass/mass*	*dimensionless*
milligrams/liter	$=$ milligrams/1000 grams	$=$ parts per million (ppm)
micrograms/liter	$=$ micrograms/1000 grams	$=$ parts per billion (ppb)
nanograms/liter	$=$ nanograms/1000 grams	$=$ parts per trillion (ppt)

The concentrations of aerosols (airborne solids or liquids) in air are expressed as the mass of a substance in a given volume (typically one cubic meter) of air. Typical expressions are milligrams of chemical X per cubic meter (m^3) of air or micrograms of chemical X per cubic meter of air. These aerosol concentrations are not equivalent to parts per million, parts per billion, and so forth.

Concentrations of gases or vapors in air are expressed as the volume of a gaseous material in a given volume of air (volume/volume) or as the mass of the material per given volume of air (mass/volume). A concentration of one liter of chemical X per one million liters of air is equivalent to a concentration of one part per million of X;

one liter of chemical Y in one billion liters of air is a Y concentration of one part per billion. The alternative expression of the air concentration of a gaseous chemical is a mass per volume, for example, the number of milligrams of X in a given volume (typically one cubic meter) of air.

Volume/volume concentrations and mass/volume concentrations of gaseous air contaminants are convertible, using the molecular weight of the gas or vapor and the density of air. For air at 25°C and a pressure of one atmosphere (United States Environmental Protection Agency standard conditions) the following conversions can be used.

$$\frac{\text{micrograms of a substance A}}{\text{cubic meter of air}} = (\text{parts per million of A}) \times (\text{molecular weight of A}) \times 40.9$$

$$\text{parts per million of A} = \frac{\text{micrograms of substance A}}{\text{cubic meter of air}} \times \frac{1}{\text{molecular weight of A}} \times \frac{1}{40.9}$$

Standard (average) Human Factors

Factor	Man	Woman	Child (3–12)
Body mass (kilograms)	70	60	15–40
Skin surface area (square meters)			
No clothing	1.8	1.6	0.9
Clothed	0.1–0.3	0.1–0.3	0.05–0.15
Respiration (liters/minute)			
Resting	7.5	6.0	5.0
Light activity	20	19	13
Air volume breathed			
(cubic meters per day)	23	21	15
Fluid intake (liters/day)			
(tap water, milk, water-based			
beverages)	2	1.4	1.0

The Chemical Elements

Element	Symbol	Proton No.	Relative atomic mass
actinium	Ac	89	[227]
aluminum	Al	13	26.9815
americium	Am	95	[243]
antimony	Sb	51	121.75
argon	Ar	18	39.948
arsenic	As	33	74.9216
astatine	At	85	[210]
barium	Ba	56	137.34
berkelium	Bk	97	[247]
beryllium	Be	4	9.0122
bismuth	Bi	83	208.98
boron	B	5	10.81
bromine	Br	35	79.904
cadmium	Cd	48	112.40
caesium	Cs	55	132.905
calcium	Ca	20	40.08

californium	Cf	98	[251]
carbon	C	6	12.011
cerium	Ce	58	140.12
chlorine	Cl	17	35.453
chromium	Cr	24	51.996
cobalt	Co	27	58.9332
copper	Cu	29	63.546
curium	Cm	96	[247]
dysprosium	Dy	66	162.50
einsteinium	Es	99	[254]
erbium	Er	68	167.26
europium	Eu	63	151.96
fermium	Fm	100	[257]
fluorine	F	9	18.9984
francium	Fr	87	[223]
gadolinium	Gd	64	157.25
gallium	Ga	31	69.72
germanium	Ge	32	72.59
gold	Au	79	196.967
hafnium	Hf	72	178.49
helium	He	2	4.0026
holmium	Ho	67	164.930
hydrogen	H	1	1.00797
indium	In	49	114.82
iodine	I	53	126.9044
iridium	Ir	77	192.2
iron	Fe	26	55.847
krypton	Kr	36	83.80
lanthanum	La	57	138.91
lawrencium	Lr	103	[257]
lead	Pb	82	207.19
lithium	Li	3	6.939
lutetium	Lu	71	174.97
magnesium	Mg	12	24.305
manganese	Mn	25	54.938
mendelevium	Md	101	[258]
mercury	Hg	80	200.59
molybdenum	Mo	42	95.94
neodymium	Nd	60	144.24
neon	Ne	10	20.179
neptunium	Np	93	[237]
nickel	Ni	28	58.71
niobium	Nb	41	92.906
nitrogen	N	7	14.0067
nobelium	No	102	[255]
osmium	Os	76	190.2
oxygen	O	8	15.9994
palladium	Pd	46	106.4
phosphorus	P	15	30.9738
platinum	Pt	78	195.09
plutonium	Pu	94	[244]
polonium	Po	84	[209]
potassium	K	19	39.102

praseodymium	Pr	59	140.907
promethium	Pm	61	[145]
protactinium	Pa	91	[231]
radium	Ra	88	[226]
radon	Rn	86	[222]
rhenium	Re	75	186.20
rhodium	Rh	45	102.905
rubidium	Rb	37	85.47
ruthenium	Ru	44	101.07
samarium	Sm	62	150.35
scandium	Sc	21	44.956
selenium	Se	34	78.96
silicon	Si	14	28.086
silver	Ag	47	107.868
sodium	Na	11	22.9898
strontium	Sr	38	87.62
sulphur	S	16	32.064
tantalum	Ta	73	180.948
technetium	Tc	43	[97]
tellurium	Te	52	127.60
terbium	Tb	65	158.924
thallium	Tl	81	204.37
thorium	Th	90	232.038
thulium	Tm	69	168.934
tin	Sn	50	118.69
titanium	Ti	22	47.90
tungsten	W	74	183.85
uranium	U	92	238.03
vanadium	V	23	50.942
wolfram (tungsten)	W	74	183.85
xenon	Xe	54	131.30
ytterbium	Yb	70	173.04
yttrium	Y	39	88.905
zinc	Zn	30	65.37
zirconium	Zr	40	91.22

The Greek Alphabet

A	α	alpha
ß	β	beta
Γ	γ	gamma
Δ	δ	delta
E	ϵ	epsilon
Z	ζ	zeta
H	η	eta
Θ	θ	theta
I	ι	iota
K	κ	kappa
Λ	λ	lambda
M	μ	mu
N	ν	nu
Ξ	ξ	xi

O	*o*	omicron
Π	π	pi
P	ρ	rho
Σ	σ	sigma
T	τ	tau
Υ	υ	upsilon
Φ	ϕ	phi
X	χ	chi
Ψ	ψ	psi
Ω	ω	omega